# Physics meets philosophy at the Planck scale

The greatest challenge in fundamental physics is how quantum mechanics and general relativity can be reconciled in a theory of 'quantum gravity'. The project suggests a profound revision of our notions of space, time, and matter, and so has become a key topic of debate and collaboration between physicists and philosophers. This timely volume collects classic and original contributions from leading experts in both fields for a provocative discussion of all the issues.

This volume contains accessible introductions to the main and less well known approaches to quantum gravity. It includes exciting topics such as the fate of spacetime in various theories, the so-called 'problem of time' in canonical quantum gravity, black hole thermodynamics, and the relationship between the interpretation of quantum theory and quantum gravity.

This book will be essential reading for anyone interested in the profound implications of trying to marry the two most important theories in physics.

Craig Callender is an Assistant Professor of Philosophy at the University of California at San Diego. Formerly a Senior Lecturer in Logic, Philosophy and Scientific Method at the London School of Economics, he has worked as Deputy Editor of the *British Journal for the Philosophy of Science* and as Associate Editor of *Mind*. His research interests include the philosophical foundations of modern physics, the problem of the direction of time, and the metaphysics of science. He has published in physics, law and philosophy journals.

Nick Huggett is an Assistant Professer of Philosophy at the University of Illinois at Chicago, who earned his Ph.D. in Philosophy from Rutgers University. He has published and lectured extensively on topics concerning the philosophical foundations of quantum mechanics, quantum field theory, and spacetime theories. He is the author of the book *Space from Zeno to Einstein: Classic Readings with a Contemporary Commentary*.

# Physics meets philosophy at the Planck scale

## Contemporary theories in quantum gravity

edited by Craig Callender

*University of California at San Diego*

Nick Huggett

*University of Illinois at Chicago*

CAMBRIDGE
UNIVERSITY PRESS

CAMBRIDGE UNIVERSITY PRESS
Cambridge, New York, Melbourne, Madrid, Cape Town, Singapore, São Paulo, Delhi

Cambridge University Press
The Edinburgh Building, Cambridge CB2 8RU, UK

Published in the United States of America by Cambridge University Press, New York

www.cambridge.org
Information on this title: www.cambridge.org/9780521664455

First published 2001

*A catalogue record for this publication is available from the British Library*

ISBN 978-0-521-66280-2 hardback
ISBN 978-0-521-66445-5 paperback

Transferred to digital printing 2009

*For Joanna and Lisa*

# Contents

vii

Contents

# Preface

Both philosophy and quantum gravity are concerned with fundamental questions regarding the nature of space, time, and matter, so it is not surprising that they have a great deal to say to each other. Yet the methods and skills that philosophers and physicists bring to bear on these problems are often very different. However, and especially in recent years, there is an increasing recognition that these two groups are indeed tackling the same issues, and moreover, that these different methods and skills may all be of use in answering these fundamental questions. Our aim in this volume is thus to introduce and explore the philosophical foundations of quantum gravity and the philosophical issues within this field.

We believe that some insight will be gained into the many deep questions raised by quantum gravity by examining these issues from a variety of perspectives. Toward this end, we collected together ten original and three previously published contributions on this topic from eminent philosophers of science, mathematicians, and physicists. Though the papers assume that the reader is scientifically literate, the majority are written with the non-specialist in philosophy or quantum gravity in mind. The brief that we gave the contributors was to write pieces that would introduce the key elements of the physics and philosophy, whilst making original contributions to the debates. The book should therefore be of interest to (and appropriate for) anyone challenged and fascinated by the deep questions facing the cutting edge of fundamental theoretical physics. We also feel that these essays will lay the foundation for a wider consideration of quantum gravity in the philosophical community, and hopefully a fruitful dialogue between physicists and philosophers. After all, many of the greatest advances in physics were inseparable from philosophical reflection on foundational questions.

We could not have completed this volume without a great deal of support and assistance, and we owe thanks to many people: to all our contributors for their efforts; to Carlo Rovelli and John Baez especially, for their encouragement and advice in the early stages; to Jeeva Anandan, Jossi Berkovitz, Jeremy Butterfield, Joy Christian, Carl Hoefer, Jeffrey Ketland, Tom Imbo, and Steve Savitt for help and comments on a number of issues; to Adam Black and Ellen Carlin, our editors, and Alan Hunt of Keyword Publishing Services Ltd. for their enthusiasm, help, and (most of all) patience. We are also grateful to Andrew Hanson for letting us use his wonderful 4-D projection of Calabi-Yau space on the cover. We also thank Damian Steer for his efforts with the bibliography and Jason Wellner for his help preparing the reprinted articles. Nick thanks the Centre for Philosophy of Natural and Social Sciences at the London School of Economics for support during a residence, and the Center for the Humanities and the Office of the Vice Chancellor for Research at the University of Illinois at Chicago for a faculty fellowship and Grants-in-Aid support. Finally, we thank Robert Weingard who taught us that these ideas existed, and of course our families, Joanna and Lisa, Ewan and Lily, without whom none of this would be worthwhile.

*Craig Callender*
*Nick Huggett*

**Introduction**

Craig Callender and Nick Huggett

In recent years it has sometimes been difficult to distinguish between articles in quantum gravity journals and articles in philosophy journals. It is not uncommon for physics journals such as *Physical Review D, General Relativity and Gravitation* and others to contain discussion of philosophers such as Parmenides, Aristotle, Leibniz, and Reichenbach; meanwhile, *Philosophy of Science, British Journal for the Philosophy of Science* and others now contain papers on the emergence of spacetime, the problem of time in quantum gravity, the meaning of general covariance, etc. At various academic conferences on quantum gravity one often finds philosophers at physicists' gatherings and physicists at philosophers' gatherings. While we exaggerate a little, there is in recent years a definite trend of increased communication (even collaboration) between physicists working in quantum gravity and philosophers of science. What explains this trend?

Part of the reason for the connection between these two fields is no doubt negative: to date, there is no recognized experimental evidence of characteristically *quantum* gravitational effects. As a consequence, physicists building a theory of quantum gravity are left without direct guidance from empirical findings. In attempting to build such a theory almost from first principles it is not surprising that physicists should turn to theoretical issues overlapping those studied by philosophers.

But there is also a more positive reason for the connection between quantum gravity and philosophy: many of the issues arising in quantum gravity are genuinely philosophical in nature. Since quantum gravity forces us to challenge some of our deepest assumptions about the physical world, all the different approaches to the subject broach questions discussed by philosophers. How should we understand general relativity's general covariance – is it a significant physical principle, or is it merely a question about the language with which one writes an equation? What is the nature of time and change? Can there be a theory of the universe's boundary conditions? Must space and time be fundamental? And so on. Physicists thinking about these issues have noticed that philosophers have investigated each of them. (Philosophers have discussed the first question for roughly 20 years; the others for at least 2,500 years.) Not surprisingly, then, some physicists have turned to the work of classic and contemporary philosophers to see what they have been saying about time, space, motion, change and so on. Some philosophers, noticing this work, have responded by studying quantum gravity. They have diverse motives: some hope that their logical skills and acquaintance with such topics may serve the physicists in their quest for a theory of quantum gravity; others hope that work in the field may shed some light on these ancient questions, in the way that modern physics has greatly clarified other traditional areas of metaphysics, and still others think of quantum gravity as an intriguing 'case study' of scientific discovery in practice. In all these regards, it is interesting to note that Rovelli (1997) explicitly and positively draws a parallel between the current interaction between physics and philosophy and that which accompanied the scientific revolution, from Galileo to Newton.

This volume explores some of the areas that philosophers and physicists have in common with respect to quantum gravity. It brings together some of the leading thinkers in contemporary physics and philosophy of science to introduce and discuss philosophical issues in the foundations of quantum gravity. In the remainder of this introduction we aim to sketch an outline of the field, introducing the basic physical ideas to philosophers, and introducing philosophical background for physicists. We are especially concerned with the questions: Why should there be a quantum theory of gravity? What are the leading approaches? And what issues might constitute the overlap between quantum gravity and philosophy?

More specifically, the plan of the Introduction is as follows. Section 1.1 sets the stage for the volume by briefly considering why one might want a quantum theory of gravity in the first place. Section 1.2 is more substantive, for it tackles the question of whether the gravitational field *must* be quantized. One often hears the idea that it is actually inconsistent with known physics to have a world wherein the gravitational field exists unquantized. But is this right? Section 1.2.1 considers an interesting argument which claims that if the world exists in a half-quantized and half-unquantized form, then either superluminal signalling will be allowed or energy–momentum will not be conserved. Section 1.2.2 then takes up the idea of so-called 'semiclassical' quantum gravity. We show that the arguments for quantizing gravity are not conclusive, but that the alternative is not particularly promising either. We feel that it is important to address this issue so that readers will understand how one is led to consider the kind of theories – with their extraordinary conceptual difficulties – discussed in the book. However, those not interested in pursuing this issue immediately are invited to skip ahead to Section 1.3, which outlines (and hints at some conceptual problems with) the two main theories of quantum gravity, superstring theory and canonical quantum gravity. Finally, Section 1.4 turns to the question of what quantum gravity and philosophy have to say to each other. Here, we discuss in the context of the papers in the volume many of the issues where philosophers and physicists have interests that overlap in quantum gravity.

A word to the wise before we begin. Because this is a book concerned with the philosophical dimensions of quantum gravity, our contributors stress philosophical discussion over accounts of state-of-the-art technical developments in physics: especially, loop quantum gravity and M-theory are treated only in passing (see Rovelli 1998 and Witten 1997 for reviews, and Major 1999 for a very accesible formal introduction). Aside from sheer constraints on space, the reasons for this emphasis are two-fold: First, developments at the leading edge of the field occur very fast and do not always endure; and second, the central philosophical themes in the field can (to a large extent) be understood and motivated by consideration of the core parts of the theory that have survived subsequent developments. We have thus aimed to provide an introduction to the philosophy of quantum gravity that will retain its relevance as the field evolves: hopefully, as answers are worked out, the papers here will still raise the important questions and outline their possible solutions. But the reader should be aware that there will have been important advances in the physics that are not reflected in this volume: we invite them to learn here what issues are philosophically interesting about quantum gravity, and then discover for themselves how more recent developments in physics relate to those issues.

## 1.1 Why quantum gravity?

We should emphasize at the outset that currently there is no quantum theory of gravity in the sense that there is, say, a quantum theory of gauge fields. 'Quantum gravity' is merely a placeholder for whatever theory or theories eventually manage to bring together our theory of the very small, quantum mechanics, with our theory of the very large, general relativity. This absence of a theory might be thought to present something of an impediment to a book supposedly on its foundations. However, there do exist many more-or-less developed approaches to the task – especially superstring theory and canonical quantum gravity (see Section 1.3) – and the assumptions of these theories and the difficulties they share can be profitably studied from a variety of philosophical perspectives.

First, though, a few words about why we ought to expect there to be a theory of quantum gravity. Since we have no unequivocal experimental evidence conflicting with either general relativity or quantum mechanics, do we really need a quantum theory of gravitation? Why can't we just leave well enough alone, as some philosophical approaches to scientific theories seem to suggest?

It might be thought that 'instrumentalists' are able to ignore quantum gravity. Instrumentalism, as commonly understood, conceives of scientific theories merely as tools for prediction. Scientific theories, on this view, are not (or ought not to be) in the business of providing an accurate picture of reality in any deeper sense. Since there are currently no observations demanding a quantum gravitational theory, it might be thought that advocates of such a position would view the endeavour as empty and misguided speculation, perhaps of formal interest, but with no physical relevance.

However, while certain thinkers may indeed feel this way, we don't think that instrumentalists can safely ignore quantum gravity. It would be unwise for them to construe instrumentalism so narrowly as to make it unnecessary. The reason is that some of the approaches to the field may well be testable in the near future. The work that won first prize in the 1999 Gravity Research Foundation Essay Competition, for instance, sketches how both photons from distant astrophysical sources and laboratory experiments on neutral kaon decays may be sensitive to quantum gravitational effects (Ellis et al. 1999). And Kane (1997) explains how possible predictions of superstring theory – if only the theory was sufficiently tractable for them to be made – could be tested with currently available technologies. We will never observe the effects of gravitational interactions between an electron and a proton in a hydrogen atom (Feynman 1995, p. 11, calculates that such interaction would change the wave function phase by a tiny 43 arcseconds in $100\,T$, where $T$ is the age of the universe!), but other effects may be directly or indirectly observable, perhaps given relatively small theoretical or experimental advances. Presumably, instrumentalists will want physics to be empirically adequate with respect to these phenomena. (We might also add the common observation that since one often doesn't know what is observable until a theory is constructed, even an instrumentalist should not restrict the scope of new theories to extant evidence.)

Another philosophical position, which we might dub the 'disunified physics' view might in this context claim that general relativity describes certain aspects of the world, quantum mechanics other distinct aspects, and that would be that. According to this view, physics (and indeed, science) need not offer a single universal theory encompassing all physical phenomena. We shall not debate the correctness of this view here, but we would like to point out that if physics aspires to provide a complete account of the world, as it traditionally has, then there must be a quantum theory of gravity. The simple reason is that general relativity and quantum mechanics cannot both be correct even in their domains of applicability.

First, general relativity and quantum mechanics cannot both be universal in scope, for the latter strictly predicts that all matter is quantum, and the former only describes the gravitational effects of classical matter: they cannot both take the whole (physical) world as their domain of applicability. But neither is the world split neatly into systems appropriately described by one and systems appropriately described by the other. For the majority of situations treated by physics, such as electrons or planets, one can indeed get by admirably using only one of these theories: for example, the gravitational effects of a hydrogen nucleus on an electron are negligible, as we noted above, and the quantum spreading of the wavepacket representing Mercury won't much affect its orbit. But in principle, the two theories govern the same systems: we cannot think of the world as divided in two, with matter fields governed by quantum mechanics evolving on a curved spacetime manifold, itself governed by general relativity. This is, of course, because general relativity, and in particular, the Einstein field equation

$$G_{\mu\nu} = 8\pi T_{\mu\nu}, \tag{1.1}$$

couples the matter–energy fields in the form of the stress–energy tensor, $T_{\mu\nu}$, with the spacetime geometry, in the form of the Einstein tensor, $G_{\mu\nu}$. Quantum fields carry energy and mass; therefore, if general relativity is true, quantum fields distort the curvature of spacetime and the curvature of spacetime affects the motion of the quantum fields. If these theories are to yield a complete account of physical phenomena, there will be no way to avoid those situations – involving very high energies – in which there are non-negligible interactions between the quantum and gravitational fields; yet we do not have a theory characterizing this interaction. Indeed, the influence of gravity on the quantum realm is an experimental fact: Peters et al. (1999) measured interference between entangled systems following different paths in the Earth's gravitational field to measure gravitational acceleration to three parts in $10^9$. Further, we do not know whether new low energy, non-perturbative, phenomena might result from a full treatment of the connection between quantum matter and spacetime. In general, the fact that gravity and quantum matter are inseparable 'in principle' will have *in practice* consequences, and we are forced to consider how the theories connect.

One natural reaction is to correct this 'oversight' and extend quantum methods to the gravitational interaction in the way that they were applied to describe the electromagnetic and nuclear interactions of matter, yielding the tremendously successful 'standard model' of quantum field theory. One way to develop this approach is to say that the spacetime metric, $g_{\mu\nu}$, be broken into two parts, $\eta_{\mu\nu} + h_{\mu\nu}$, representing a flat background spacetime and a gravitational disturbance respectively; and that we look for a quantum field theory of $h_{\mu\nu}$ propagating in a flat spacetime described by $\eta_{\mu\nu}$. However, in contrast to the other known forces, it turns out that all unitary local quantum field theories for gravity are non-renormalizable. That is, the coupling strength parameter has the dimensions of a negative power of mass, and so standard arguments imply that the divergences that appear in perturbative calculations of physical quantities cannot be cancelled by rescaling a finite number of physical parameters: ultimately the theory depends on an infinite number of quantities that would need to be fixed empirically. More troubling is the strong suggestion from study of the 'renormalization group' that such non-renormalizable theories become pathological at short distances (e.g. Weinberg 1983) – perhaps not too surprising a result for a theory which attempts in some sense to 'quantize distance'.

Thus the approach that worked so well for the other forces of nature does not seem applicable to gravity. Some new strategy seems in order if we are to marry quantum theory and relativity. The different programmes – both the two main ones, canonical quantum gravity and superstring theory, and alternatives such as twistor theory, the holographic hypothesis, non-commutative geometry, topological quantum field theory, etc. – all explore different avenues of attack. What goes, of course, is the picture of gravity as just another quantum field on a flat classical spacetime – again, not too surprising if one considers that there is no proper distinction between gravity and spacetime in general relativity. But what is to be expected, if gravity will not fit neatly into our standard quantum picture of the world, is that developing quantum gravity will require technical and philosophical revolutions in our conceptions of space and time.

## 1.2 Must the gravitational field be quantized?

### 1.2.1 No-go theorems?

Although a theory of quantum gravity may be unavoidable, this does not automatically mean that we must *quantize* the classical gravitational field of general relativity. A theory is clearly needed to characterize systems subject to strong quantum and gravitational effects, but it does not follow that the correct thing to do is to take classical relativistic objects such as the Riemann tensor or metric field and quantize them: that is, make them operators subject to non-vanishing commutation relations. All that follows from Section 1.1 is that a new theory is needed – nothing about the nature of this new theory was assumed. Nevertheless, there are arguments in the literature to the effect that it is inconsistent to have quantized fields interact with non-quantized fields: the world cannot be half-quantized-and-half-classical. If correct, given the (apparent) necessity of quantizing matter fields, it would follow that we must also quantize the gravitational field. We would like to comment briefly on this type of argument, for we believe that they are interesting, even if they fall short of strict no-go theorems for any half-and-half theory of quantum gravity.

We are aware of two different arguments for the necessity of quantizing fields that interact with quantum matter. One is an argument (e.g. DeWitt 1962) based on a famous paper by Bohr and Rosenfeld (1933) that analysed a semiclassical theory of the electromagnetic field in which 'quantum disturbances' spread into the classical field. These papers argue that the quantization of a given system implies the quantization of any system to which it can be coupled, since the uncertainty relations of the quantized field 'infect' the coupled non-quantized field. Thus, since quantum matter fields interact with the gravitational field, these arguments, if correct, would prove that the gravitational field must also be quantized. We will not discuss this argument here, since Brown and Redhead (1981) contains a sound critique of the 'disturbance' view of the uncertainty principle underpinning these arguments.

Interestingly, Rosenfeld (1963) actually denied that the 1933 paper showed any inconsistency in semiclassical approaches. He felt that empirical evidence, not logic, forced us to quantize fields; in the absence of such evidence 'this temptation [to quantize] must be resisted' (1963, p. 354). Emphasizing this point, Rosenfeld ends his paper with the remark, 'Even the legendary Chicago machine cannot deliver sausages if it is not supplied with hogs' (1963, p. 356). This encapsulates the point of view we would like to defend here.

The second argument, which we will consider, is due to Eppley and Hannah (1977) (but see also Page and Geilker 1981 and Unruh 1984). The argument – modified in places by us – goes like this. Suppose that the gravitational field were relativistic (Lorentzian) and classical: not quantized, not subject to uncertainty relations, and not allowing gravitational states to superpose in a way that makes the classical field indeterminate. The contrast is exactly like that between a classical and quantum particle.[1] Let us also momentarily assume the standard interpretation of quantum mechanics, whereby a measurement interaction instantaneously collapses the wave function into an eigenstate of the relevant observable. (See, for example Aharonov and Albert 1981, for a discussion of the plausibility of this interpretation in the relativistic context.)

Now we ask how this classical field interacts with quantized matter, for the moment keeping all possibilities on the table. Eppley and Hannah (1977) see two (supposedly) exhaustive cases: gravitational interactions either collapse or do not collapse quantum states.

Take the first horn of the dilemma: suppose the gravitational field *does not* collapse the quantum state of a piece of matter with which it interacts. Then we can send superluminal signals, in violation of relativity, as conventionally understood. Eppley and Hannah (1977) (and Pearle and Squires 1996) suggest some simple ways in which this can be accomplished using a pair of entangled particles, but we will use a modification of Einstein's 'electron in a box' thought experiment. However, the key to these examples is the (seemingly unavoidable) claim that if a gravitational interaction does not collapse a quantum state, then the dynamics of the interaction depend on the state. In particular, the way a classical gravitational wave scatters off a quantum object would depend on the spatial wave function of the object, much as it would depend on a classical mass distribution. Thus, scattering experiments are at least sensitive to changes in the wave function, and at best will allow one to determine the form of the wave function – without collapsing it. It is not hard to see how this postulate, together with the usual interpretation of quantum measurements, allows superluminal signalling.

We start with a rectangular box containing a single electron (or perhaps a microscopic black hole), in a quantum state that makes it equally likely that the electron will be found in either half of the box. We then introduce a barrier between the two halves and separate them, leaving the electron in a superposition of states corresponding to being in the left box and being in the right box. If the probabilities of being in each box are equal, then the state of the particle will be:

$$\psi(x) = 1/\sqrt{2}(\psi_L(x) + \psi_R(x)),\tag{1.2}$$

where $\psi_L(x)$ and $\psi_R(x)$ are wave functions of identical shape but with supports inside the left and right boxes respectively.

Next we give the boxes to two friends Lefty and Righty, who carry them far apart (without ever looking in them of course). In Einstein's original version (in a letter discussed by Fine 1986, p. 35–39, which is a clarification of the EPR argument in its published form), when Lefty looked inside her box – and say found it empty – an element of reality was instantaneously present in Righty's box – the presence of the electron – even though the boxes were spacelike separated. Assuming the collapse postulate, when Lefty looks in her box a state transition,

$$1/\sqrt{2}(\psi_L(x) + \psi_R(x)) \rightarrow \psi_R(x)\tag{1.3}$$

occurs. In the familiar way, either some kind of spooky non-local 'action' occurs or the electron was always in Righty's box and quantum mechanics is incomplete, since $\psi(x)$ is indeterminate between the boxes. Of course, this experiment does not allow signalling, for if Righty now looks in his box and sees the electron, he could just as well conclude that he was the first to look in the box, collapsing the superposition. And the long run statistics generated by repeated measurements that Righty observes will be 50 : 50, electron : empty, whatever Lefty does – they can only determine the correlation by examining the joint probability distribution, to which Righty, at his wing, does not have access.

In the present case the situation is far more dire, for Righty can use our non-collapsing gravitational field to 'see' what the wave function in his box is without collapsing it. We simply imagine that the right-hand box is equipped with apertures that allow gravitational waves in and out, and that Righty arranges a gravitational wave source at one of them and detectors at the others.[2]

Since the scattering depends on the form of the wave function in the box, any changes in the wave function will show up as changes in the scattering pattern registered by the detectors. Hence, when Lefty now looks in her box – and suppose this time she finds the electron – Righty's apparatus will register the collapse instantaneously; there will be no scattering source at all, and the waves will pass straight through Righty's box. That is, before Lefty looks, the electron wave function is $\psi(x)$ and Righty's gravity wave scatters off $\psi_R(x)$; after Lefty collapses the electron, its state is $\psi_L(x) + 0$ and so Righty's gravity wave has no scattering source. And since we make the usual assumption that the collapse is instantaneous, the effect of looking in the left box is registered on the right box superluminally. So, if Righty and Lefty have a prior agreement that if Lefty performs the measurement then she fancies a drink after work, otherwise she wants to go to the movies, then the apparatus provides Righty with information about Lefty's intentions at a spacelike separated location.[3]

It is crucial to understand that this experiment is *not* a variant of 'Wigner's friend'. One should absolutely not think that scattering the gravity wave off the electron wave function leads to an entangled state in which the gravity wave is in a quantum superposition, which is itself collapsed when measured by the detectors, producing a consequent collapse in the electron wave function. Of course, such things might occur in a theory of quantum gravity, but they cannot occur in the kind of theory that we are presently discussing: a theory with a *classical* gravitational field, which just means a theory in which there are no quantum superpositions of the gravitational field. There is in this theory no way of avoiding signalling by introducing quantum collapses of the gravitational field, since there is nothing to collapse.

It is also important to see how the argument depends on the interpretation of quantum mechanics. On the one hand it does not strictly require the standard interpretation of quantum mechanics, but can be made somewhat more general. In our example, the component of the wave function with support on Righty's box went from $\psi_R(x)$ to 0, which is a very sharp change. But the argument doesn't need a sharp change, it just needs a detectable change, to $\epsilon\psi_R(x)$, say. On the other hand, it is necessary for the argument that normal measurements can produce effects at spacelike separated regions. For then the gravitational waves provide an abnormal way of watching a wave function without collapsing it, to see when such effects occur. Thus, an interpretation of quantum mechanics that admits a dynamics which prevents superluminal propagation of any disturbance in the wave function will escape this argument. Any no-collapse theory whose wave function is governed at all times by a relativistic wave equation will be of this type.

The conclusion of this horn of the dilemma is then the following. If one adopts the standard interpretation of quantum mechanics, and one claims that the world is divided into classical (gravitational) and quantum (matter) parts, and one models quantum–classical interactions without collapse, then one must accept the possibility of superluminal signalling. And further, though practical difficulties may prevent one from ever building a useful signalling device, the usual understanding of relativity prohibits superluminal signalling, even in principle. Of course, this interpretation of relativity is a subtle matter in a number of ways, for instance concerning the possibility of Lorentz-invariant signalling (Maudlin 1994) and even the possibility of time travel (see, e.g. Earman 1995a). And of course, given the practical difficulty of performing such an experiment, we do not have definitive empirical grounds for ruling out such signalling. But since the kind of signalling described here could pick out a preferred foliation of spacetime – on which the collapse occurs – it does violate relativity in an important sense. Thus, someone who advocates a standard interpretation of quantum mechanics, a half-and-half view of the world and a no-collapse theory of classical–quantum interactions must deny relativity as commonly understood. They would need a very different theory that could accommodate the kind of superluminal signalling demonstrated, but that also approximates the causal structure of general relativity in all the extant experiments. (Note that this conclusion is in line with our earlier, more general, argument for the existence of a theory of quantum gravity; and note that the present argument really only demonstrates the need for a new theory – it does not show that quantizing the metric field is the only way to escape this problem.)

Of course, as mentioned, one *might* be able to avoid this horn of the dilemma by opting for a no-collapse interpretation of quantum mechanics, e.g. some version of Bohmian mechanics, or Everettian theories. We are not aware of any actual proposal for a half-and-half world that exploits this possibility (e.g. Bohmian quantum gravity – see below – aims to quantize the gravitational field). But the space may exist in the logical geography. In Bohm's theory, however measurements can have non-local effects on particle positions. Signalling could therefore occur if scattering at the gravitational field depended on the particle configuration and not only the wave function.

Let's turn to the other horn of the dilemma, where now we suppose that gravitational interactions can collapse quantum states of matter. Interestingly enough, there are a number of concrete suggestions that gravity should be thus implicated in the measurement problem, so it is perhaps not too surprising that attempts to close off this horn are, if anything, even less secure.

Eppley and Hannah's (1977) argument against a collapsing half-and-half theory is that it entails a violation of energy–momentum conservation. First, we assume that when our classical gravitational wave scatters off a quantum particle its wave function collapses, to a narrow Gaussian say. Second, we assume that the gravity wave scatters off the collapsed wave function as if there were a point particle localized at the collapse site. Then the argument is straightforward: take a quantum particle with sharp momentum but uncertain position, and scatter a gravity wave off it. The wave function collapses, producing a localized particle (whose position is determined by observing the scattered wave), but with uncertain momentum according to the uncertainty relations. Making the initial particle slow and measuring the scattered gravity wave with sufficient accuracy, one can pinpoint the final location sharply enough to ensure that the uncertainty in final momentum is far greater than the sharp value of the initial momentum. Eppley and Hannah conclude that we have a case of momentum non-conservation, at least on the grounds that a subsequent momentum measurement could lead to a far greater value than the initial momentum. (Or perhaps, if we envision performing the experiment on an ensemble of such particles, we have no reason to think that the momentum expectation value after will be the same as before.)

As with the first argument, the first thing that strikes one about this second argument is that it does not obviously depend on the fact that it is an interaction with the gravitational field that produces collapse. Identical reasoning could be applied to any sufficiently high resolution particle detector, given the standard collapse interpretation of measurement. Since this problem for the collapse interpretation is rather obvious, we should ask whether it has any standard response. It seems that it does: as long as the momentum associated with the measuring device is much greater than the uncertainty it produces, then we can sweep the problem under the rug. The non-conservation is just not relevant to the measurement undertaken. If this response works for generic measurements, then we can apply it in particular to gravitationally induced collapse, leaving Eppley and Hannah's argument inconclusive.

But how satisfactory is this response in the generic case? Just as satisfactory as the basic collapse interpretation: not terribly, we would say. Without rehearsing the familiar arguments, 'sweeping quantities under the rug' in this way seems troublingly *ad hoc*, pointing to some missing piece of the quantum puzzle: hidden variables perhaps or, as we shall consider here, a precise theory of collapse. Without some such addition to quantum mechanics it is hard to evaluate whether such momentum non-conservation should be taken seriously or not, but with a more detailed collapse theory it is possible to pose some determinate questions. Take, as an important example, the 'spontaneous localization' approaches of Ghirardi, Rimini, and Weber (1986) or, more particularly here, of Pearle and Squires (1996). In their models, energy is indeed not conserved in collapse, but with suitable tuning (essentially smearing matter over a fundamental scale), the effect can be made to shrink below anything that might have been detected to date.[4]

Whether such an answer to non-conservation is satisfactory depends on whether we must take the postulate of momentum conservation as a fundamental or experimental fact, which in turn depends on our reasons for holding the postulate. In quantum mechanics, the reasons are of course that the spacetime symmetries imply that the self-adjoint generators of temporal and spatial translations commute, $[\hat{H}, \hat{P}] = 0$, and the considerations that lead us to identify the generator of spatial transformations with momentum (cf., e.g. Jordan 1969). The conservation law, $d\langle \hat{P} \rangle / dt = 0$ then follows simply. But of course, implicit in the assumption that there is a self-adjoint generator for temporal translations, $\hat{H}$, is the assumption that the evolution operator, $\hat{U}(t) = e^{-i\hat{H}t/\hbar}$, is unitary. But in a collapse, it is exactly this assumption that breaks down: so what Eppley and Hannah in fact show is only that in a collapse our fundamental reasons for expecting momentum conservation fail. But if all that remains are our empirical reasons, then the spontaneous localization approaches are satisfactory on this issue, as are other collapse models that hide momentum non-conservation below the limits of observation. Thus, the incompleteness problem aside, sweeping momentum uncertainty under the rug need not do any harm.[5] In this respect, it is worth noting that if gravitational waves cause quantum jumps, then the effect must depend in some way on the strength of the waves. The evidence for this assertion is the terrestrial success of quantum mechanics despite the constant presence on Earth of gravity waves from deep space sources (and indeed from the motions of local objects). If collapse into states sharp in position were occurring at a significant rate then, for instance, we would not expect matter to be stable, since energy eigenstates are typically not sharp in position, nor would we observe electron diffraction, since electrons with sharp positions move as localized particles, without interfering.[6]

Since the size of the collapse effect must depend on the gravitational field in some way, one could look for a theory in which the momentum of a gravitational wave was always much larger than the uncertainty in any collapse it causes. Then any momentum non-conserving effect would be undetectable, and Eppley and Hannah's qualms could again be swept under the rug.

Indeed, the spontaneous localization model developed by Pearle and Squires (1996) has these features to some degree. They do not explicitly model a collapse caused by a gravitational wave, but rather use a gravitational field whose source is a collection of point sources, which punctuate independently in and out of existence. In their model, the rate at which collapse occurs depends (directly) on the mass of the sources and (as a square root) on the probability for source creation at any time, so that a stronger source field produces a stronger gravitational field and a greater rate of collapse. And the amount of energy produced by collapse is undetectable by present instruments, given a suitable fixing of constants, sweeping it under the rug. (Pearle and Squires also argue that the collapse rate is great enough to prevent signalling.)

Finally, Roger Penrose also links gravity to collapse (see Chapter 13). He in fact advocates a model in which gravity is quantized, but this is not crucial to the measure of collapse rate he offers. He proposes that the time rate of collapse of a superposition of two separated wave packets, $T$, is determined by the (Newtonian) gravitational self-energy of the difference of the two packets, $E_\Delta$: $T \sim \hbar/E_\Delta$. Penrose also proposes a test for this model, which Joy Christian criticizes and refines in Chapter 14.

As we said at the start of this section, arguments in the style of Eppley and Hannah do not constitute no-go theorems against half-and-half theories of quantum gravity. There are ways to evade both horns of the dilemma: adopting a no-collapse theory could preclude superluminal signalling in the first horn, and allowing for unobservable momentum non-conservation makes the result of the second horn something one could also live with. This particular argument, which is often repeated in the literature and in conversation, fails. Though we have not shown it, we would like to here register our skepticism that any argument in the style of Eppley and Hannah's could prove that it is inconsistent to have a world that is part quantum and part classical. Rosenfeld (1963) is right. Empirical considerations must create the necessity, if there is any, of quantizing the gravitational field.

Even so, the mere possibility of half-and-half theories does not make them attractive, and aside from their serious attention to the measurement problem, it is important to emphasize that they have not yielded the kind of powerful new insights that attract large research communities. Note too that most physicists arguing for the necessity of quantum gravity do not take the above argument as the main reason for quantizing the gravitational field. Rather, they usually point to a list of what one might call methodological points in favour of quantum gravity; see Chapter 2 for one such list. These points typically include various perceived weaknesses in contemporary theory, and find these sufficiently suggestive of the need for a theory wherein gravity is quantized. We have no qualms with this kind of argument, so long as it is recognized that the need for such a theory is not one of logical or (yet) empirical necessity.

### 1.2.2 The semiclassical theory

Finally, we should discuss a specific suggestion for a half-and-half theory due to Møller (1962) and Rosenfeld (1963), which appears – often as a foil – in the literature from time to time (see Chapters 2 and 13). This theory, 'semiclassical quantum gravity' (though any half-and-half theory is in some sense semiclassical), postulates first that the spacetime geometry couples to the expectation value of the stress–energy tensor:

$$G_{\mu\nu} = 8\pi \langle \hat{T}_{\mu\nu} \rangle_{\Psi}.$$  (1.4)

$G_{\mu\nu}$ is the *classical* Einstein tensor and $\langle \hat{T}_{\mu\nu} \rangle_{\Psi}$ is the expectation value for the stress–energy operator given that the *quantum* state of the matter fields is $\Psi$. Clearly this is the most obvious equation to write down given the Einstein field equation of classical general relativity, and given quantum rather than classical matter: $\langle T_{\mu\nu} \rangle_{\Psi}$ is the most obvious 'classical' quantity that can be coupled to $G_{\mu\nu}$.

Now, eqn. 1.4 differs importantly from the classical equation (eqn. 1.1), in that the latter is supposed to be 'complete', coding all matter–space and matter–matter dynamics: in principle, no other dynamical equations are required. Equation 1.4 cannot be complete in this sense, for it only imposes a relation between the spacetime geometry and the *expectation value* of the matter density, but typically many quantum states share any given expectation value. For example, being given the energy expectation value of some system as a function of time does not, by itself, determine the evolution of the quantum state (one also needs the standard connection between the Hamiltonian and the dynamics). So the semiclassical theory requires a separate specification of the quantum evolution of the matter fields: a Schrödinger equation on a curved spacetime. But this means that semiclassical quantum gravity is governed by an unpleasantly complicated dynamics: one for which we must seek 'self-consistent' solutions to two disparate of motions. Finding a model typically proceeds by first picking a spacetime – say a Schwarzschild black hole – and solving the Schrödinger equation for the matter fields on the spacetime. But this ignores the effect of the field on the spacetime, so next one wants to find the stress–energy tensor for this solution, and plug that into the semiclassical equation, to find corrections to the original spacetime. But then the assumption of a Schwarzschild solution no longer holds, and the Schrödinger equation must be solved for the new geometry, giving a new stress–energy expectation value to be fed back into the semiclassical equation, and so on and so on. What one of course hopes is that this process converges on a spacetime and matter field that satisfy both equations, but in the absence of such solutions it is not even clear that the equations are mutually consistent.

This lack of unity is one reason that physicists by-and-large do not take the semiclassical theory seriously as a 'fundamental' theory. The idea instead is that it can be used as an heuristic guide to some suggestive results in the absence of a real theory of quantum gravity; in the famous case mentioned, 'Hawking radiation' is produced by quantum fields in a Schwarzschild spacetime, and if one could feed the 'back reaction' into the semiclassical equation one would expect to find the black hole radius decreasing as it evaporated. Unfortunately, a number of technical problems face even this example, and though research is active and understanding is increasing, there are still no known interesting solutions with evolving spacetimes in four dimensions (see Wald 1994 for a discussion).

Things look even worse when one considers the collapse dynamics for quantum mechanics. On the standard interpretation, the unitary dynamics for a system must be supplemented with a collapse during certain 'measurement' interactions. If the semiclassical theory were complete, then not only would the unitary evolution of quantum matter need to be contained in eqn. 1.4, but so would the collapse. But – turning around an argument of Unruh's (1984) – it is impossible for eqn. 1.4 to contain a sharp collapse: the Einstein tensor is necessarily conserved, $G_{\mu;\nu}{}^{\nu} = 0$, but a collapse would lead to a discontinuity in the stress–energy expectation value, $\langle \hat{T}_{\mu}{}^{\nu} \rangle_{;\nu} \neq 0$. Nor does it seem plausible that eqn. 1.4 contains a smoother collapse: why should measuring events invariably produce the appropriate change in $G_{\mu\nu}$? In that case, the theory needs to be supplemented with a collapse dynamics, perhaps along the lines suggested by Pearle and Squires, but certainly constrained by the arguments considered in Section 1.2.1. In this context one point is worth noting: if the collapse mechanism is sharp then Unruh's argument shows either that eqn. 1.4 is incorrect, or that the LHS is not a tensor field everywhere, but only on local patches of spacetime, with 'jumps' in the field outside the patches.[7]

Our assessment of the semiclassical theory *as a candidate fundamental theory* is much the same as for half-and-half theories in general: while they are not impossible, if one weighs the insights they offer against the epicycles they require for their maintenance, then they do not appear to be terribly progressive. Certainly they take seriously the measurement problem, and so address arguments by, for instance, Penrose (1989) that gravity and collapse are interrelated. And certainly, the semiclassical theory has provided an invaluable and revealing tool for exploring the boundary between general relativity and quantum mechanics, yielding a picture of what phenomena – such as black hole thermodynamics – might be expected of a theory of quantum gravity. But on the other hand, arguing that half-and-half theories are fundamental involves more negotiating pitfalls than producing positive results. Thus it is not surprising that, although half-and-half theories have not been shown to be inconsistent, they are not the focus of most work in quantum gravity.

### 1.3 Approaches to quantum gravity

As we mentioned earlier, there currently is no quantum theory of gravity. There are, however, some more-or-less developed approaches to the field, of which the most actively researched fall into one of two broad classes, superstring theory and canonical quantum gravity. Correspondingly, most of the chapters in this book that deal explicitly with current research in the field discuss one or the other of these programmes; thus it will be useful to give here preparatory sketches of both classes.

(A warning: more space is devoted here to the canonical approach than to string theory. This does not mirror their relative popularities in the contemporary physics community, but reflects the greater development of philosophical discussion within the canonical programme.)

### 1.3.1 Superstrings

First, superstring theory, which is discussed in Chapters 5, 6, and 7. Superstring theory seeks to provide a unified quantum theory of all interactions, in which the elementary entities are one-dimensional extended objects (strings), not point particles. Superstring theory arose from work on the strong interaction in the late 1960s and early 1970s when it was shown that all the properties of a certain interesting model of the strong interaction (Veneziano's model) could be duplicated by a Lagrangian theory of a relativistic string. Though interesting, this idea did not really take off. In the mid-1970s, however, it was demonstrated that graviton–graviton (quanta of the gravitational field) scattering amplitudes were the same as the amplitudes of a certain type of closed string. This fact led to the idea that superstring theory is a theory of all forces, not only the strong interaction.

Superstrings, consequently, are not meant to represent single particles or single interactions, but rather they represent the entire spectrum of particles through their vibrations. In this way, superstring theory promises a novel and attractive 'ontological unification'. That is, unlike electrons, protons and neutrons – which together compose atoms – and unlike quarks and gluons – which together compose protons – strings would not be merely the smallest object in the universe, one from which other types of matter are composed. Strings would not be merely *constituents* of electrons or protons in the same manner as these entities are constituents of atoms. Rather, strings would be all that there is: electrons, quarks, and so on would simply be different vibrational modes of a string. In the mid-1980s this idea was taken up by a number of researchers, who developed a unitary quantum theory of gravity in ten dimensions.

A sketch of the basic idea is as follows (see Chapters 5 and 6 for more extensive treatments). Consider a classical one-dimensional string propagating in a relativistic spacetime, sweeping out a two-dimensional worldsheet, which we can treat as a manifold with 'internal' co-ordinates $(\sigma, \tau)$. One can give a classical treatment of this system in which the canonical variables are $X^\mu(\sigma, \tau)$, where $X^\mu$ ($\mu = 0, 1, 2, 3$) are the spacetime co-ordinates of points on the string worldsheet. When the theory is quantized, this embedding function is treated as a quantum field theory of excitations *on the string*, $\hat{X}^\mu(\sigma, \tau)$. A number of fascinating and suggestive properties arise as necessary consequences of such quantization.

- For instance, it was found that every consistent interacting quantized superstring theory necessarily includes gravity. That is, closed strings all have gravitons (massless spin-2 particles) in their excitation spectrum and open strings contain them as intermediate states.
- In addition, the classical spacetime metric on the background spacetime must satisfy a version of Einstein's field equations (plus small perturbations) if we (plausibly) demand that $X^\mu(\sigma, \tau)$ be conformally invariant on the string. Thus, arguably, general relativity follows from string theory in the appropriate limit.

• It is also necessary that the theory be supersymmetric (a symmetry allowing transformations between bosons and fermions), a property independently attractive to physicists seeking unified theories.

So part of the motivation for string theory comes from the feeling that it is almost too good to be a coincidence that the mere requirement of quantizing a classical string automatically brings with it gravity and supersymmetry. Of course, another notorious consequence of quantizing a string is that spacetime must have a dimension $n$, where $n > 4$ ($n = 26$ for the bosonic string, $n = 10$ when fermions are added). But even here it can be claimed that the extra spacetime dimensions do not arise by being put in artificially (as is the case, arguably, in Kaluza–Klein theory). Rather, they again arise as a necessary condition for consistent quantization.

More recently, since the mid-1990s, string theorists have explored various symmetries known as 'dualities' in order to find clues to the non-perturbative aspects of the theory. These dualities, they believe, hint that the five different existing classes of string theories may in fact just be aspects of an underlying theory sometimes called 'M-theory', where 'M' = 'magic', 'mystery', 'membrane', or we might add, 'maybe'. (In a sense, superstring theory can be seen as returning to its early foundations, since the ideas of Veneziano were based upon a [different] kind of 'duality' in his model of the strong interaction.) There have been many exciting results along these lines during the 1990s. These are briefly described in Chapters 2 and 5 (see also Witten 1997). In addition, important and impressive results concerning the thermodynamics of black holes have also been derived from the perspective of string theory, and these are described and discussed in Chapter 7.

It is perhaps legitimate to view the difference between the perturbative superstring theory of the mid-1980s and the non-perturbative M-theory of 1994–present as the difference between whether superstring theory is a new fundamental theory or not. The older superstring theory is, in a sense, not fundamental; quantum mechanics still was. Superstring theory was 'just' quantum mechanics applied to classical strings. (Of course there was no 'just' about it as regards the mathematical and physical insight needed to devise the theory!) But with today's string field theory, we see intimations of a wholly new fundamental theory in the various novel dualities and non-perturbative results. However, it is still troubling that so much of the success of string theory derives from the enormous mathematical power and elegance of the theory, rather than from empirical input. In his article, Weingard draws attention to this point, arguing that, unlike other theories which turned out to be successful, such as general relativity, string theory is not based on any obvious physical 'clues'. This article was written before the recent developments in M-theory, so we leave it to the reader to consider whether the situation has changed.

Another issue of philosophical significance, discussed in Chapters 5 and 6, concerns the nature of spacetime according to string theory. The original formulation of string theory was envisaged as an extension of perturbative quantum field theory from point particles to strings; thus, as we described earlier, strings were taken to carry the gravitational field on a flat background spacetime. (Part of the promise of this approach was that renormalization difficulties are at least rendered more tractable, and at best do not occur at all.) On this view, then, spacetime appears very much as it did classically: we have matter and forces evolving on an 'absolute', non-dynamical manifold. One disadvantage of this approach is easily overcome: as Witten describes, it is simple to generalize from the assumption of a flat background by inserting your metric of choice into the Lagrangian for the string field. This observation, plus the fact that conformal invariance for the field on the string demands that the Einstein equation be satisfied by the spacetime metric, leads Witten to propose that we should not see spacetime as an absolute background in string theory after all. Since spacetime is captured by the field theory on the string, 'one does not have to have spacetime any more, except to the extent that one can extract it from a two-dimensional field theory'. (In Chapter 6, Weingard makes a similar point in his discussion of the second quantization of string theory.) To support his claim Witten also describes a duality symmetry of string theory that identifies small circles with larger circles. The idea is that no circle can be shrunk beneath a certain scale, and so there is a minimum – quantum – size in spacetime, and so no absolute background continuum. More on this claim in the final section.

### 1.3.2 Canonical quantum gravity

In contrast to superstring theory, canonical quantum gravity seeks a non-perturbative quantum theory of only the gravitational field. It aims for consistency between quantum mechanics and gravity, not unification of all the different fields. The main idea is to apply standard quantization procedures to the general theory of relativity. To apply these procedures, it is necessary to cast general relativity into canonical (Hamiltonian) form, and then quantize in the usual way. This was (partially) successfully done by Dirac (1964) and (differently) by Arnowitt, Deser, and Misner (1962). Since it puts relativity into a more familiar form, it makes an otherwise daunting task seem hard but manageable. In the remainder of this section we will give an intuitive sketch of the steps involved in this process, but be aware that many (unsolved) difficulties lie in the way of its successful completion. The reader should also be warned that we introduce these ideas with an out-of-date formulation of the theory, namely, the geometrodynamical formulation. More sophisticated and successful formulations exist – notably, the Ashtekar variable and loop variable approaches – but as these do not by themselves significantly affect the philosophical issues facing canonical quantum gravity, we here confine ourselves to the simpler and more intuitive picture.

In the standard Hamiltonian formulation (all this material is covered more fully in Chapter 10), one starts with canonical variables – say the position, $x$, and momentum, $p$, of a particle – which define the appropriate phase space for the system. Given a point in the phase space – say the instantaneous position and momentum of the particle – the Hamiltonian, $H(x, p)$, for the system will generate a unique trajectory with respect to a time parameter. So, to apply this approach to general relativity the first job is to define the relevant variables. The intuitive picture sought is one in which a three-dimensional spatial manifold, $\Sigma$, evolves through an arbitrary time parameter $\tau$, so the natural thing is to decompose spacetime into space and time. In the geometrodynamical formulation, a spatial 3-metric $h_{ab}(x)$ on $\Sigma$ plays the role of the canonical position, and a canonically conjugate momentum $p^{ab}(x)$ is also defined (it is closely related to the extrinsic curvature of $\Sigma$ in the spacetime). The phase space for this system is thus the space of all possible 3-spaces and conjugate momenta, so the pair $(h, p)$ fix an instantaneous state of $\Sigma$. (It is worth noting that the foliation of spacetime implied by this procedure is at odds with the central tenets of general relativity and will be a source of difficulties further down the line.) Finally, one gives a Hamiltonian that generates trajectories through phase space in agreement with relativity: trajectories such that the stack of 3-spaces form a model of Einstein's field equation.

Canonical quantization of such a system then means (in the first place) finding an operator representation of the canonical variables obeying the canonical commutation relations, say, $[\hat{x}, \hat{p}] = i\hbar$, or in our case $[\hat{h}, \hat{p}] = i\hbar$. Usually one hopes that smooth (wave) functions, $\psi(x)$, on the canonical position space (configuration space) will carry such a representation (since $\hat{x} = x$ and $\hat{p} = i\hbar\,\partial/\partial x$ are operators on such functions satisfying the commutation relations). One finally obtains a quantum Hamiltonian operator by replacing the canonical variables in the classical Hamiltonian with their operator representations: $H(x, p) \rightarrow \hat{H}(\hat{x}, \hat{p})$. States of the quantum system are of course represented by the wave functions, and evolutions are generated by the quantized Hamiltonian via the Schrödinger equation: $\hat{H}\psi = i\hbar\,\partial\psi/\partial t$. If all this went through for general relativity, then we would expect states of quantum gravity to be wave functions $\psi(h)$ over the configuration space of possible – Riemannian – 3-metrics, $Riem(3)$, and physical quantities including the canonical variables to be represented as operators on this space.

The general relativistic Hamiltonian generates evolutions in the direction of the time that parameterizes the stack of 3-spaces, but there is no reason why this should be normal to any given point on a 3-space. So, if $T$ is a vector (field) in the direction of increasing stack time in the spacetime, and $n$ is a unit vector (field) everywhere normal to the 3-spaces, we can decompose $T$ into normal and tangential components, $T = Nn + \vec{N}$. $N$ is the 'lapse' function, coding the normal component of $T$, and $\vec{N}$ is the 'shift' vector (field), which is always tangent to the 3-space. Not surprisingly, when one works through the problem (writing down a Lagrangian, then using Hamilton's equations), the Hamiltonian for the system can be split into parts generating evolutions normal and tangent to the 3-spaces. Added together, these parts generate transformations in the direction of the time parameter. Thus, a standard Hamiltonian takes the form (for certain functions, $C^\mu$, of the canonical variables):

$$H = \int d^3x\,(C \cdot N + N^i C_i). \tag{1.5}$$

Variation of $h$ and $p$ yields six of the Einstein equation's ten equations of motion, and variation of the $N$ and $\vec{N}$ produces the so-called 'Hamiltonian' and 'momentum' constraints' (which hold at each point in spacetime):

$$C = C_i = 0. \tag{1.6}$$

In other words, not every point of our phase space actually corresponds to a (hypersurface of a) solution of general relativity, but only those for which $(h, p)$

satisfy eqn. 1.6: in the formulation general relativity is a *constrained Hamiltonian system*. As a consequence of eqn. 1.6, $H = 0$, though in the classical case at least, this does not mean that there is no dynamics: the first six equations of motion ensure that the 3-space geometry varies with time. It does however lead to some deep problems in the theory, both in the classical and quantum contexts.

The idea of a constraint is of course fairly straightforward: consider, for example, a free particle moving on a plane with freely specifiable values of position, $x$ and $y$, and of momentum, $p_x$ and $p_y$: there are four degrees of freedom. However, if confined to a circle, $x^2 + y^2 = a$, the particle must satisfy the constraint $xp_x + yp_y = 0$, which allows us to solve for one of the variables in terms of the others: the constrained system has only three degrees of freedom. Pictorially, the unconstrained particle's state may be represented anywhere in the four-dimensional phase space spanned by the two position and two momenta axes, but the constrained particle can only 'live' on the three-dimensional subspace – the 'constraint hypersurface' – on which the constraint holds. In this model the simplest way to approach the motion is to reparameterize the system to three variables in which the constraint is automatically satisfied: effectively making the constraint hypersurface the phase space.

Now, finding a constraint for a Hamiltonian system often (though not in the previous toy example) indicates that we are dealing with a gauge theory: there is some symmetry transformation between states in the phase space that leaves all dynamical parameters unchanged. The usual understanding is that since any physical quantities must 'make a difference' dynamically, all observables (physically real quantities) must be gauge invariant. (Note that this is a much stronger notion than a covariant symmetry, the idea that transformed quantities, though distinguishable, obey the same equations of motion.) Such systems are of course fundamental to contemporary field theory, since imposing local gauge invariance on a field requires introducing a 'connection', $A$, which allows comparisons of values of the field at infinitesimally separated points (working as the affine connection to allow differentiation of fields over spacetime). $A$ acts as a second field in the gauge invariant field equation, mediating interactions; in quantum field theory, the original field represents 'matter' and the connection field represents exchange particles, such as photons (see Redhead 1983). Note however that $A$ is an example of a gauge *non*-invariant quantity, so although it is crucial for understanding interactions, it contains unphysical degrees of freedom. This apparent paradox shows the subtleties involved in understanding gauge theories.[8]

We can apply these lessons to the Hamiltonian formulation of general relativity. The constraint appears to be connected to a symmetry, this time the general covariance of the theory, understood as diffeomorphism invariance. That is, if $\langle M, g, T \rangle$ represents a spacetime of the theory (where $M$ is a manifold, $g$ represents the metric field and $T$ the matter fields), and $D*$ is a smooth invertible mapping on $M$, then $\langle M, D * g, D * T \rangle$ represents *the very same* spacetime. Crudely, smooth differences in how the fields are arranged over the manifold are not physically significant.

It is vital to note that we have already reached the point at which controversial philosophical stances must be taken. As Belot and Earman explain, to understand diffeomorphism this way, as a gauge symmetry, is to take a stance on Einstein's infamous 'hole argument', and hence on various issues concerning the nature of spacetime; in turn, these issues will bear on the 'problem of time', introduced below. We will maintain the gauge understanding at this point, since it is fairly conventional among physicists (as several articles here testify).

Though the matter is subtle, if intuitively plausible, the momentum and Hamiltonian constraints (eqn. 1.6) are believed to capture the invariance of general relativity under spacelike and timelike diffeomorphisms respectively (e.g. Unruh and Wald 1989). As it happens (unlike the toy case) the constraints cause the Hamiltonian to vanish. This is not atypical of generally covariant Hamiltonian systems (see Belot and Earman's discussion of a parametrized free particle in Chapter 10 for an example), but what is atypical in this theory is that the Hamiltonian is *entirely* composed of constraints.

In fact, since things have been chosen nicely so that the momentum and Hamiltonian constraints are associated with diffeomorphisms tangent and normal to the 3-space, $\Sigma$, respectively, satisfying the former with a reparameterization is easy: instead of counting every $h$ on $\Sigma$ as a distinct possibility, we take only equivalence classes of 3-spaces related by diffeomorphisms to represent distinct states. This move turns our earlier configuration space into 'superspace', so we now want quantum states to be wave functions over superspace.

Once this move is made, all that is left of the Hamiltonian is the Hamiltonian constraint. But now a reparameterization seems out of the question, for the Hamiltonian constraint is related to diffeomorphisms in the time direction. If we try to form a state space in which all states related by temporal diffeomorphisms are counted as the same, then we have no choice but to treat whole spacetimes, not just 3-spaces, as states. Otherwise it just makes no sense to pose the question of whether the two states are related by the diffeomorphism. But in this case the state space consists, not of 3-geometries, but of full solutions of general relativity. And in this case there are no trajectories of evolving solutions, and the Hamiltonian and Schrödinger pictures no longer apply.

Despite this difficulty, it is still possible to quantize our system by (more or less) following Dirac's quantization scheme and requiring that the constraint equations be satisfied as operator equations, heuristically writing:

$$\hat{C}\Psi = \hat{C}_i\Psi = 0. \tag{1.7}$$

Since the momentum constraints are automatically satisfied in superspace, the focus in canonical quantum gravity is on

$$\hat{C}\Psi = 0, \tag{1.8}$$

the so-called *Wheeler–DeWitt equation*. This equation's interpretation has generated much controversy, as we shall describe in the next section. For now, be aware that very significant conceptual and formal difficulties confront the intuitive picture we have sketched: not least that there are no known solutions to the problem as constructed so far![9] There are, however, some solutions to this equation when one writes the theory in terms of Ashtekar's so-called 'new variables'. These variables have overcome many technical obstacles to the older canonical theory, and have greatly reinvigorated the canonical programme in the past decade (see Rovelli 1998b for a review).

Finally, and extremely speculatively, we note that both superstring theory and the canonical programme have evolved greatly from their initial formulations, and, as far as we are aware, it is possible that they are converging in some way: for example, perhaps some descendent of the Ashtekar formulation will turn out to be a realization of M-theory. That is, the future may reveal an analogy between the historical developments of quantum gravity and the Schrödinger and Heisenberg formulations of quantum mechanics. Were this to be the case, then one would expect fruitful new insights into all the issues raised here.

## 1.4 What quantum gravity and philosophy have to say to each other

Quantum gravity raises a multitude of issues interesting to philosophically minded thinkers. Physicists working in the field challenge some of our deepest assumptions about the world, and the philosophical tradition has a strong interest in many of these assumptions. In this final section we will describe a variety of topics in the subject that are of mutual concern to physicists and philosophers: some issues that bear on historically philosophical questions, and some new foundational issues. Our aim is to give philosophers a good outline of the problems that define the field, and to give physicists a sketch of how philosophers have investigated such issues – and of course to show how the arguments of our contributors fit into a broad dialogue concerning the foundations of quantum gravity. (Note that the papers are not exactly organized according to the following scheme, because many of them address several distinct issues.)

### 1.4.1 The demise of classical spacetime

- A number of the contributors make comments relevant to the 'fate of spacetime' in the quantum regime. Since we have already mentioned Witten's views, we will start there. He claims that, despite the original idea of strings propagating on a fixed background spacetime, spacetime arises entirely from the more fundamental two-dimensional conformal field on the string (an argument supported by duality symmetry). If correct (and if string theory is correct), then this view constitutes a considerable advance on our philosophical understanding of the nature of space. Views on the nature of space are as old as the idea of a general account of motion (see Huggett 1999): Plato and Descartes believed that matter and space were identical; Aristotle and other plenists often denied the existence of space by denying the vacuum of the atomists; and of course Newton and Leibniz were in famous opposition on the question of whether space was 'absolute' or 'relative'. The recent philosophical tradition (Friedman 1983 is especially influential here) has divided this question up in a number of ways: whether spacetime is dynamical or not; whether there is literally a manifold of points distinct from matter; and the meaning of 'relativity principles' as symmetries. Naturally, postgeneral relativity the answer to the first question is affirmative (though not without subtleties), but the other two topics are addressed by quantum gravity; the nature of the symmetries will come up later, for now we see the claim that in string theory spacetime is not distinct from matter (i.e. strings) but derives from it. The physical argument seems pretty straightforward, and it seems to us that philosophers should pick up this challenge: does Witten offer a sound answer to the problem of the nature of space, and how does it fit with historical proposals?

One oddity with this view is worth mentioning at once: the spacetime metric field that appears in the stringy Lagrangian must apparently be defined everywhere, not just on those points which the string occupies. Witten's spacetime seems to exist where matter and hence the fundamental two-dimensional field does not, so does it truely fade away?

- Canonical quantum gravity also has implications for the nature of spacetime, raised here by John Baez and Carlo Rovelli in Chapters 8 and 4 respectively. At first, it might seem that this approach should diverge from that of Witten by postulating a manifold as distinct from matter as that of general relativity, though subject to quantum effects: quantum spacetime should be to quantum matter as classical spacetime is to classical matter. However, as we shall mention below, it seems unlikely that canonical quantum gravity can be formulated in the space of 3-metrics as suggested earlier, and the significant advances in the theory of the last fifteen years have come from the 'loop' formulation due, *inter alia*, to Ashtekar, Rovelli, and Smolin, introduced here by Rovelli as a logical development of the insights of general relativity and quantum mechanics. The idea is that quantum gravity has as a basis (of quantum states) networks with spins values ($\frac{1}{2}, 1, \frac{3}{2} \ldots$) associated with the vertices. The picture may not seem intuitive, but if the nodes represent quantized regions, then suitable operators for geometrical quantities, such as volume, can be found, and so an understanding of the spacetime represented by such a state attained. (What is unknown at present is how the theory relates to general relativity in a classical limit.) If the loop basis is not unitarily equivalent to the 3-metric basis then (if it is correct) it too offers a picture in which spacetime is not fundamental, but a result of a more basic reality: in this case the spin network. Rovelli, in Chapter 4, claims that this too is a form of relationism.

- A closely related question is raised in Chapter 8 by Baez in his discussion of 'topological quantum field theory'. This is an approach to quantum gravity that gets away from a background spacetime by utilizing an analogy – brought out in 'category theory' – between the topological properties of a space and the quantum formalism. This enables one to construct models of quantum evolutions, involving topological change, satisfying a set of appropriate axioms, without worrying about the details of the dynamics of (quantum) geometry. The analogy is very suggestive, and may be a clue to uniting quantum mechanics and general relativity, but it comes at a price: the spacetimes in the theory have no local properties, such as a metric or causal structure, and so the theory cannot be the whole story. As Baez mentions, the ideas have been useful in canonical quantum gravity to study the dynamics in analogy to the Feynman approach of quantum field theory: one calculates sums over ways in which surfaces can interact by branching and joining – 'spin foams'.

An important philosophical question brought up here is how the causal structure of spacetime is to be built into this kind of theory. Causation is a perennial topic for philosophers: from Aristotle, to Hume and his sceptical descendants, who claim that causation is just 'constant conjunction' of some kind; to contemporary accounts in subjunctive terms – 'A caused B just in case if A *had not* happened then neither *would* B have happened' – or based on statistical considerations; to the understanding of causal structure in relativity. Apparently this is not a topic that is well understood in quantum gravity, but it is one that philosophers should address. Another question of philosophical interest here concerns the topology change that is the basis of the theory. Philosophers have rarely considered the exotic possibility of the topology of space changing with time, but it is a possibility with relevance to some traditional philosophical issues, for example Aristotle's, Descartes', and Kant's claimed 'unity of space' (see Callender and Weingard 2000).

- As we have just indicated, the fate of spacetime in many approaches to the subject is that classical spacetime structure (loosely, a semi-Riemannian metric on a continuous manifold) breaks down. In Chapter 2, Butterfield and Isham treat at greater length the consequences of the various programmes and speculations for the notion of spacetime. Common to these programmes and speculations is talk of the gravitational field in quantum gravity 'fluctuating'. But can this really make sense? In Chapter 3, Weinstein (a philosopher) takes this naive idea seriously, to see whether it can really hold water. He argues that it cannot without the theory failing to capture all observable gravitational phenomena. The main idea behind some of his critique is that fluctuations in the gravitational field imply fluctuations in the spatiotemporal, and hence causal, structure of the world. But it is hard to see how one can make sense of canonical commutation relations and hence quantize anything in the absence of a stable causal structure.

### 1.4.2 The nature of time

- Next, consider what is possibly the deepest of all philosophical puzzles, the nature of time (see, e.g. Le Poidevin and MacBeath 1993). Not only do certain programmes imply the breakdown of classical spacetime structure, but they also threaten to say something 'special' about time. Indeed, quantum gravity in all of its formulations seems forced to say something novel about this subject, for it must reconcile a conflict in the understanding of time between quantum theory and general relativity. Canonical quantum mechanics, since it is based on the Hamiltonian formulation, describes systems evolving with respect to a time parameter: either the preferred foliation of Galilean spacetime or – covariantly – the instantaneous hypersurfaces of an inertial frame. But general relativity is famously hostile to any such time parametrization (except in very special cases with nice symmetries). First of all, there is no such thing as time, *simpliciter*, in the theory, but rather a variety of time variables. There is the completely arbitrary co-ordinate time which, unlike time in Minkowski spacetime, has no metrical properties. There is also the proper time of an observer, but this cannot be extrapolated out to be the unique measure of time for all observers. And then there are various 'cosmic time' variables, such as Weyl's and Milne's definitions of cosmic time, though these are usually dependent upon special distributions of matter–energy or on special geometrical properties of the spacetime holding. In general relativity, there are plenty of cosmological solutions that do not even allow the possibility of spacetime being foliated by global spacelike hypersurfaces. Finally, the spacetime metric is dynamical in general relativity but non-dynamical in quantum mechanics. Thus, the two conceptions of time are very different. Quantum gravity, therefore, must say something about time, even if only that one of these two conceptions is fundamentally correct and the other only approximately right. Its verdict on this issue and others like it will be of interest to philosophers and physicists alike. In Chapter 2, Butterfield and Isham survey this problem (among others) and explain what some of the major programmes suggest is right.

- Some more specific problems connected with time arise in canonical quantum gravity. First, there is the notorious 'problem of time'. Since Belot and Earman discuss this in detail in Chapter 10, we will only briefly mention the problem. The problem is obvious enough: the Wheeler–DeWitt equation has no time dependence! Like a particle in a (non-degenerate) eigenstate of zero energy, the quantum state of the universe does not change, contrary to experience. (Note that this problem does not affect the classical theory: in that context a spacetime can evolve even if its Hamiltonian vanishes. It is only when we follow Dirac's prescription for the quantum interpretation of the constraints that trouble strikes.) Further, since the Wheeler–DeWitt equation must hold at all times, it holds before and after measurements. Thus there is no way to encode the information gained from a measurement back into the equation, which aggravates the first problem.

- Then, there is the related problem of observables. As we explained above, the natural interpretation of a gauge theory is that the gauge degrees of freedom do not correspond to physical transformations, and thus one concludes that all observable quantities must be gauge invariant, unchanged under the action of the gauge transformations (actually this is what is usually meant by a gauge symmetry). In formal terms this means classically that the Poisson bracket of any constraint and observable vanishes, $\{C, O\} = 0$; on quantizing according to the Dirac scheme, the operators corresponding to the constraint and observable must commute, $[\hat{C}, \hat{O}] = 0$. Now in general relativity we find that the Hamiltonian, $H$, which is supposed to generate evolutions is itself a constraint, and so commutes with any observable; but $\{H, O\} = 0$ and $[\hat{H}, \hat{O}] = 0$ mean that any observable quantity (its value or expectation value) is a constant of the motion. Equivalently, since the two parts of the Hamiltonian generate diffeomorphisms, all physical quantities must be diffeomorphism invariant. As a rule, constants of the motion tend to be pretty dull physical quantities. In fact, if $\Sigma$ is compact, the system has no known observables; if $\Sigma$ is open, then trivial quantities may be defined, but they are generally acknowledged to be useless to quantum gravity. In particular, since the spatial 3-metric presumably changes with time, it cannot be an observable (which leaves one wondering what the point is of promoting it to an operator). Of course this flies in the face of the conventional understanding of general relativity, which holds that the metric is observable.

Note that although the problem described afflicts both classical and quantum versions of the theory, it raises an additional conflict between quantum general relativity and the usual interpretation of quantum mechanics. Any quantum mechanical observable with time-dependent values will satisfy $[\hat{H}, \hat{Q}] \neq 0$, which means that there are no simultaneous eigenstates of $\hat{H}$ and $\hat{Q}$. But since $H$ is a constraint, the fundamental postulate of Dirac's quantization is that every state of the system is an eigenstate of $\hat{H}$: $\hat{H}\Psi = 0$. But then no possible states are eigenstates of such a $\hat{Q}$, and so the usual account of quantum measurement breaks down: there simply cannot be a collapse into a state of definite $Q$ if no such states exist!

Proposed solutions to these problems are compared and evaluated in detail by Belot and Earman in Chapter 10, and we think they are only half-joking when they divide them into 'Parmenidean' and 'Hericlitean' kinds. Parmenides (later followed by the more familiar Zeno) argued that all change was illusion, and the view associated with Rovelli (intimated in Chapter 4, but outlined more fully in Chapter 10 by Belot and Earman) has this character. According to Rovelli, all physical quantities are irreducibly relational, and hence timeless: for example, 'the clock read one as the mouse ran down' is a physical property, but it is not analysable into '(at t the clock read one) and (at t the mouse ran down)', since readings and positions at t are not physical. The idea of course is to bite the bullet, and accept that only diffeomorphism invariant quantities are physical, so there are only timeless truths.

Heraclites, on the other hand, argued for permanent flux, and the views proposed by Kuchař (e.g. 1992), among others, have this character. In the more extreme formulations they go against the spirit of general relativity and assume a preferred foliation, which gives physical significance to observable quantities that are time-specific. Kuchař's proposal is more subtle, seeking to find a meaningful time-parameter without violating the spirit of general relativity, but strong enough to deny that the scalar constraint has the same force as the vector constraint: quantities should be invariant under spatial diffeomorphisms, but not under timelike diffeomorphisms.

We would speculate that the problem of change is perhaps the oldest philosophical subject, and its solutions are the source of many metaphysical problems. The oldest version asks how change is possible at all: if A changes, then it is no longer the same, and hence no longer A, so A has not changed, but ceased to exist! In recent years the problem has been most focussed on the nature of identity – especially the nature of persistence through time – and the meaning of the spacetime view of the world: are all truths 'tenseless', or is some sense to be made of a 'specious present' (see Le Poidevin and MacBeath 1993 and references therein; Williams 1951 is an enjoyable classic article on this topic). Clearly, the views of Rovelli and Kuchař have crucial bearing on these arguments, and must be taken seriously by philosophers of time, though, as Belot and Earman point out in Chapter 10 , which (if either) of their insights is correct is something to be determined in part by the success of their approaches in solving physical problems.

### 1.4.3 The interpretation of general relativity

- Since diffeomorphism invariance is the symmetry underlying general relativity's general covariance, the interpretation of general covariance – a hotly disputed topic since the theory's inception – may be relevant to solving some of the above problems. Indeed, as many of the chapters in this volume make abundantly clear (especially Chapters 9 and 10), there most certainly is a connection between issues in quantum gravity and the interpretation of general covariance, and in particular,

Einstein's famous hole argument, which itself bears on the relational–substantival debate concerning space. In recent years this argument has attracted a great deal of attention among philosophers. Earman and Norton (1987) argued that the example shows that the 'manifold substantivalist' – one committed to the literal existence of a spacetime manifold distinct from matter – had to accept radical indeterminism: there are infinitely many models of general relativity that agree outside the hole, but disagree inside. Put another way (as Belot and Earman suggest in Chapter 10), the substantivalist must accept that the points of two manifolds can be compared, independently of the various fields on them; that is the meaning of their distinctness from matter. Substantivalists have replied in a variety of ways, some by arguing that while the comparison can be made without regard to matter fields, it cannot be made without regard for the metric field, others use the developments in our understanding of the logic of possibility to argue that diffeomorphism invariance is compatible with substantivalism. Since one's preferred solution to the problem of time bears on the meaning of diffeomorphism invariance, it is only natural to expect that it will have bearing on the hole argument and hence the nature of spacetime. (Looked at from another angle, as Belot and Earman again point out, the issues here are a very special case of the philosophical issues that arise in understanding gauge degrees of freedom, which are at once unphysical, but also seemingly essential for modern field theories. See also Redhead 1975.)

- Another issue, raised by Penrose in Chapter 13, suggests that there are problems with approaches in which there are quantum superpositions of spacetime, as, for instance, there would be in the intuitive approach to canonical quantum gravity we sketched in Section 1.3 (though it is less clear what to say about the loop approach). Penrose points out that general relativity's principle of general covariance seems at odds with the quantum mechanical principle of superposition. For suppose $|\alpha\rangle = |\phi(x)\rangle|\Psi\rangle$ represents a state in which a particle is localized in a spacetime with a sharp metric, and $|\beta\rangle = |\phi(x + a)\rangle|\Psi'\rangle$ represents the particle shifted and the appropriate new metric eigenstate. Now, we can imagine the state $|\alpha\rangle + |\beta\rangle$ representing an entangled system involving particle and gravitational field superpositions. Certainly such a state is distinct from either $|\alpha\rangle$ or $|\beta\rangle$, but how can these two states be distinct? Since the particles only differ by a displacement, we can suppose the metrics to as well, and so the spacetimes involved are diffeomorphic: given (one reading of) general covariance the two spacetimes and hence quantum states are one and the same.

  In Chapter 9, Julian Barbour tackles this issue (and the preceding one), arguing that there is a canonical way to identify points between slightly differing spacetimes; the key insight, he thinks, comes from a method of deducing the 'best match' between one relative configuration of particles and another. In the case of 3-geometries he claims this method is simply Hilbert's variational principle. This idea traces its origins to work in mechanics by Lagrange, Lange, and Mach, among others, and it has relevance to Machianism about spacetime and the hole argument.

- It is often thought that only quantum mechanics has a problem of interpretation. However, the above issues and others make it plausible that even general relativity has a problem of interpretation. The interpretation of general relativity is of

25

obvious importance to quantum gravity: it is important to understand properly the assumptions that go into relativity so that one can better appreciate what one can and cannot give up when creating a new theory. In this spirit, in Chapter 11 Harvey Brown and Oliver Pooley examine John S. Bell's (1976) paper 'How to Teach Special Relativity', and apply the lesson they learn from it regarding special relativity to general relativity. Bell's paper explains the Lorentz transformations dynamically, à la Lorentz: by showing how they can be derived from the structure of matter (in particular from electrodynamics) in relative motion. Einstein related this explanation of relativity to his own account, as kinetic theory explanations (so-called 'constructive theory' explanations) relate to thermodynamical explanations (so-called 'principle theory' explanations). Like Bell, Einstein claimed that Lorentz's constructive explanation was necessary for a proper understanding of special relativity. Brown and Pooley seek to extend this point to general relativity. They claim it clarifies the role of kinematics and dynamics in special and general relativity, as well as the role of rods and clocks in the two theories.

### 1.4.4 The interpretation of quantum mechanics

- Another connection between philosophy and quantum gravity involves the notorious measurement problem in quantum mechanics. Physicists and philosophers of science have both devoted much time and energy to discussing this topic. Is the measurement problem and the interpretation of quantum mechanics relevant to quantum gravity? In this volume we are fortunate enough to have sharply divided answers to this controversial question. In Chapter 4, Rovelli emphatically claims that there is no connection between the two, whereas Sheldon Goldstein and Stefan Teufel in Chapter 12 claim that the connection makes all the difference in the world, and that the failure to acknowledge it is responsible for many of the conceptual problems in the field, such as the problem of time.
- The Bohmian approach, here advocated by Goldstein and Teufel, also falls under this heading. In the case of particle quantum mechanics, Bohm's theory describes point bodies – 'beables' – whose definite motions are determined by their collective locations (not momentum) and the wave function, which is itself determined by the ordinary Schrödinger equation (not the particle locations). Goldstein and Teufel explain in Chapter 12 how this picture can be carried over to canonical quantum gravity, where the evolution of a definite 3-geometry – the beable – is determined by a wave function, which is itself determined by the Wheeler–DeWitt equation. This time, since the wave function is not representing the physical state but, as it were, driving the 3-geometry, the problem of time does not arise: there is no inconsistency in a stationary wave function leading to an evolving spacetime in this theory. Further, in Bohmian mechanics the usual quantum formalism for observables is not taken as a matter of fundamental postulate, but rather should emerge from analysis of experiments within the framework. In this case the problem of observables cannot get off the ground: whatever quantities can be shown to be measurable in experiment are observable. Finally, since physical quantities come directly from the beables, not the wave function, and since in a model of Bohmian quantum gravity there is only one stationary wave function, it also seems

that the need for an inner product is moot: the theory simply sidesteps one of the hardest technical problems of canonical quantum gravity (explained below). On the downside, despite many interesting applications of Bohmian quantum gravity to conceptual and physical problems – for example to the problem of time by Holland (1993), Callender and Weingard (1994, 1996), and many others; to black holes by Kenmoku et al. (1997); to the initial singularity by Callender and Weingard (1995); and to various cosmological models by Blaut and Kowalski-Glikman (1996) – Bohmian quantum gravity has not been developed as seriously as some other approaches, so it is difficult to say whether or not it has hard problems of its own.

- In Chapters 13 and 14, Roger Penrose and Joy Christian also see a connection between gravity and the measurement problem. Penrose develops the attractive idea that the gravitational field can be drafted in to help answer the measurement problem. The idea is that the gravitational field is the 'trigger' that stimulates collapses of quantum superpositions before they become macroscopic. In principle, this interpretation should give slightly (though currently unobserved) different results than standard quantum mechanics. We therefore have the exciting possibility of testing this collapse theory against the standard theory. Penrose, here and elsewhere, proposes some such experiments. But will they really succeed in testing his theory? Christian claims that one type of experiment will not work and proposes some others in its place, as well as greatly elaborating the conceptual position of Penrose's model within quantum gravity.

- Issues analogous to those concerning collapses arise from Hawking's (1974) celebrated results concerning the 'evaporation' of black holes (the most important achievement of the semiclassical theory). In a non-unitary wave function collapse, information about the prior state must be erased: one cannot invert the dynamics to reconstruct the original wave function. Similarly, information about a system's state falls into a black hole with the system, but – on the standard treatment – is neither contained in the thermal radiation given off as the black hole evaporates away nor in the black hole itself. Like measurement, it is hard to reconcile this 'information loss' with the unitary evolution required by quantum mechanics.

In Chapter 7, Unruh explains the ideas of black hole thermodynamics, and especially the significance of black hole entropy. He also uses an ingenious parallel between light transmission in spacetime and sound transmission in moving water (if a black hole is a region from which light cannot escape, then a 'dumb hole' is a region from which sound cannot escape, for instance because the water inside is moving in the opposite direction supersonically) to show that Hawking's result is insensitive to short length scale physics. However, it is such short length physics, and in particular string theory, that may offer a solution to the information loss problem according to recent proposals.

First of all, one can (in some very special cases) calculate a 'traditional' entropy for a black hole by counting the number of corresponding (in a loose sense) string states and taking their logarithm; the result is (in those special cases) in agreement with the Hawking-style calculation. This raises the possibility that the information about systems which fall into the black hole is in fact contained in the strings.

That is, knowing the details of the thermal radiation and the exact string state might suffice to reconstruct the in-falling state, as knowledge of the radiation and internal state of a hot poker would allow resconstruction of the heating process. Unruh criticizes this proposal, arguing that if the strings are inside the black hole then they cannot influence the thermal radiation in the appropriate way, but if they are outside the black hole then they cannot be suitably affected by the in-falling system. The only possiblity seems to be some kind of non-local interaction between strings inside and outside the hole, but this scenario is deeply unappealing.

### 1.4.5 The status of the wave function

- Another topic of interest to both physicists and philosophers concerns the status of quantum cosmology. Quantum cosmology, in contrast to quantum gravity, aims to provide a rationale for a particular choice of boundary conditions for our universe. The most familiar schemes of this kind are the famous Hartle and Hawking (1983) No-Boundary Proposal and the Vilenkin (1982) initial wave function of the universe. Some physicists conceive of quantum cosmology as a prescriptive enterprise: they believe that there are laws of quantum cosmology. Hawking, Hartle, and Vilenkin are all engaged in what we might call 'cosmogenic' theories. They are trying to find laws that uniquely determine the initial conditions of the universe. But is this search scientifically respectable? What possible justification could there be for the choice of a particular boundary condition – aside from the fact that it works, i.e. that it leads to what we observe? Any inductive inference from a single case is unwarranted, so how can we scientifically justify talk of laws and causes for the universe as a whole? While none of the contributors addresses this question directly, Goldstein and Teufel in Chapter 12 do speculate about the meaning of the universal wave function. And these considerations do raise the question of whether the usual probabilistic interpretation of the wave function can be carried over to the case of the universal wave function.
- Related problems arise in particular approaches to quantum gravity. To give an example, consider the problem of the interpretation of the Wheeler–DeWitt equation (eqn. 1.8). The naive suggestion is that it receive the same interpretation as does quantum mechanics. (By interpretation we here mean only rules for extracting predictions, not a solution to the measurement problem; until we know how to extract predictions from the theory it doesn't even have the *luxury* of having a measurement problem!) The naive interpretation would be to think of $\Psi(h, p)$ as a probability, which when squared yields the probability that an observer will measure the values $h$ and $p$. Even putting aside the question of where this observer of the whole universe is, we know this scheme cannot work (at least straightforwardly). This can be seen by comparing the Wheeler–DeWitt equation to the more familiar Klein–Gordon equation. If we impose the Klein–Gordon Hamiltonian as a restriction on the space of physical states, then the analogy between the resulting equation and the Wheeler–DeWitt equation is very strong. In fact, when the Wheeler–DeWitt equation is reduced to two degrees of freedom, it *is* this resulting equation. Now recall that the Klein–Gordon equation suffers from a very serious problem: its inner product $\langle \Psi_1 | \Psi_2 \rangle$ is not positive definite and therefore cannot be

used to define a probability. Thus we have a problem with defining a sensible inner product for the position representation (see, e.g. Teller 1995). The same problem threatens here, and is one motivation for loop quantization, for one can define an inner product on the space of 3-metrics in that approach.

The first sections of this introduction sought to clarify the motivations for the search for a theory of quantum gravity and give a useful outline of the two major programmes. In this final section we provided a sketch of many (but surely not all) of the major philosophical dimensions of quantum gravity. We hope that the reader, equipped with these ideas will now have sufficient context to tackle the chapters of this book, seeing how they relate to the broad physical and philosophical issues that surround the topic; if so, he or she will surely find them as exciting and illuminating as we have. We look forward to the debates that they will spark!

### Notes

Many thanks to John Baez, Jossi Berkovitz, Jeremy Butterfield, Carl Hoefer, Tom Imbo, Jeffrey Ketland and Carlo Rovelli for indispensable comments on this introduction. Portions of the Introduction are based on 'Why Quantize Gravity (or Any Other Field for that Matter)?', presented at the Philosophy of Science Association Conference (Callender and Huggett 2001).

1. In fact, it is these assumptions of classicality that do the work in the argument; the fact that the field of interest is not just classical, but also gravitational, does not play a role.
2. Obviously this thought experiment relies on some extreme idealizations, since we in effect postulate that the electron is screened off from all other fields so that no correlations are lost. But this does not detract from the point of the example: superluminal signalling that picks out a preferred foliation of spacetime must be impossible in principle, not just practice, if we take relativity seriously and literally.
3. A couple of other comments: (a) If Lefty found her box empty then Righty would still measure an effect, providing that the scattering is sensitive to the amplitude of the wave function (if not, one could still arrange signalling either by using enough boxes to ensure that Lefty will find an electron in one, or by using one of the other schemes mentioned above); (b) As Aharanov and Vaidman (1993) point out, this kind of arrangement will not permit signalling with their 'protective observations', though they are similar in allowing measurements of the wave function without collapse.
4. Of course, one might be concerned that such *ad hoc* fine-tuning of parameters is an indication that we are reaching a 'degenerating' phase of the spontaneous localization program, adding epicycles to save the theory. But this judgement may be premature: Pearle and Squires (1996) suggest how some such parameters may be derived.
5. Alternately, one might try to maintain momentum conservation *on average* (as Brown and Redhead 1981, footnote 21, suggest) in collapses. That is, one could seek to complete the collapse dynamics of QM in such a way that the expectation value for momentum was always conserved.
6. If gravitational waves do cause quantum jumps, then a wave empinging on the Earth could clearly have disastrous consequences if it were sufficiently powerful to collapse every piece of matter!
7. One issue that has captured a great deal of attention in the semiclassical theory is the so-called 'loss of information' problem (e.g. Belot, Earman, and Ruetsche 1999): as the black hole evaporates away, there is a transition from a pure to mixed state for the matter fields, reminiscent of the collapse in measurement. While this has worried many, it is not really so surprising given what we have been saying: no unitary evolution can produce such a transition, but in the model one is effectively invoking eqn. 1.4 as a second – non-unitary – equation of motion. It is really just another reflection of the point that in a half-and-half approach, one must be careful about how to include collapses.

8. So too does comparison with our toy example of a constraint. For instance, if one takes electromagnetism written in terms of the gauge potential $A$, and attempts to remove the unphysical degrees of freedom by reparameterizing with gauge invariant quantities, such as the magnetic field, $B = \nabla \times A$, one introduces non-locality into the theory (as shown by the notorious Aharonov–Bohm effect).

9. We should point out that there are reasons to wonder whether this intuitive picture of 3-metrics evolving on superspace is anything other than a metaphor. The 3-metric does not weave its path between Cauchy hypersurfaces like point particles in classical dynamics do. Wheeler's so-called 'Thick Sandwich' conjecture is false: that given any two 3-metrics in superspace there exists a spacetime between them such that they arise as induced metrics on two disjoint Cauchy surfaces. Different foliations between the two surfaces changes the curve between them. And even the Thin Sandwich conjecture – that given a point and tangent vector in superspace, there is a unique spacetime realizing the initial condition – has only limited applicability. See Bartnik and Fodor (1993) for more.

On the positive side, note that Christian (1997) shows how to exactly quantize a simpler spacetime theory – Newton–Cartan theory – along these lines.

# Part I

# Theories of Quantum Gravity and their Philosophical Dimensions

# 2 Spacetime and the philosophical challenge of quantum gravity

Jeremy Butterfield and Christopher Isham

## 2.1 Introduction

### 2.1.1 Prologue

Any branch of physics will pose various philosophical questions: for example about its concepts and general framework, and the comparison of these with analogous structures in other branches of physics. Indeed, a thoughtful consideration of any field of science leads naturally to questions within the philosophy of science.

However, in the case of quantum gravity we rapidly encounter fundamental issues that go well beyond questions within the philosophy of science in general. To explain this point, we should first note that by 'quantum gravity' we mean any approach to the problem of combining (or in some way 'reconciling') quantum theory with general relativity.[1] An immense amount of effort has been devoted during the past forty years to combining these two pillars of modern physics. Yet although a great deal has been learned in the course of this endeavour, there is still no satisfactory theory: rather, there are several competing approaches, each of which faces severe problems, both technical and conceptual.

This situation means that there are three broad ways in which quantum gravity raises philosophical questions beyond the philosophy of science in general.

(1) Each of the 'ingredient theories' – quantum theory and general relativity – poses significant conceptual problems in its own right. Since several of these problems are relevant for various topics in quantum gravity (as Chapter 8 of this volume bears witness), we must discuss them, albeit briefly. We will do this in Section 2.2, to help set the stage for our study of quantum gravity. But since these problems are familiar from the literature in the philosophy, and foundations, of physics, we shall be as brief as possible.

(2) The fundamentally disparate bases of the two ingredient theories generate major new problems when any attempt is made to combine them. This will be the main

focus of this chapter. But even to summarize these new conceptual problems is a complicated and controversial task: complicated because these problems are closely related to one another, and to the technical problems; and controversial because what the problems are, and how they are related to one another, depends in part on problematic matters in the interpretation of the ingredient theories. Accordingly, the main task of this chapter will be just to give such a summary; or rather, part of such a summary – roughly speaking, the part that relates to the treatment of spacetime.[2] We undertake this task in Section 2.3 and the following sections.

(3) The contrast between our lacking a satisfactory theory of quantum gravity and our having supremely successful ingredient theories, raises questions about the nature and function of philosophical discussion of quantum gravity. It clearly cannot 'take the theory as given' in the way that most philosophy of physics does; so how should it proceed? Though we are cautious about the value of pursuing meta-philosophical questions, we think this question deserves to be addressed. We do so in the course of Section 2.1.2 (where we also emphasize how unusual quantum gravity is, as a branch of physics).

So much by way of gesturing at the entire scope of 'philosophy of quantum gravity'. In the rest of this subsection, we shall make some brief general comments about the source of the conceptual problems of quantum gravity, and thereby lead up to a more detailed prospectus.

Despite the variety of programmes, and of controversies, in quantum gravity, most workers would agree on the following, admittedly very general, diagnosis of what is at the root of most of the conceptual problems of quantum gravity. Namely: general relativity is not just a theory of gravity – in an appropriate sense, it is also a theory of spacetime itself; and hence a theory of quantum gravity must have something to say about the quantum nature of space and time. But though the phrase 'the quantum nature of space and time' is portentous, it is also very obscure, and opens up a Pandora's box of challenging notions.

To disentangle these notions, the first thing to stress is that – despite the portentous phrase – there is, in fact, a great deal *in common* between the treatments of space and time given by the ingredient theories, quantum theory and general relativity. Specifically, they both treat space and time as aspects of spacetime, which is represented as a four-dimensional differentiable manifold, while the metrical structure of spacetime is represented by a Lorentzian metric on this manifold.

In view of this, one naturally expects that a theory of quantum gravity will itself adopt this common treatment, or at least its main ingredient, the manifold conception of space and time. Indeed, we will see in Section 2.4 that the three main research programmes in quantum gravity do accept this conception. So returning to our phrase, 'the quantum nature of space and time': although 'quantum' might suggest 'discrete', the 'quantum nature of space and time' need not mean abandoning a manifold conception of space and time at the most fundamental level. And consequently, the clash mentioned above – between the disparate bases of the ingredient theories, quantum theory and general relativity – need not be so straightforward as the contradiction between discreteness and continuity. As we shall see in Section 2.3 and following sections, the clash is in fact both subtle and multi-faceted.

But we will also see in Section 2.4 that in various ways, and for various reasons, these programmes do not accept all of the common treatment, especially as regards the dimensionality and metric structure of spacetime: the main difference being their use of some type of quantized metric. Furthermore, the two main *current* programmes even suggest, in various ways, that the manifold conception of spacetime is inapplicable on the minuscule length–scales characteristic of quantum gravity.

So the situation is curious: although the ingredient theories have much in common in their treatments of space and time, this common treatment is threatened by their attempted unification. This situation prompts the idea of departing more radically from the common treatment. Besides, quite apart from the challenges to the manifold conception that come from the programmes in Section 2.4, other quantum gravity programmes reject this conception from the outset. We explore these ideas in Section 2.5.

Here it is helpful to distinguish two general strategies for going beyond the common treatment, the first more specific than the second. First, one can quantize a classical structure which is part of that treatment, and then recover it as some sort of classical limit of the ensuing quantum theory. Second, and more generally, one can regard such a structure as phenomenological, in the physicists' sense of being an approximation, valid only in regimes where quantum gravity effects can be neglected, to some other theory (not necessarily a quantum theory). In more philosophical jargon, this second strategy is to regard the classical structure as emergent from the other theory – though here 'emergence' must of course be understood as a relation between theories, not as a temporal process. As we will see, both these strategies can be applied either to metrical structure (as in the programmes in Section 2.4), or to structures required just by the manifold conception, such as topology (Section 2.5).

But whichever of these two strategies one adopts, there are clearly two rather different ways of thinking about the relation between the familiar treatment of space and time – in common between the ingredient theories – and the treatment given by the, as yet unknown, theory of quantum gravity (with or without a manifold at the fundamental level). First, one can emphasize the emergence of the familiar treatment: its being 'good enough' in certain regimes. We adopt this perspective in a complementary essay.[3] Second, one can emphasize instead 'the error of its ways'. That is, one can emphasize how quantum gravity suggests limitations of the familiar treatment (though the conclusions of the examination will of course be tentative, since we have no satisfactory theory of quantum gravity). That is the perspective of this paper.[4]

In more detail, our plan is as follows. First we make some general orienting remarks about philosophy of quantum gravity (Section 2.1.2) and realism (Section 2.1.3). We then briefly review the bearing on quantum gravity of the conceptual problems of the ingredient theories: quantum theory (Section 2.2.1), and more briefly, general relativity (Section 2.2.2). In Section 2.3, we introduce the enterprise of quantum gravity proper. We state some of the main approaches to the subject; summarize some of the main motivations for studying it (Section 2.3.1); and introduce some of the conceptual aspects that relate closely to spacetime: the role of diffeomorphisms (Section 2.3.2) and the problem of time (Section 2.3.3).

In Section 2.4, we first set up the four topics in terms of which we will survey three well-developed research programmes in quantum gravity. The topics are: (i) the extent to which the programme uses standard quantum theory; (ii) the extent to which it uses standard spacetime concepts; (iii) how it treats spacetime diffeomorphisms; and (iv) how it treats the problem of time. After some historical remarks (Section 2.4.2), we turn to our three research programmes: the old particle-physics approach (Section 2.4.3), superstring theory (Section 2.4.4), and canonical quantum gravity (Section 2.4.5). (We also briefly treat a distinctive version of the latter, the Euclidean programme.) By and large, these three programmes are 'conservative' in the sense that although they in some way quantize metric structure, they use standard quantum theory, and they treat the spacetime manifold in a standard way. But in various ways, they also suggest that this treatment of spacetime is phenomenological. To mention two obvious examples: in the superstring programme, there are suggestions that many *different* manifolds play a role. And in the main current version of canonical quantum gravity, there are suggestions that quantities such as area and volume are quantized; and that the underlying structure of space or spacetime may be more like a combinatorial network than a standard continuum manifold.

This situation prompts the idea of a quantum gravity programme that *ab initio* goes beyond the standard treatments of spacetime and/or quantum theory. So in Section 2.5, we discuss some of these more radical ideas: quantizing spacetime structures other than the metric, or regarding such structures as phenomenological – where this could involve abandoning the manifold conception in favour of a variety of novel mathematical structures.

Finally, a caveat about the scope of the chapter. Even for the well-established research programmes, we shall omit – or mention only in passing – many of the more recent ideas. This is partly because of lack of space, and partly because technical ideas need to acquire a certain degree of maturity before philosophical reflection becomes appropriate. Similarly, we will not mention many of the more speculative ideas that have emerged over the years. Instead, we have deliberately limited ourselves to some of the well-established ideas in a few of the well-established research programmes, and to just a few of the more speculative ideas.

### 2.1.2 No data, no theory, no philosophy?

Before embarking on this detailed analysis of the conceptual challenge of quantum gravity, it is important to consider briefly the general idea of philosophical discussion of quantum gravity: more pointedly, to defend the idea, in the face of the fact that there is far from being a universally agreed theory of quantum gravity! So in this subsection, we will point out the peculiarity of quantum gravity in comparison with other branches of theoretical physics, and discuss how this affects the way in which one can write about the subject from a philosophical perspective. This will lead to some discussion of the philosophical positions we intend to adopt, especially as regards realism.

The most obvious peculiarity is a dire lack of data. That is, there are no phenomena that can be identified unequivocally as the result of an interplay between general relativity and quantum theory[5] – a feature that arguably challenges the right of quantum gravity to be considered as a genuine branch of science at all!

This lack of obvious empirical data results from a simple dimensional argument that quantum gravity has a natural length scale – the Planck length defined using dimensional analysis as[6] $L_P := (G\hbar/c^3)^{1/2}$ – and this is extremely small: namely $10^{-35}$ m. By comparison: the diameters of an atom, proton, and quark are, respectively, about $10^{-10}$ m, $10^{-15}$ m, and $10^{-18}$ m. So the Planck length is as many orders of magnitude from the (upper limit for) the diameter of a quark, as that diameter is from our familiar scale of a metre! The other so-called 'Planck scales', such as the Planck energy and the Planck time, are equally extreme: the Planck energy $E_P$ has a value of $10^{22}$ MeV, which is far beyond the range of any foreseeable laboratory-based experiments; and the Planck time (defined as $T_P := L_P/c$) has a value of about $10^{-42}$ s.

These values suggest that the only physical regime where effects of quantum gravity might be studied directly – in the sense that *something* very specific can be expected – is in the immediate post big-bang era of the universe – which is not the easiest thing to probe experimentally! This problem is compounded by the fact that, modulo certain technical niceties, *any* Lorentz-invariant theory of interacting spin-2 gravitons with a conserved energy–momentum tensor will yield the same low energy scattering amplitudes as those obtained from a perturbative expansion of the Einstein Lagrangian. Thus different quantum gravity theories might only reveal their differences empirically at very high energies.

This lack of data has implications for both physics and philosophy. For physics, the main consequence is simply that it becomes very difficult to build a theory: witness the fact that a fully satisfactory quantum theory of gravity remains elusive after forty years of intense effort. Of course, a great deal has been learnt about the features that such a theory might possess, albeit partly by eliminating features that are now known not to work. A good example is the perturbative unrenormalizability of the particle-physics approach to quantum gravity (see Section 2.4.3).

But there is more to the difficulty of theory-construction than just the lack of data: and this 'more' relates to philosophy in two ways. Firstly, this difficulty is partly due to conceptual problems which clearly bear on philosophical discussions of concepts such as space, time, and matter. Here we have in mind both kinds of conceptual problem, listed as (1) and (2) in Section 2.1.1 (see pp. 33–34): those arising from the disparateness of the bases of general relativity and quantum theory, and also problems about each of these theories in themselves. To take an obvious example of the latter, quantum gravity is usually taken to include quantum cosmology; and here, the idea of a 'quantum state of the universe' immediately confronts conceptual problems about quantum theory such as the meaning of probability, and the interpretation of the quantum state of a closed system – in this case, the universe in its entirety.[7] And as we shall see in Section 2.2.2, such examples do not only come from quantum theory, with its notorious conceptual problems. They also come from general relativity – whose foundations are murkier than philosophers commonly take them to be. To sum up: the difficulty of theory-construction is partly due to conceptual problems (and is thereby related to philosophy), not just lack of data.

Secondly, theory-construction is difficult because there is not even agreement on what *sorts* of data a quantum theory of gravity would yield, if only we could get access to them! More precisely, the dimensional argument discussed above suggests that only phenomena at these very small distances, or high energies, would exhibit quantum gravity effects; which implies that the main application of the quantum theory of gravity may be to the physics of the very early universe. However, this argument rests on the assumption that any physically measurable quantity can be expressed as a power series in a number like $E/E_P$ (where $E$ is some characteristic energy scale for an experiment), so that the quantity's predicted value is tiny in any experiment that probes energy-scales that are much lower than $E_P$. But experience elsewhere in quantum field theory suggests that 'non-perturbative effects' could also occur, in which the predicted values of certain measurable quantities are not analytic functions of the coupling constant, and this totally changes the argument: for example, if $x$ is a very small number, then $|\log x|$ is very large! But, in the case of quantum gravity it is anybody's guess what *sort* of effects would be exhibited – and therefore, what sort of predictions the envisaged quantum theory of gravity is meant to give. For example, for all we know, it might predict the masses of all the elementary particles.

This uncertainty puts us in a 'double bind'. On the one hand, it is ferociously difficult to find the theory without the help of data, or even an agreed conception of the *sort* of data that would be relevant. On the other hand, we can only apply our present theories to (and get evidence from) regimes well way from those determined by the Planck scale; and we cannot judge what phenomena might be relevant to a theory of quantum gravity, until we know what the theory is.

In this predicament, theory-construction inevitably becomes much more strongly influenced by broad theoretical considerations, than in mainstream areas of physics. More precisely, it tends to be based on various *prima facie* views about what the theory *should* look like – these being grounded partly on the philosophical prejudices of the researcher concerned, and partly on the existence of mathematical techniques that have been successful in what are deemed (perhaps erroneously) to be closely related areas of theoretical physics, such as non-abelian gauge theories. In such circumstances, the goal of a research programme tends towards the construction of abstract theoretical schemes that are compatible with some preconceived conceptual framework, and are internally consistent in a mathematical sense.

This situation does not just result in an extreme 'underdetermination of theory by data', in which many theories or schemes, not just a unique one, are presented for philosophical assessment. More problematically, it tends to produce schemes based on a wide range of philosophical motivations, which (since they are rarely articulated) might be presumed to be unconscious projections of the chthonic psyche of the individual researcher – and might be dismissed as such! Indeed, practitioners of a given research programme frequently have difficulty in understanding, or ascribing validity to, what members of a rival programme are trying to do. This is one reason why it is important to uncover as many as possible of the assumptions that lie behind each approach: one person's 'deep' problem may seem irrelevant to another, simply because the starting positions are so different.

This situation also underlines the importance of trying to find some area of physics in which any putative theory could be tested directly. A particularly important question in this context is whether the dimensional argument discussed above can be overcome, i.e. whether there are measurable quantum gravity effects well below the Planck scales; presumably arising from some sort of non-perturbative effect. However, the existence of such effects, and the kind of phenomena which they predict, are themselves likely to be strongly theory-dependent.

It follows from all this that the subject of quantum gravity does not present the philosopher with a conceptually or methodologically unified branch of physics, let alone a well-defined theory; but instead with a wide and disparate range of approaches. We must turn now to discussing what this situation implies about the scope and possible topics of philosophical discussion.

As we conceive it, philosophy of physics is usually concerned with either: (i) meta-physical/ontological issues, such as the nature of space, or physical probability; or (ii) epistemological/methodological issues, such as underdetermination of theories, or scientific realism, with special reference to physics. And in both areas (i) and (ii), the discussion is usually held in focus by restricting itself to a reasonably well-defined and well-established physical theory; or at worst, to a small and homogeneous set of reasonably well-defined rival theories. Hence the image of philosophy of physics as a Greek chorus, commenting *post facto* on the dramatic action taking place on a well-defined stage. But, as we have seen, this kind of role is not available in quantum gravity, where there are no reasonably well-defined, let alone well-established, theories. So should philosophical discussion simply wait until the subject of quantum gravity is much better established?

We think not. For there are at least three (albeit overlapping) ways in which to pursue the subject 'philosophy of quantum gravity'; each of them valuable, and each encompassing many possible projects. The first two are straightforward in that all will agree that they are coherent endeavours; the third is more problematic. This chapter will be an example of the second way. Broadly speaking, they are, in increasing order of radicalism:

(1) One can undertake the 'normal' sort of philosophical analysis of some sufficiently specific and well-defined piece of research in quantum gravity; even though the price of it being well-defined may be that it is too specific/narrow to warrant the name 'theory' – so that, in particular, it would be unwise to give much credence to the ontological or interpretative claims it suggests. An example might be one of the specific perturbative approaches to string theory – for example, the type II-B superstring – which recent research in the area suggests might merely be one particular perturbative regime of an underlying theory ('M-theory') whose mathematical and conceptual structures are quite different from those of a quantized loop propagating in a spacetime manifold. This sort of analysis is certainly valuable, even if the limitations of any specific example must make one wary of any ontological or interpretative suggestions which arise. For examples of such analyses, see Chapters 8–11.

(2) One can try to relate a *range* of conceptual problems about general relativity, quantum theory, and their having disparate bases, to a *range* of approaches or research programmes in quantum gravity.

To make this endeavour manageable, one must inevitably operate at a less detailed level than in (1) above. The hope is that despite the loss of detail, there will be a compensating value in seeing the overall pattern of relationships between conceptual problems and mathematical/physical approaches to quantum gravity. Indeed, one can hope that such a pattern will be illuminating, precisely because it is not tied to details of some specific programme that may be on the proverbial hiding to nowhere.

Such a pattern of relationships can be envisioned in two ways, according to which 'side' one thinks of as constraining the other. One can think of problems and ideas about general relativity and quantum theory as giving constraints on – or heuristic guides to our search for – quantum gravity. (Of course, we harbour no illusions that such conceptual discussion is a *prerequisite* for successful theory-construction – of that, only time will tell.) And vice versa, one can think of quantum gravity as constraining those problems and ideas, and even as suggesting possible changes to the foundations of these constituent theories. In what follows, we shall see examples of both kinds of constraint.

(3) One can try to study quantum gravity in the context of some traditional philosophical ideas that have nothing to do with the interpretation of general relativity and quantum theory per se – for example, traditional concepts of substance and attribute. Again, one can think of such a relation in two ways: the philosophical idea giving constraints on quantum gravity; and vice versa, quantum gravity reflecting back on the philosophical idea.

We admit to finding this endeavour alluring. But, again, we harbour no illusions that such traditional philosophical ideas are *likely* to be heuristically helpful, let alone a prerequisite, for theory-construction. Similarly, one must be tentative about constraints in the opposite direction; i.e. about the idea that a traditional philosophical position could be 'knocked out' by a quantum gravity proposal, where 'knocked out' means that the position is shown to be, if not false, at least 'merely' phenomenological, or approximately true, in ways that philosophers tend not to realize is on the cards. Since no quantum gravity proposal is well-established, any such 'knock-out' is tentative.

This chapter will exemplify type (2) above, although we emphasize that we are sympathetic to type (3). More generally, we are inclined to think that in the search for a satisfactory theory of quantum gravity, a fundamental reappraisal of our standard concepts of space, time, and matter may well be a necessary preliminary. Thus we are sceptical of the widespread idea that at the present stage of quantum gravity research, it is better to try first to construct an internally consistent mathematical model and only then to worry about what it 'means'. But such a reappraisal is fiercely hard to undertake; and accordingly, this chapter adopts 'the middle way' – type (2).

A final remark about the various ways to pursue 'philosophy of quantum gravity'. Although type (2) involves, by definition, surveying ideas and approaches, this by no means implies that it encompasses a single project, or even that it encompasses only one project focussing on the nature of space and time. For example, here is an alternative project, relating to traditional positions in the philosophy of geometry.

The immense developments in pure and physical geometry from Riemann's *habilitationsschrift* of 1854 to the establishment of general relativity, led to a transformation of the philosophy of geometry beyond recognition. In particular, Kant's apriorism fell by the wayside, to be replaced by empiricism and conventionalism of various stripes. With this transformation, the idea that at very small length–scales, space might have a non-manifold structure became a 'live option' in a way that it could not have been while Kant's influence held sway in its original form. Yet, in fact, this idea has had only a small role in the philosophy of geometry of the last

150 years – for the perfectly good reason that no significant physical theory took it up. (Its main role is via the view – endorsed by, for example, Grünbaum – that in a discrete space, but not a manifold, the metric is, or can be, intrinsic, and thereby non-conventional.)

But nowadays, there are several 'unconventional' approaches to quantum gravity that postulate a non-manifold structure for spacetime; and even in the more conventional approaches, which do model space or spacetime with a differentiable manifold, there are often hints of a discrete structure that lies beneath the continuum picture with which one starts. For example, in the canonical programme, area and volume variables become discrete; and in superstring theory there are strong indications that there is a minimal size for length.

These proposals prompt many questions for philosophers of geometry; the obvious main one being, how well can the traditional positions – the various versions of empiricism and conventionalism – accommodate such proposals? We will not take up such questions here, though we like to think that this chapter's survey of issues will help philosophers to address them.

### 2.1.3 Realism?

Finally, we should briefly discuss the bearing of our discussion on the fundamental questions of realism. Thus it is natural to ask us (as one might any authors in the philosophy of physics): Does the discussion count for or against realism, in particular scientific realism; or does it perhaps presuppose realism, or instead its falsity?

Our answer to this is broadly as follows. We will write as if we take proposals in quantum gravity realistically, but in fact our discussion will not count in favour of scientific realism – nor indeed, against it. This lack of commitment is hardly surprising, if only because, as emphasized in Section 2.1.2, quantum gravity is too problematical as a scientific field, to be a reliable test-bed for scientific realism. But we will fill out this answer in the rest of this subsection. In short, we will claim that: (i) we are not committed to scientific realism; and (ii) there is a specific reason to be wary of reifying the mathematical objects postulated by the mathematical models of theoretical physics. We shall also make a comment relating to transcendental idealism.

#### 2.1.3.1 *Beware scientific realism*

Scientific realism says, roughly speaking, that the theoretical claims of a successful, or a mature, scientific theory are true or approximately true, in a correspondence sense, of a reality independent of us. So it is a conjunctive thesis, with an ontological conjunct about the notion of truth as correspondence, and an independent reality; and an epistemic conjunct of 'optimism' – about our mature theories 'living up' to the first conjunct's notions.

Obviously, discussions of quantum gravity (of any of the three types of Section 2.1.2) need not be committed to such a doctrine, simply because, whatever exactly 'successful' and 'mature' mean, quantum gravity hardly supplies us with such theories! That is: even if one endorses the first conjunct of scientific realism, the second conjunct does not apply to quantum gravity. So there is no commitment, whatever one's view of the first conjunct.

But there is another way in which we might seem to be committed to scientific realism: namely, through our treatment of the 'ingredient theories', quantum theory and general relativity. In Section 2.2 and following sections we will often write about the interpretation of these theories, in an ontological (rather than epistemological or methodological) sense; as does much current work in the philosophy of physics. For example, we will mention the so-called 'interpretations' of quantum theory (Copenhagen, Everettian, pilot-wave, etc.), which are in fact ontologies, or world-views, which the philosopher of quantum theory elaborates and evaluates. Similarly, as regards general relativity for example, we will sometimes write about the existence of spacetime points as if they were objects.

But it should not be inferred from our writing about these topics in this way that we are committed to some form of scientific realism. There is no such entailment; for two reasons, one relating to each of scientific realism's two conjuncts. The first, obvious reason concerns the epistemic optimism of scientific realism. Clearly, elaborating and evaluating ontologies suggested by scientific theories is quite compatible with denying this optimism.

The second reason is perhaps less obvious. We maintain that such elaboration and evaluation of ontologies involves no commitment to a correspondence notion of truth, or approximate truth, characteristic of realism. This can seem surprising since for philosophers of physics, 'electron' and 'spacetime point' come as trippingly off the tongue, as 'chair' and other words for Austin's 'medium-sized dry goods' come off all our tongues, in everyday life. And this suggests that these philosophers' account of reference and truth about such topics as electrons is as realist, as is the account by the so-called 'common-sense realist' of reference to chairs, and of the truth of propositions about chairs. But the suggestion is clearly false. Whatever general arguments (for example, about ontological relativity) can be given against realist accounts of reference and truth in regard to chairs (and of course, rabbits and cats – Quine's and Putnam's 'medium-sized wet goods'!) can no doubt also be applied to electrons and spacetime points. Indeed, if there is to be a difference, one expects them to apply with greater, not lesser, force; not least because (at least, in the case of the electron) of the notorious difficulty in understanding a quantum 'thing' in any simple realist way.

Here we should add that in our experience, philosophers of physics do *in fact* tend to endorse realist accounts of reference and truth. We suspect that the main cause of this is the powerful psychological tendency to take there to be real physical objects, corresponding in their properties and relations to the mathematical objects in mathematical models, especially when those models are very successful. But this tendency is a cause, not a reason; i.e. it does not support the suggestion we denied above, that elaborating a physical theory's ontology implies commitment to realism. Whitehead had a vivid phrase for this tendency to reification: 'the fallacy of misplaced concreteness'.

For this chapter, the main example of this psychological urge will be the tendency to reify spacetime points, which we shall discuss in more detail in Section 2.2.2. For now, we want just to make three general points about this tendency to reifi-cation. The third is more substantial, and so we devote the next subsection to it.

First, such reification is of course a common syndrome in the praxis of physics, and indeed the rest of science; carried over, no doubt, from an excessive zeal for realism about say, chairs, in everyday life. Certainly, in so far as they take a view on these matters, the great majority of physicists tend to be straightforward realists when referring to electrons, or even such exotic entities as quarks.

Second, reification is not *just* a psychological tendency, or a pedagogic crutch. It can also be heuristically fruitful, as shown by successful physical prediction based on the mathematics of a theory; for example, Dirac's prediction of the positron as a 'hole' in his negative energy 'sea' (though he at first identified the holes with protons!).

### 2.1.3.2 *The fragility of ontology in physics*

Setting aside our general cautiousness about scientific realism, there is a specific reason to be wary of misplaced concreteness in theoretical physics. We cannot develop it fully here, but we must state it; for it applies in particular to such putative objects as spacetime points, which will of course be centre-stage in this chapter.

The reason arises from the idea that physics aims to supply a complete description of its subject matter. It does not matter how exactly this idea is made precise: for example, what exactly 'complete description' means, and whether this aim is part of what we mean by 'physics'. The rough idea of physics aiming to be complete is enough. For it entails that in physics, or at least theoretical physics, a change of doctrine *about* a subject matter is more plausibly construed as a change of subject matter *itself*, than is the case in other sciences. So in physics (at least theoretical physics), old ontologies are more liable to be rejected in the light of new doctrine.

We can make the point with a common-sense example. Consider some body of common-sense doctrine, say about a specific table, or tables in general. Not only is it fallible – it might get the colour of the specific table wrong, or it might falsely say that all tables have four legs – it is also bound to be incomplete, since there will be many facts that it does not include – facts which it is the business of the special sciences, or other disciplines, to investigate; for example, the material science of wood, or the history of the table(s). Similarly for a body of doctrine, not from common sense, but from a science or discipline such as chemistry or history, about any subject matter, be it wood or the Napoleonic wars. There are always further facts about the subject matter, not included in the doctrine. Indeed this is so, even if the body of doctrine is the conjunction of all the facts about the subject matter expressible in the taxonomy (vocabulary) of the discipline concerned. Even for such a giant conjunction, no enthusiast of such a science or discipline is mad enough, or imperialist enough, to believe that it gives a complete description of its subject matter. There are always other conjuncts to be had, typically from other disciplines.

Not so, we submit, for physics – or at least theoretical physics. Whether it be madness, imperialism, or part of what we mean by 'physics', physics *does* aspire to give just this sort of complete description of its subject matter. And this implies that when 'other conjuncts arrive' – i.e. new facts come to light in addition to those given by the most complete available physical description – it is more reasonable to construe the new facts as involving a change of subject matter, rather than as an additional piece of doctrine about the old subject matter.

Note that we do *not* say that the first construal is always more reasonable than the second – by no means! Only that it is usually more reasonable than in other sciences, simply because of physics' aspiration to completeness. To take an obvious example: the very fact that a quantum-theoretic description of the electron before the discovery of spin aspired to be complete, makes it more reasonable to construe the discovery of the magnetic moment of the electron as a change of subject matter – the replacement of the 'old' ontology comprising the spinless electron, by one with a spinning electron – rather than as just additional doctrine about the old subject matter, the 'old electron'. To sum up this point: the fact that physics aspires to give a complete description of its subject matter gives a specific reason to be wary of reifying the objects postulated by physical theories.[8]

### 2.1.3.3 *The question of transcendental idealism*

Any discussion about realism – even one mainly concerned, as we are, with scientific realism – raises the issue of 'transcendental idealism'. By this we mean the issue of whether a distinction can, or should, be made between 'appearances' – i.e. in modern terms, the results obtainable by scientific enquiry (obtainable perhaps in principle, or in ideal conditions, if not in fact) – and 'things-in-themselves', i.e. the world 'in itself'. This issue comes to mind all the more readily in a discussion of realism concerning space and time, since it was in connection with these notions that Kant forged his transcendental idealism.

Of course we have no space here to address this enormous issue.[9] Instead, we confine ourselves to two short remarks. First, like most authors in the philosophy of physics (and almost all theoretical physicists),[10] we will write for convenience and brevity 'as if' there is nothing beyond 'appearances': i.e. the distinction above cannot be made.

Second, a remark specific to quantum gravity that relates to the minuscule size of the Planck length, emphasized at the start of Section 2.1.2. Namely, it is *so* minuscule as to suggest that some aspects of reality that underly a theory of quantum gravity do not deserve such names as 'appearance', 'phenomenon', or 'empirical'. Agreed, there is no hint in the writings of Kant or other Kantians that one should restrict the word 'appearance' to what is *practically* accessible. And one naturally thinks that an 'item' (event, state of affairs, call it what you will) that is localized in spacetime, or that somehow has aspects localized in spacetime, is *ipso facto* an appearance, part of empirical reality – be it, or its aspects, ever so small. But our present suggestion is that one should resist this, and consider taking the inaccessibility of these scales of length, energy, etc. to be *so* extreme as to be truly 'in principle'. To put the point in terms of 'empirical': the suggestion is that these items, or their localized aspects, are not empirical, though one might still call them 'real' and 'actual'.[11]

If this is right, one could perhaps reconcile various Kantian claims that space and time must have some features – for example, being continua – as an *a priori* matter with the claims of those quantum gravity programmes that deny space and time those features. The apparent contradiction would be an artefact of an ambiguity in 'space and time': the quantum gravity programmes would *not* be about space and time in the Kantian sense. Finally, we should emphasize that in envisaging such a reconciliation, we are not trying to defend specific Kantian claims, such as its being an *a priori* matter that space and time are continua; or that geometry is Euclidean. Indeed, we join most physicists in being sceptical of such specific claims, not least because the history of physics gives remarkable examples of the creative, albeit fallible, forging of new concepts. But we *are* sympathetic to the broader Kantian idea that human understanding of reality must, as an *a priori* matter, involve certain notions of space and time.

## 2.2 Conceptual problems of quantum theory and general relativity

As discussed in Section 2.1.1, the overarching question of this chapter is: what part (if any) of the ingredient theories' common treatment of spacetime – i.e. as a differentiable manifold with a Lorentzian metric – needs to be given up in quantum gravity? It is already clear (sad to say!) that there is no agreement about the answer to this question. As we shall see in more detail in Sections 3 and following sections, there is a wide variety of different quantum gravity programmes, giving different answers. And more confusingly, these different answers do not always represent simple disagreements between the programmes: sometimes two programmes are aiming to do such very different things, that their different answers need not contradict each other. Hence this chapter's project of undertaking a survey.

But as we shall also see, this variety of programmes (and of aims) is due in part to the fact that significant conceptual problems about the ingredient theories are still unsolved: both problems about the nature of quantum reality, and problems about the nature of space and time – in part, traditional philosophical problems, though of course modified in the light of general relativity and quantum theory. So it will help to set the stage for our survey, to devote this section to describing such problems. Of course, we cannot give a thorough discussion, or even an agreed complete *list*, of these problems.[12] We confine ourselves to briefly discussing some issues that are specifically related to quantum gravity.

In this discussion, we will place the emphasis on problems of quantum theory, for two reasons; only the second of which concerns quantum gravity. First, we agree with the 'folklore' in the philosophy of physics that quantum theory faces more, and worse, conceptual problems than does general relativity. In a nutshell, general relativity is a classical field theory (of gravitation); and broadly speaking, such theories are not mysterious, and their interpretation is not controversial. On the other hand, quantum theory *is* mysterious, and its interpretation *is* controversial. This is attested not only by the struggles of its founding fathers, but also by the ongoing struggles with issues such as the non-Boolean structure of the set of properties of a physical system, the lack of values for quantities associated with superpositions, the phenomenon of quantum entanglement – and of course, the 'confluence' of these three issues in the 'measurement problem'. Indeed, thanks to these struggles, the issues are not nearly so intractable as they were seventy, or even forty, years ago. Though mystery remains, there are nowadays several flourishing schools of thought about how to interpret quantum theory: we will mention some of them in Section 2.2.1 below.

Second, as we shall see in Section 2.4, despite general relativity's 'merits' of interpretative clarity over quantum theory, the main quantum gravity programmes tend to put much more pressure on the framework of standard general relativity, and thus on spacetime concepts, than they do on quantum theory. Like most other research programmes using quantum theory, they simply use the standard quantum theoretic formalism, and do not address its conceptual problems. It is this disparity that motivates this chapter's choice of spacetime as the main topic of its survey. But arguably, this acceptance of standard quantum theory is a mistake, for two reasons. First, in general, it would seem wise for a research programme that aims to combine (or somehow reconcile) two theories, to rely more heavily on the clearer ingredient theory, than on the mistier one! Second, as we shall see in Section 2.2.1, it turns out that in various ways the search for a quantum theory of gravity raises the conceptual problems of quantum theory in a particularly acute form – and even puts some pressure on its mathematical formalism. In any case – whether or not this acceptance of quantum theory is a mistake – in this section we will briefly 'redress the balance'. That is to say, we will emphasize the pressure that quantum gravity puts on quantum theory.[13]

Accordingly, our plan will be to discuss first (in Section 2.2.1) the conceptual problems of quantum theory, especially in relation to quantum gravity; and then the conceptual problems of general relativity (in Section 2.2.2).

### 2.2.1 Interpreting quantum theory

In this subsection, our strategy will be to distinguish four main approaches to interpreting quantum theory, in order of increasing radicalism; and to show how each relates to topics, or even specific approaches, in quantum gravity. The first two approaches (discussed in Sections 2.2.1.1 and 2.2.1.2) are both conservative about the quantum formalism – they introduce no new equations. But they differ as to whether they are 'cautious' or 'enthusiastic' about the interpretative peculiarities of quantum theory. The third and fourth approaches (discussed in Sections 2.2.1.3 and 2.2.1.4) do introduce new equations, but in various (different) ways remain close enough to standard quantum theory to be called 'interpretations' of it.

We stress that although our catalogue of four approaches is by no means maverick, we make no claim that it is the best, let alone the only, way to classify the various, and complexly interrelated, interpretations of quantum theory. But of course we believe that with other such classifications, we could make much the same points about the connections between interpreting quantum theory and quantum gravity.

On the other hand, we cannot consider the details of individual interpretations within each approach. And in view of the chapter's overall project, we shall emphasize how the interpretations we do mention relate to space and time – at the expense of other aspects of the interpretation. For example, we will not mention even such basic aspects as whether the interpretation is deterministic – except in so far as such aspects relate to space and time.

### 2.2.1.1 *Instrumentalism*

We dub our first approach to interpreting quantum theory, 'instrumentalism'. We intend it as a broad church. It includes views that apply to quantum theory some general instrumentalism about all scientific theories; and views that advocate instrumentalism only about quantum theory, based on special considerations about that subject. We will not comment on the first group, since we see no special connections with quantum gravity. Or more precisely, we see no connections other than those which we already adumbrated from another perspective, that of realism, in Section 2.1, especially Section 2.1.3.

On the other hand, some views in the second group do have connections with quantum gravity, albeit 'negative' ones. Thus consider the Copenhagen interpretation of quantum theory: understood, not just as the minimal statistical interpretation of the quantum formalism in terms of frequencies of measurement results, but as insisting on a classical realm external to the quantum system, with a firm 'cut' between them, and with no quantum description of the former. In so far as this classical realm is normally taken to include classical space and time,[14] this suggests that, in talking about 'quantum gravity', we are making a category error by trying to apply quantum theory to something that forms part of the classical background of that theory: 'what God has put assunder, let no man bring together'. We shall say more later about the view that a quantum theory of gravity should, or can, be avoided (Section 2.3.1.2). But for the most part we will accept that serious attempts should be made to construct a 'quantum theory of space and time' (or, at least, of certain aspects of space and time); with the understanding that, in doing so, it may be necessary radically to change the interpretation – and, perhaps, the mathematical formalism – of quantum theory itself.

In endeavouring to interpret quantum theory, regardless of quantum gravity, this second group of views is notoriously problematic. It is not just a matter of it being difficult to understand (or to defend) Bohr's own views, or views similar to his. There are quite general problems, as follows. Any view that counts as 'instrumentalism specifically about quantum theory' (i.e. any view in this group) must presumably do either or both of the following:

(1) Deny that the quantum state describes individual systems, at least between measurements; or in some similar way, it must be very cautious about the quantum description of such systems.
(2) Postulate a 'non-quantum' realm, whose description can be taken literally (i.e. not instrumentalistically, as in (1)); usually this realm is postulated to be 'the classical realm', understood as macroscopic, and/or the domain of 'measurement results', and/or described by classical physics.

But recent successful applications of quantum theory to individual microphysical systems (such as atoms in a trap), and to mesoscopic systems (such as SQUIDs) have made both (1) and (2) problematic. This suggests in particular that we should seek an interpretation in which no fundamental role is ascribed to 'measurement', understood as an operation external to the domain of the formalism (see the next subsection).

### 2.2.1.2 *Literalism*

Like instrumentalism, we intend this approach to be a broad church. The idea is to make the interpretation of quantum theory as 'close' as possible to the quantum formalism. (Hence the name 'literalism'; 'realism' would also be a good name, were it not for its applying equally well to our third and fourth approaches.) In particular, one rejects the use of a primitive notion of measurement, and associated ideas such as a special 'classical realm', or 'external observer' that is denied a quantum-theoretic description. Rather, one 'cuts the interpretation to suit the cloth of the formalism'; revising, if necessary, traditional philosophical opinions, in order to do so. Hence our remark at the start of Section 2.2.1 that this approach is 'enthusiastic' about the interpretative peculiarities of quantum theory, while instrumentalism is 'cautious'.

As we see it, there are two main types of literalist view: Everettian views, and those based on quantum logic. Of these types, the first has been much discussed in connection with quantum gravity (especially quantum cosmology); but the second, hardly at all in this connection. Accordingly, we shall only treat the first.[15]

As usually presented, the main aim of an Everettian view (or, as it is sometimes called these days, a 'post-Everett' view) is to solve the 'measurement problem': i.e. the threat that at the end of a measurement, macroscopic objects (such as an instrument pointer) will have no definite values for familiar quantities like position – contrary to our experience. More specifically, the aim is to solve this problem without invoking a collapse of the state vector, or an external observer. This involves: (i) resolving the state vector of a closed system as a superposition of eigenstates of a 'preferred quantity'; (ii) interpreting each of the components as representing definite positions for pointers and other macroscopic objects; and then (iii) arguing that, although there is no collapse, you will only 'see' one component in the superposition.

This summary description leaves open some crucial questions. For example:

(a) How is the preferred quantity to be chosen? Should it be in terms of familiar quantities such as position of macroscopic objects, so that each summand secures a definite macroscopic realm ('many worlds'); or should it involve arcane quantities concerning brains, whose eigenstates correspond to experiences of a definite macroscopic realm ('many minds')?

(b) Should one say that for each component there is a physically real 'branch', not just the possibility of one?

(c) How should one justify the claim that you will not 'see the other components': by some process of 'splitting of the branches', or by appeal to decoherence making the interference terms that are characteristic of the presence of other components, negligibly small?

We do not need to discuss these issues here, which have been much debated in the philosophy of quantum theory.[16] For our purposes, it suffices to note the four main connections of Everettian views with quantum gravity, specifically quantum cosmology.

The first connection has been evident from the earliest discussions of Everettian views. Namely, whatever the exact aims of a theory of quantum cosmology, in so far as it posits a 'quantum state of the universe', the Everettian promise to make sense of the quantum state of a closed system makes this interpretation particularly attractive.

The second connection concerns the more extreme Everettian view in which the universe is deemed literally to 'split'. In so far as this might involve some transformation of the topology of space, one naturally imagines implementing this with the aid of ideas from quantum gravity.

The third connection relates to decoherence, mentioned in question (c) above. Much recent work has shown decoherence to be a very efficient and ubiquitous mechanism for making interference terms small (and so for securing an apparent reduction of a quantum system's state vector); essentially by having the correlational information that these terms represent 'leak out' to the system's environment. Though this work in no way relies on Everettian views, Everettians can, and do, appeal to it in answering question (c). Furthermore, the work has been adapted to the discussion within quantum cosmology of how we 'see' a single classical space or spacetimes, despite the fact that in quantum cosmological models the quantum state of the universe assigns non-zero amplitude to many such spaces. The idea is that decoherence destroys the interference terms, 'hiding all but one'. (Typically, inhomogeneous modes of the gravitational field act as the environment of the homogeneous modes, which form the system.) (For more discussion, see Ridderbos 1999.)

The fourth connection relates to time. Presumably, any Everettian view must specify not only probabilities for values of its preferred quantity at each time; it must also specify joint ('conjunctive') probabilities for values at sequences of times, i.e. a rule for the temporal evolution of these values. Traditionally, Everettians tended not to give such a rule, but recently they have done so, often in the context of the 'consistent-histories' approach to quantum theory.

There is a specific reason for quantum cosmologists to focus on the consistent-histories formalism, apart from the general need to specify a rule for the evolution of values. As we shall see in more detail in Section 2.3.3, quantum gravity, and thereby quantum cosmology, is beset by 'the problem of time'. One response to this severe problem is to seek a new type of quantum theory in which time does not play the central role that it does in the standard approach. And precisely because the consistent-histories approach concerns many times, it suggests various ways in which the formalism of quantum theory can be generalized to be less dependent on the classical concept of time (Hartle 1995, Isham and Linden 1994).

This last point gives an example of an important, more general idea, which goes beyond the discussion of Everettian views. Namely, it is an example of how issues in quantum gravity can put pressure on the actual formalism of quantum theory – not just on some traditional interpretative views of it, such as the Copenhagen interpretation.

### 2.2.1.3 *Extra values*

Again, we intend this approach to be a broad church. Like the Everettian views discussed above, it aims to interpret quantum theory – in particular, to solve the measurement problem – without invoking a collapse of the state vector. And it aims to do this by postulating values for some 'preferred quantity' or quantities, in addition to those given by the orthodox 'eigenvalue–eigenstate link',[17] together with a rule for the evolution of such values.

But there are two differences from the Everettian views. First, 'extra values' makes no suggestion that there is a physically real 'branch' for every component in the resolution of the state vector in terms of the preferred quantity. (So there is no suggestion that 'branches splitting' prevents the detection of interference terms.) Second, 'extra values' aspires to be more precise from the outset about which quantity is preferred, and the dynamics of its values.[18]

The best-known examples of this approach are the deBroglie–Bohm 'pilot-wave' or 'causal' interpretation of quantum theory (Valentini 2000), and the various kinds of modal interpretation (Bub 1997). Thus the pilot-wave interpretation of quantum mechanics postulates a definite value for the position of each point-particle, evolving according to a deterministic guidance equation. The corresponding interpretation of quantum field theory postulates a definite field configuration, and again a deterministic guidance equation. On the other hand, modal interpretations postulate that which quantity is 'preferred' depends on the state, and they consider various stochastic dynamics for values.

Within this approach, only the pilot-wave interpretation has been discussed in connection with quantum gravity. The main idea is to 'make a virtue of necessity', as follows.[19] On the one hand, the guidance equations (at least as developed so far) require an absolute time structure, with respect to which the positions or field configurations evolve. (So for familiar quantum theories on Minkowski spacetime, the relativity of simultaneity, and the Lorentz-invariance of the theory, is lost at the fundamental level – but recovered at a phenomenological level.) On the other hand: in quantum gravity, one response to the problem of time is to 'blame' it on general relativity's allowing arbitrary foliations of spacetime; and then to postulate a preferred foliation of spacetime with respect to which quantum theory should be written. Most general relativists feel this response is too radical to countenance: they regard foliation-independence as an undeniable insight of relativity. But an advocate of the pilot-wave interpretation will reply that the virtues of that interpretation show that sacrificing fundamental Lorentz-invariance is a price worth paying in the context of flat spacetime; so why not also 'make a virtue of necessity' in the context of curved spacetime, i.e. general relativity?

Indeed, this suggestion has been developed in connection with one main approach to quantum gravity; namely, the quantum geometrodynamics version of the canonical quantum gravity programme. We will discuss this in more detail in Section 2.4.5. For the moment, we just note that the main idea of the pilot-wave interpretation of quantum geometrodynamics is to proceed by analogy with the interpretation of quantum field theories such as electrodynamics on flat spacetime. Specifically, a wave function defined on 3-geometries (belonging to the three-dimensional slices of a preferred foliation) evolves in time, and deterministically guides the evolution of a definite 3-geometry.

To sum up: 'extra values' preserves the usual unitary dynamics (the Schrödinger equation) of quantum theory, but adds equations describing the temporal evolution of its extra values. And the best developed version of 'extra values' – the pilot-wave interpretation – has been applied only to the quantum gravity programme based on quantum geometrodynamics.

### 2.2.1.4 New dynamics

This approach is more radical than 'extra values'. Instead of adding to the usual unitary dynamics of quantum theory, it replaces that dynamics, the motivation being to solve the measurement problem by dynamically suppressing the threatened superpositions of macroscopically distinguishable states. During the past fifteen years, there has been considerable development of this approach, especially in the wake of the 'spontaneous localization' theories of Ghirardi, Rimini, and Weber (1986), and Pearle (1989).

This approach has natural links with quantum gravity. Indeed, from the point of view of physical theory itself, rather than its interpretation, it is a closer connection than those reviewed in the previous subsections. For it is natural to suggest that the proposed deviation from the usual dynamics be induced by gravity (rather than being truly 'spontaneous'). This is natural for at least two reasons: (i) gravity is the only *universal* force we know, and hence the only force that can be guaranteed to be present in all physical interactions; and (ii) gravitational effects grow with the size of the objects concerned – and it is in the context of macroscopic objects that superpositions are particularly problematic.

We emphasize that this idea – that gravity is involved in the reduction of the state vector – is different from, and more radical than, the idea in Section 2.2.1.2 that some modes of the gravitational field might act as the environment of a system for a decoherence process that yields an apparent state-vector reduction. Here, there is no invocation of an environment; i.e. there is reduction for a strictly isolated system.

This idea has been pursued in various ways for several decades. In particular, adapting the idea to *quantum* gravity: since general relativity treats gravity as space-time curvature, the most straightforward implementation of the idea will require that a quantum superposition of two spacetime geometries, corresponding to two macroscopically different distributions of mass-energy, should be suppressed after a very short time. Penrose has been particularly active in advocating this idea. More specific implementations of the idea involve variants of the spontaneous localization theories; for example, Pearle and Squires (1996), which also contains a good bibliography.[20]

### 2.2.2 Interpreting general relativity

We turn now to consider the conceptual problems of general relativity, especially those related to quantum gravity. However, our discussion will be briefer than that of Section 2.2.1, for the two reasons given at the start of the section. First, general relativity is essentially a classical field theory, and its interpretation is less mysterious and controversial than that of quantum theory. Second, subsequent sections will give ample discussion of the pressure that quantum gravity puts on general relativity.

Specifically, we shall confine ourselves to brief remarks about one aspect of the grand debate between 'absolute' *versus* 'relational' conceptions of space and time: namely, the question in what sense, if any, spacetime points are objects.[21] (We shall discuss spacetime points, but most of the discussion could be straightforwardly rephrased as about whether regions in spacetime are objects; or, indeed – in a canonical approach – points or regions in 3-space, or in time.) For this question bears directly on the discussion in subsequent sections of the treatment of spacetime in quantum gravity.[22]

The debate between 'absolute' and 'relational' conceptions of space and time has many strands. Nowadays, philosophers separate them (at least in part) by distinguishing various senses. For example, does a spatio-temporal structure being 'absolute' mean that it is 'non-dynamical', i.e. unaffected by material events; or that

it is not determined by (supervenient on) the spatio-temporal relations of material bodies?[23] And what spatio-temporal structure does the 'absolutist' take to be
absolute: space (as by Newton), or the four-dimensional metric of spacetime, or
the connection? Once these senses are distinguished, it becomes clear that general
relativity supports 'relationism' in the senses that: (i) its four-dimensional metric
and connection are dynamical; and (ii) in its generic models, no space, i.e. no foliation of spacetime, is preferred (whether dynamically or non-dynamically). On the
other hand, it supports absolutism in the sense that the *presence* in the theory of
the metric and connection is not determined by the spatio-temporal relations of
material bodies.

But the consensus on these issues about relatively technical senses of 'absolute'
leaves outstanding the question whether we should interpret general relativity as
commited to the existence of spacetime points (or regions) as physical objects.
We are wary of the 'Yes' answer to this question (which became popular in the
1960s, with the rise of scientific realism). But this is not just because we are wary of
scientific realism; and in particular, of reifying the objects and structures postulated
by physical theories (as discussed in Section 2.1.3). We also have two more specific
reasons. The first contains a more general moral about reification; the second is
specific to spacetime points:

(1) To explain the first reason, we should begin by admitting that it is especially
tempting to take spacetime points as the fundamental physical objects of both our
'ingredient theories' – general relativity and quantum theory – and not just as points
in mathematical models. There are two specific factors prompting this reification.

Roughly speaking, the first factor is this. As usually formulated, the theories agree
with one another in using such points, endowed with the (highly sophisticated)
structure of a differentiable manifold. But to be precise, one needs to respect the
distinction between a (putative) physical spacetime point, and an (undeniably postulated!) point in a mathematical model of spacetime based on standard set theory.
So one should express this first factor by saying that, as usually formulated, the
theories agree in using the latter points, i.e. those in the mathematical models. Even
the most ardent realist must allow this distinction in principle, if he or she is to
avoid begging the question; though he or she may well go on to suggest that as a
realist, they can take (and perhaps prefers to take) the physical points in which they
believe, as elements of the mathematical model – say as the bottom-level elements
in a set-theoretic definition of a manifold equipped with a Lorentzian metric and
some matter fields.

The second factor is that, as usually formulated, the theories postulate the points
(in the sense of using set theory) initially, i.e. at the beginning of their formalism;
the rest of physical reality being represented by mathematical structures (vector,
tensor, and operator fields, etc.) defined over the sets of points.[24] Here, as in the
first factor, to be precise – and to avoid begging the question – one must take these
postulated points to be those in the mathematical models, not the putative physical
points. The other structures representing fields, etc., then become properties and
relations among these postulated points; or, more generally, higher-order properties
and relations; or in a formal formulation of the theory, set-theoretic surrogates for
such properties and relations. And again, to avoid begging the question one must

understand 'represent' as not committing one to the represented fields *really* being properties and relations. For that would commit one to there being objects which instantiated them; and these would no doubt be spacetime points and $n$-tuples of them.

However, notwithstanding these cautionary remarks, most people who bother to think about such matters succumb to Whitehead's fallacy of misplaced concreteness, by positing a one-to-one correspondence between what is undeniably real in the Platonic realm of mathematical form, and what is, more problematically, 'real' in the world of physical 'stuff'.

But tempting though this reification is, it is very questionable: not least because it overlooks the fact that these theories can be formulated in *other* (usually less well-known) ways, so as to postulate initial structures that, from the usual viewpoint, are complex structures defined *on* the points.

More precisely, the theories can be formulated so as to postulate initially not: (i) mathematical objects that represent spacetime points (again, understanding 'represent' as not committing one to spacetime points being genuine physical objects); but rather (ii) mathematical objects that represent (again understood non-commitally) fields, and similar items – items that in the usual formulations are represented by complex mathematical structures (formally, set-theoretic constructions) defined over the initially postulated representatives of spacetime points.

To give the flavour of such formulations, here is a standard example from the simpler setting of topological spaces, rather than differentiable manifolds. Consider a compact Hausdorff space $X$ and let $\mathcal{A}$ denote the ring of real-valued functions on $X$. Then it is a famous theorem in topology that both the set $X$ (i.e. its points) and the topology of $X$ can be uniquely reconstructed from just the *algebraic* structure of $\mathcal{A}$. Specifically, the closed subsets of $X$ are in one-to-one correspondence with the (closed) ideals in the commutative ring (actually, $C^*$-algebra) $\mathcal{A}$; and the points of $X$ correspond to maximal ideals in $\mathcal{A}$.

The implication of this result is that, from a mathematical perspective, a theory based on such a topological space – modelling, say, physical space – can be formulated in such a way that the fundamental mathematical entity is not the set $X$ of spatial points – on which *fields* are then defined – but rather a commutative ring, on which spatial *points* are then defined: viz. as maximal ideals. Put in graphic terms, rather than writing $\phi(x)$, one writes $x(\phi)$!

We emphasize that this idea is by no means esoteric from the perspective of theoretical physics. For example, the subject of 'non-commutative' geometry starts from precisely this situation and then posits a non-commutative extension of $\mathcal{A}$. In this case the algebra remains, but the points go in the sense that the algebra can no longer be written as an algebra of functions on anything.

(2) There is also another reason for wariness about the existence of spacetime points as physical objects; a reason relating to symmetry transformations. The idea goes back to Leibniz; but in modern terms, it is as follows: given that a model of a theory represents a physical possibility, the model obtained by applying a global symmetry transformation to it describes *the same* physical possibility. In the context of spacetime theories, this idea means that taking points to be physical objects

involves a distinction without a difference. So in particular, the existence of translation invariance in Newtonian or Minkowski spacetime shows that points should not be taken as physical objects. We think that most physicists would concur with this idea.

In the context of general relativity, such considerations become perhaps yet more convincing, in view of Einstein's 'hole argument'. We cannot enter into details about this argument,[25] which has interesting historical and physical, as well as philosophical, aspects. Suffice it to say that: (i) the argument applies not just to general relativity, but to any generally covariant theory postulating a spacetime manifold; and (ii) according to the argument, general covariance (that is: the diffeomorphism–invariance of the theory), together with spacetime points being physical objects, implies a radical indeterminism: and such indeterminism is unacceptable – so that we should conclude that points are not physical objects. That is, the points occurring in the base-sets of differentiable manifolds with which general relativity models spacetime should not be reified as physically real.[26] We shall take up this theme again in more detail in Section 2.3.2.

## 2.3 Introducing quantum gravity

We turn now to our main project: surveying how quantum gravity suggests fundamental limitations in the familiar treatment of space and time that is common to the 'ingredient theories' – quantum theory and general relativity. In this section, we first give some details about the variety of approaches to quantum gravity (Section 2.3.1). We then give a more detailed discussion of two conceptual aspects relating specifically to spacetime: viz. the role of diffeomorphisms (Section 2.3.2) and the problem of time (Section 2.3.3). This will set the stage for the discussion in Section 2.4 of the treatment of spacetime in three of the main research programmes in quantum gravity.

### 2.3.1 Approaches to quantum gravity

In this section, we begin to give a more detailed picture of quantum gravity research. We will first survey some motivations for studying quantum gravity (Section 2.3.1.1). Then we will consider, but reject, the view that quantum gravity can be avoided (Section 2.3.1.2). We will then describe four broad approaches to quantum gravity (Section 2.3.1.3).

#### 2.3.1.1 Motivations for studying quantum gravity

In surveying quantum gravity, it is useful to begin with the various motivations for studying the subject. For as we have seen, quantum gravity does not have a well-established body of 'facts' against which theories can be tested in the traditional way. In consequence, although some people's motivations refer to potential observations or experiments – particularly in the area of cosmology – most motivations are of a more internal nature: namely, the search for mathematical consistency, or the implementation of various quasi-philosophical views on the nature of space and time. And, since these different motivations have had a strong influence on researchers' technical approaches to the subject, it is important to appreciate them in order to understand what people have done in the past, and to be able to judge if they succeeded in their endeavours: since to be adjudged 'successful' a theory must presumably either point beyond itself to new or existing 'facts' in the world, or else achieve some of its own internal goals.

It is useful pedagogically to classify motivations for studying quantum gravity according to whether they pertain to the perspective of elementary particle physics and quantum field theory, or to the perspective of general relativity. As we shall see, this divide substantially affects one's approach to the subject, in terms of both the goals of research and the techniques employed.

*Motivations from the perspective of elementary particle physics and quantum field theory:*

(i) Matter is built from elementary particles that *are* described in quantum theoretical terms, and that certainly interact with each other gravitationally. Hence it is necessary to say *something* about the interface between quantum theory and general relativity, even if it is only to claim that, 'for all practical purposes', the subject can be ignored; (see Section 2.3.1.2 for further discussion).

(ii) Relativistic quantum field theory might only make proper sense if gravity is included from the outset. In particular, the short-distance divergences present in most such theories – including those that are renormalizable, but not truly finite – might be removed by a fundamental cut-off at the Planck energy. Superstring theory (see Section 2.4.4) is arguably the latest claimant to implement this idea.

(iii) A related claim is that considerations about quantum gravity will be a necessary ingredient in any fully consistent theory of the unification of the three *non*-gravitational forces of nature.[27] The underlying idea here is as follows.

The mark of unification in a field theory is the equality of the coupling constants that determine the strengths of the different forces. However, in the quantum version these coupling 'constants' are energy dependent (they are said to 'run' with the energy) and therefore forces that are not unified at one energy may become so at a different one. It turns out that the running constants of the electromagnetic, weak and strong nuclear forces can be shown to 'meet', i.e. to be equal or at least approximately equal, at around $10^{20}$ MeV. The fact that $10^{20}$ MeV is 'quite close to' the Planck energy (viz. only two orders of magnitude less) then suggests that quantum gravity may have a role to play in this unification of forces.

This line of thought also prompts the more specific suggestion that a successful theory of quantum gravity *must* involve the unification of all four fundamental forces. As we shall see in Section 2.4, one of the key differences between the two most currently active research programmes, superstring theory and canonical quantum gravity, is that the former adopts this suggestion – it aims to provide a scheme that encompasses all the forces – while the latter asserts that a quantum theory of pure gravity *is* possible.

*Motivations from the perspective of a general relativist:*

(i) Spacetime singularities arise inevitably in general relativity if the energy and momentum of any matter that is present satisfies certain, physically well-motivated, positivity conditions. It has long been hoped that the prediction of such pathological behaviour can be avoided by the correct introduction of quantum effects.

(ii) A related point is that, once quantum mechanical effects are included, black holes produce Hawking radiation and, in the process, slowly lose their mass. But the nature of the final state of such a system is unclear, and much debated, providing another reason for studying quantum gravity.

(iii) Quantum gravity should play a vital role in the physics of the very early universe, not least because, in standard classical cosmology, the 'initial event' is an example of a spacetime singularity. Possible applications include:

(a) finding an explanation of why spacetime has a macroscopic dimension of four;[28]

(b) accounting for the origin of the inflationary evolution that is believed by many cosmologists to describe the universe as it expanded after the initial big-bang.

(iv) Yet more ambitiously, one can hope that a theory of quantum gravity will provide a quantum cosmology, and thereby an understanding of the very origin of the universe in the big-bang as some type of quantum 'event'.

However, special problems are posed by quantum cosmology, for example about the interpretation of the quantum state (see Section 2.2.1); so one might well take the view that quantum gravity research should not get distracted by debating these problems. Certainly it would be a signal achievement to have a theory that successfully handled quantum theory and general relativity 'in the small', even if it could not be applied to the 'universe in its totality' – a problematic concept on any view! In any case, we shall from now on largely set aside quantum cosmology, and its special problems.[29]

### 2.3.1.2 *Can quantum gravity be avoided?*

The argument is sometimes put forward that the Planck length $L_P := (G\hbar/c^3)^{1/2} \simeq 10^{-35}$ m is so small that there is no need to worry about quantum gravity except, perhaps, in recherché considerations of the extremely early universe – i.e. within a Planck time ($\simeq 10^{-42}$ s) of the big-bang. However, as we hinted in the first motivation listed in Section 2.3.1.1:

- Such a claim is only really meaningful if a theory exists within whose framework genuine perturbative expansions in $L_P/L$ can be performed, where $L$ is the length scale at which the system is probed: one can then legitimately argue that quantum effects are ignorable if $L_P/L \ll 1$. So we must try to find a viable theory, even if we promptly declare it to be irrelevant for anything other than the physics of the very early universe.

- The argument concerning the size of $L_P$ neglects the possibility of *non*-perturbative effects – an idea that has often been associated with the claim that quantum gravity produces an intrinsic cut-off in quantum field theory.

A very different, and less radical, view is that – although we presumably need some sort of theory of quantum gravity for the types of reason listed in Section 2.3.1.1 – it is *wrong* to try to construct this theory by quantizing the gravitational field, i.e. by applying a quantization algorithm to general relativity (or to any other classical theory of gravity). We shall develop this distinction between the general idea of a theory of quantum gravity, and the more specific idea of quantized version of general relativity, immediately below (Section 2.3.1.3). For the moment, we mention some reasons advanced in support of this view.

- The metric tensor should not be viewed as a 'fundamental' field in physics, but rather as a phenomenological description of gravitational effects that applies only in regimes well away from those characterized by the Planck scale. Again, diverse reasons can be given for this viewpoint: we cite three. One example is superstring theory. Here, the basic quantum entities are very different from those in classical general relativity, which is nevertheless recovered as a phenomenological description. Another (very different) example is Jacobson's re-derivation of the Einstein field equations as an equation of state (Jacobson 1995), which (presumably) it would be no more appropriate to 'quantize' than it would the equations of fluid dynamics.[30] Yet a third example is Brown's view of the metric, even in special relativity, as phenomenological (see Chapter 11).

- The gravitational field is concerned with the structure of space and time – and these are, par excellence, fundamentally classical in nature and mode of functioning. As we mentioned before, this might be defended from the viewpoint of (a version of) the Copenhagen interpretation (Section 2.2.1) – or even from a Kantian perspective (Section 2.1.3).

If it is indeed wrong to quantize the gravitational field (for whichever of the above reasons), it becomes an urgent question how matter – which presumably *is* subject to the laws of quantum theory – should be incorporated in the overall scheme. To discuss this, we shall focus on the so-called 'semiclassical quantum gravity' approach. Here, one replaces the right-hand side of Einstein's field equations by a quantum expectation value, so as to couple a classical spacetime metric $\gamma$ to quantized matter by an equation of the form

$$G_{\mu\nu}(\gamma) = \langle\psi|T_{\mu\nu}(g,\hat{\phi})|\psi\rangle, \tag{2.1}$$

where $|\psi\rangle$ is some state in the Hilbert space of the quantized matter variables $\hat{\phi}$. Thus the source for the gravitational field, i.e. the right-hand side of eqn. 2.1 – is the expectation value of the energy–momentum tensor $T_{\mu\nu}$ of the quantized matter variables. In this context, we note the following:

- In the case of electromagnetism, the well-known analysis by Bohr and Rosenfeld (1933) of the analogue of eqn. 2.1 concluded that the electromagnetic field *had* to be quantized to be consistent with the quantized nature of the matter to which it couples. However, the analogous argument for general relativity does not go through (Rosenfeld 1963), and – in spite of much discussion since then (for example, see Page and Geilker 1981) – there is arguably still no definitive proof that general relativity *has* to be quantized in some way.
- The right-hand side of eqn. 2.1 generates a number of technical problems. For example, the expectation value has the familiar 'ultra-violet' divergences that come from the mathematically ill-defined short-distance behaviour of quantum fields. Regularization methods only yield an unambiguous expression when the spacetime metric $\gamma$ is time-independent – but there is no reason why a semiclassical metric should have this property.[31] In addition, there have been several arguments implying that solutions to eqn. 2.1 are likely to be unstable against small perturbations, and – therefore – physically unacceptable.
- It is not at all clear how the state $|\psi\rangle$ is to be chosen. In addition, if $|\psi_1\rangle$ and $|\psi_2\rangle$ are associated with a pair of solutions $\gamma_1$ and $\gamma_2$ to eqn. 2.1, there is no obvious connection between $\gamma_1$ and $\gamma_2$ and any solution associated with a linear combination of $|\psi_1\rangle$ and $|\psi_2\rangle$. Thus the quantum sector of the theory has curious non-linear features, and these generate many new problems of both a technical and a conceptual nature.

So much by way of reviewing the reasons one might give for not quantizing general relativity. We make no claim that our 'replies' to these reasons – for example, our last two 'bullet-points' – are definitive. But we will from now on accept that *some type* of theory of quantum gravity should be sought.

### 2.3.1.3 *Four types of approach to quantum gravity*

In seeking such a theory, there are four broad types of approach one can adopt. We shall introduce them as answers to a series of questions (questions which develop Section 2.1.1's contrast between the two strategies, quantization and emergence). Broadly speaking, these questions will place them in an order of increasing radicalism. So, let us ask: should we adopt a diorthotic scheme in which general relativity is regarded as 'just another classical field theory' to be quantized in a more-or-less standard way? Or should we instead expect the theory of quantum gravity to look quite different from quantized general relativity, but nevertheless have general relativity emerge from it as some sort of low-energy (large-length) limit? This option itself breaks down into two alternatives, according to whether the theory is a quantization of *some* classical theory; or something that is constructed with no prior reference at all to a classical system. A fourth alternative is, should both quantum theory and general relativity emerge from a theory that looks quite different from both?

We will now develop the contrast between these four alternatives. Our survey in later sections will not need to decide between them, but of the three programmes that we discuss in Section 2.4, two adopt the first alternative, and indeed, their implications for the treatment of spacetime are better understood than those of programmes adopting the other alternatives. Aspects of the third and fourth alternatives will be taken up in Section 2.5.

1. *Quantize general relativity.* The idea is to start with the classical theory of general relativity and then to apply some type of quantization algorithm. This is intended to be analogous to the way in which the classical theory of, for example, an atom bound by the Coulomb potential is 'quantized' by replacing certain classical observables with self-adjoint operators on a Hilbert space; or, to take another example, the way in which classical electromagnetism is quantized to yield quantum electrodynamics.[32]

In the context of quantum gravity, the task is usually taken to be quantization of the metric tensor regarded as a special type of field. In practice, the techniques that have been adopted fall into two classes: (i) those based on a spacetime approach to quantum field theory – in which the operator fields are defined on a four-dimensional manifold representing spacetime; and (ii) those based on a canonical approach – in which the operator fields are defined on a three-dimensional manifold representing physical space. We shall discuss (i) and (ii) in more detail in Sections 2.4.3 and 2.4.5, respectively.

2. *General relativity as the low-energy limit of a quantization of a different classical theory.* If a quantization algorithm is applied to some classical theory, then that theory is naturally recovered as a classical limit of the ensuing quantum theory. In particular, this procedure provides a natural interpretation of the physical variables that arise in the quantum theory as the result of 'quantizing' the corresponding classical theory.

But there are various senses of 'classical limit': it can refer to special states whose evolution over time follows classical laws, or to certain quantum quantities taking values in a range where classical theory is successful. So given a quantization of a classical theory, some *other* classical theory might also be a classical limit of it, in some good sense.

Hence the idea in the context of quantum gravity, that general relativity might emerge as a low-energy (large-distance) classical limit of a quantum theory, that is given to us as a quantization of a different classical theory. Of course, in view of our lack of data about quantum gravity, one expects it will be very hard to guess the correct classical theory from which to start.

Despite this difficulty, this type of approach is exemplified by the main current research programme: superstring theory, which quantizes a classical 'string theory' and yet has general relativity as a low-energy limit. We shall discuss this programme in Section 2.4.4.[33] For the moment, suffice it to say that the dimensional nature of the basic Planck units lends support to the idea of a theory that could reproduce standard general relativity in regimes whose scales are well away from that of the Planck time, length, energy, etc. This remark is reinforced by a well-known body of work to the effect that, with appropriate caveats, general relativity is necessarily recovered as the low-energy limit of *any* interacting theory of massless spin-2 particles propagating on a Minkowski background, in which the energy and momentum are conserved (Boulware and Deser 1975). The most notable example of this type is the theory of closed superstrings which has a natural massless, spin-2 excitation.

However, superstring theory is by no means the only example of this type of approach. For it is conservative, in that the classical 'string theory' that it quantizes assumes the classical concept of a manifold. Roughly speaking, in perturbative superstring theory, the quantum variables are the functions that embed the string in a continuum spacetime manifold. But there have been more radical attempts to quantize aspects of space, or spacetime, itself. For example, there have been several attempts to construct a quantum theory of topology; and there have been attempts to quantize causal structures in which the underlying set is discrete. However, recovering general relativity as a classical limit of theories of this type is by no means trivial, since the implication is that the *differentiable manifold* structure of spacetime – not just its metric tensor – must be understood in some phenomenological sense. We will postpone to Section 2.5 discussion of these more radical attempts to quantize 'spacetime itself'.

Finally, a general point about this type of approach. Given the general scenario where we obtain a classical theory as a limit of a quantum theory, it is natural to wonder what would happen if one tried to quantize this derivative classical theory (in the case of interest to us, general relativity). Generally speaking, this does not give back the initial quantum theory. That is unsurprising, in view of our comments above about the variety of classical limits. But we should also make two more specific remarks. First, one reason why one does not get back the initial quantum theory may be that the classical limit is non-renormalizable: this is well known to be the case for general relativity (as we will discuss in Section 2.4.3). Second, this feature does not render the 're-quantization' procedure completely useless. Indeed, genuine quantum predictions can be obtained by empirically fixing the appropriate number of renormalization constants, where what is 'appropriate' is determined by the energy at which the theory is to be employed. Theories of this type are called 'effective field theories', and are a valuable tool in modern theoretical physics. For a recent review in the context of general relativity see Donoghue (1998).

3. *General relativity as the low-energy limit of a quantum theory that is not a quantization of a classical theory.* The procedure of going from classical to quantum has become so ubiquitous (for example, look at the content of a typical undergraduate lecture course on quantum theory!) that one might be tempted to assume that all quantum theories necessarily arise in this way. However, there is no good reason why this should be so. So it is certainly reasonable to consider the construction of a quantum theory *ab initio* with no fundamental reference to an underlying classical theory – for example as a representation of some group or algebra. The question then arises whether a quantum theory of this type may have a classical limit of some sort, even though it is *not* obtained by the quantization of such. A good example of a quantum theory of this kind was the 'current algebra' approach to strong-interaction physics that was intensely studied in the 1960s.

Of course, one might well fear that in quantum gravity, with its dire lack of data, this type of approach will be at least as hard to implement successfully as is the previous one: the correct group or algebra might be as hard to guess, as is the correct classical system to quantize in the previous approach. However, recent developments in understanding the non-perturbative aspects of superstring theory suggest that this type of approach may well come to the fore in that programme (see Section 2.4.4).

4. *Start* ab initio *with a radically new theory.* The idea here is that both classical general relativity *and* standard quantum theory emerge from a theory that looks very different from both. Such a theory would indeed be radically new. Recall that we classified as examples of the second type of approach above, quantizations of spatial or spatio-temporal structure other than the metric; for example, quantizations of topology or causal structure. So the kind of theory envisaged here would somehow be still more radical than that; presumably by not being a quantum theory, even in a broad sense – for example, in the sense of states giving amplitudes to the values of quantities, whose norms squared give probabilities.

Of course, very little is known about potential schemes of this type, let alone whether it is necessary to adopt such an iconoclastic position in order to solve the problem of quantum gravity. We shall mention some possible clues in Section 2.5. For the moment, we want just to emphasize the philosophical interest of this type of approach, for it is often motivated by the view that the basic ideas behind general relativity and quantum theory are so fundamentally incompatible that any complete reconciliation will necessitate a total rethinking of the central categories of space, time, and matter. And as we mentioned in Section 2.1.2 (item 3), we like to think that philosophy could have a role in that enterprise!

As mentioned above, all four types of approach have been followed in the past (albeit in a very limited way in regard to the third and fourth types). Until fifteen years ago, the bulk of the effort was devoted to the first – the active quantization of classical general relativity, so that two of the three programmes reviewed in Section 2.4 are of this type. But nowadays the dominant programme, viz. superstrings, is of the second type; although the second most dominant programme – canonical quantum gravity – is of the first type; and both these programmes have touches of the third type. In short, it remains a matter of vigorous debate which of these types of approach will ultimately prove to be the most fruitful.

### 2.3.2 The role of diffeomorphisms

To set the stage for Section 2.4, we devote the rest of this section to discussing two conceptual aspects that relate specifically to spacetime: viz. the role of diffeomorphisms (this subsection); and the problem of time (Section 2.3.3).

#### 2.3.2.1 *Spacetime diffeomorphisms in classical general relativity*

The group of spacetime diffeomorphisms $\mathcal{D}$ plays a key role in classical general relativity; and its status in quantum gravity raises some major conceptual issues.[34]

In considering these matters, it is important to distinguish between the pseudo-group of local co-ordinate transformations and the genuine group $\mathcal{D}$ of global diffeomorphisms. Compatibility with the former can be taken to imply that the theory should be written using tensorial objects on spacetime. On the other hand, diffeomorphisms are active transformations of spacetime, and invariance under $\mathcal{D}$ implies, we take it, that the points in spacetime have no direct physical significance (see the discussion of realism, and the hole argument in Sections 2.1.3 and 2.2.2.) Of course, this is also true in special relativity, but it is mitigated there by the existence of inertial reference frames that can be transformed into each other by the Poincaré group of isometries of the Minkowski metric.

Put somewhat differently, the action of $\mathcal{D}$ induces an action on the space of spacetime fields, and the only thing that has immediate physical meaning is the space of equivalence classes under this action, i.e. two field configurations are regarded as physically equivalent if they are connected by a diffeomorphism transformation. Technically, this is analogous in certain respects to the situation in electromagnetism whereby a vector potential $A_\mu$ is equivalent to $A_\mu + \partial_\mu f$ for all functions $f$. However, there is an important difference between electromagnetism and general relativity. Electromagnetic gauge transformations occur at a fixed spacetime point $X$, and the physical configurations can be identified with the values of the field tensor $F_{\mu\nu}(X)$, which depends *locally* on points of $\mathcal{M}$. On the other hand, a diffeomorphism maps one spacetime point to another, and therefore one obvious way of constructing a diffeomorphism-invariant object is to take a scalar function of spacetime fields and integrate it over the whole of spacetime, which gives something that is very *non-local*. The idea that 'physical observables' are naturally non-local is an important ingredient in some approaches to quantum gravity.

#### 2.3.2.2 *Diffeomorphisms in quantum gravity*

The role of diffeomorphisms in quantum gravity depends strongly on the approach taken to the subject. For example, if the structure of classical relativity is expected to appear only in a low-energy limit – as, for example, is the case for superstring theory – there is no strong reason to suppose that the group of spacetime diffeomorphisms $\mathcal{D}$ will play any fundamental role in the quantum theory. On the other hand, in schemes which involve the active quantization of the classical gravitational field, $\mathcal{D}$ is likely to be a key ingredient in forcing the quantum theory to comply with the demands of general relativity. However, it should be noted that the situation in 'canonical' quantum gravity is less clear-cut: this programme is based on a prior decomposition into space plus time, and this is bound to obscure the role of spacetime diffeomorphisms.

In general terms, there are at least three[35] ways in which $\mathcal{D}$ could appear in the quantum theory; which will be exemplified in the programmes surveyed in Sections 2.4 and following sections:

(i) as an *exact* covariance/invariance group;[36]
(ii) as a *subgroup* of a bigger group;
(iii) as a limited concept associated with a phenomenological view of spacetime (or space).

In the first two options one could say that the diffeomorphisms form a 'precise' concept since the mathematical object that occurs in the formalism is exactly the classical group $\mathcal{D}$. The third option, (iii), is somewhat different and flows naturally from the view that spacetime is a phenomenological concept of limited applicability: the same would then be expected for the diffeomorphisms of the manifold that models spacetime in this limited sense. We shall say more about this in Section 2.5.

The idea that $\mathcal{D}$ is an exact covariance/invariance group (option (i) above) plays a key role in several approaches to quantum gravity. For example, (i) is one of the central properties of so-called 'topological quantum field theory', which seems to have potential applications in quantum gravity. And we will see in Section 2.4 that (i) also plays a major role in the particle-physics programme (Section 2.4.3), albeit with the qualification mentioned in note 36 [to option (i)]; and in a less clear-cut way, in canonical quantum gravity.

On the other hand, the idea that $\mathcal{D}$ is a subgroup of a bigger covariance group (option (ii) above) is endorsed by the perturbative approach to superstring theory. In short, the idea is that the extra fields associated with supersymmetry lead to a much larger covariance group; more details are provided in Section 2.4.4.

### 2.3.3 The problem of time

Closely related to the role of diffeomorphisms is the infamous 'problem of time'. This problem is central in any approach to quantum gravity that assigns a significant *prima facie* role to classical general relativity (unlike, say, superstring theory). For the problem arises from the very different roles played by the concept of time in quantum theory and in general relativity; and the problem lies at the heart of many of the deepest conceptual issues in such approaches to quantum gravity. To present the problem, we will consider the roles of time, first in quantum theory, and then in general relativity.[37]

#### 2.3.3.1 *Time in quantum theory*

In quantum theory, time is not a physical quantity in the normal sense, since it is not represented by an operator. Rather, it is treated as a *background* parameter which, as in classical physics, is used to mark the evolution of the system; witness the parameter $t$ in the time-dependent Schrödinger equation.[38]

Besides, the idea of an event happening at a given time plays a crucial role in the technical and conceptual foundations of quantum theory:

- One of the central requirements of the *scalar product* on the Hilbert space of states is that it is conserved under the time evolution given by the Schrödinger equation. This is closely connected to the unitarity requirement that probabilities always sum to one.
- More generally, a key ingredient in the construction of the Hilbert space for a quantum system is the selection of a complete set of quantities that are required to commute at a fixed value of time.
- Conceptually, the notion of *measuring* a quantity at a given time, to find its value at that time, is a fundamental ingredient of both the minimal statistical interpretation of the theory, and the Copenhagen interpretation (see Section 2.2.1.1).

Furthermore, all these ideas can be extended to systems that are compatible with special relativity: the unique time system of Newtonian physics is simply replaced with the set of relativistic inertial reference frames. The quantum theory can be made independent of a choice of frame, provided that the theory carries a unitary representation of the Poincaré group of isometries of the metric of Minkowski spacetime. In the case of a relativistic quantum field theory, the existence of such a representation is closely related to the microcausality requirement that fields evaluated at spacelike-separated points must commute. For example, a scalar quantum field $\hat{\phi}(X)$ satisfies the commutation relations

$$[\hat{\phi}(X), \hat{\phi}(Y)] = 0 \tag{2.2}$$

whenever the spacetime points $X$ and $Y$ are spacelike separated.

Finally, we note that this background time is truly an abstraction in the sense that according to quantum theory, no *physical* clock can provide a precise measure of it (Unruh and Wald 1989): there is always a small probability that a real clock will sometimes run backwards with respect to it.

### 2.3.3.2 *Time in general relativity; and the problem of time*

When we turn to classical general relativity, the treatment of time is very different. Time is not treated as a background parameter, even in the liberal sense used in special relativity, viz. as an aspect of a fixed, background spacetime structure. Rather, what counts as a choice of a time (i.e of a timelike direction) is influenced by what matter is present (as is, of course, the spatial metrical structure). The existence of many such times is reflected in the fact that if the spacetime manifold has a topology such that it can be foliated as a one-parameter family of spacelike surfaces, this can generally be done in many ways – without any subset of foliations being singled out in the way families of inertial reference frames are singled out in special relativity. From one perspective, each such parameter might be regarded as a legitimate definition of (global) time. However, in general, there is no way of selecting a particular foliation, or a special family of such, that is 'natural' within the context of the theory alone. In particular, these definitions of time are in general unphysical, in that they provide no hint as to how their time might be measured or registered.

But the main problem about time in general relativity arises when we turn to quantum gravity, where the disparate nature of the treatments of time in quantum theory and in general relativity becomes of paramount significance. We shall see various more specific versions of this problem in each of the research programmes reviewed in Section 2.4. But for the moment, we introduce the problem in general terms.

General relativity accustoms us to the ideas that: (i) the causal structure of spacetime depends on the metric structure, represented by the metric tensor $\gamma$; and (ii) the metric and causal structures are influenced by matter, and so vary from one model of the theory to another. In general relativity, these ideas are 'kept under control' in the sense that in each model, there is of course a *single* metric tensor $\gamma$, representing a single metric and causal structure. But once we embark on constructing a quantum theory of gravity, we expect some sort of quantum fluctuations in the metric, and so also in the causal structure. But in that case, how are we to formulate a quantum theory with a fluctuating causal structure?

This general statement of the problem is clearly relevant if one proposes a spacetime-oriented approach to formulating the quantum theory; since then one's prototype quantum theories will emphasize a fixed background causal structure. But the same statement of the problem arises on various other approaches to quantum gravity. For example, if one takes the view that the spacetime metric is only a coarse-grained, phenomenological construct of some type, then so is the causal structure. And again, the question arises how we are to formulate a quantum theory with such a causal structure.[39]

Though this problem is at bottom conceptual, it has clear technical aspects. In particular, a probabilistic causal structure poses severe technical problems for relativistic quantum field theory, whose standard formulation presupposes a fixed causal structure. For example, a quantum scalar field satisfies the microcausal commutation relations in eqn. 2.2, whereby fields evaluated at spacelike separated spacetime points commute. However, the concept of two points being spacelike separated has no meaning if the spacetime metric is probabilistic or phenomenological. In the former case, the most likely scenario is that the right-hand side of the commutator in eqn. 2.2 never vanishes, thereby removing one of the foundations of conventional quantum field theory.

In practice, the techniques that have been used to address the problem of time fall into one of the following three strategies, to all of which we shall return in Section 2.4:

(1) Use a fixed background metric – often chosen to be that of Minkowski spacetime – to define a fiducial causal structure with respect to which standard quantum field theoretical techniques can be employed. This is the strategy adopted by the old particle-physics programme, using Minkowski spacetime (Section 2.4.3).

This strategy raises questions about how the background structure is to be chosen. Of course, Minkowski spacetime seems very natural from the perspective of standard quantum field theory, but it is rather arbitrary when seen in the context of general relativity. One possibility is that the background structure could come from a *contingent* feature of the actual universe; for example, the $3°$ K microwave background radiation. However, structure of this type is approximate and is therefore applicable only if fine details are ignored. In addition, the problem of rigorously constructing (even free) quantum fields has only been solved for a very small number of background spacetimes; certainly there is no reason to suppose that well-defined quantum field theories exist on a generic spacetime manifold. Also, there is a general matter of principle: should we require a theory of quantum gravity to work for 'all possible' universes (however that is made precise), or can it depend on special features of the actual one in which we live?

(2) Accept the fact that there is no background spacetime reference system and attempt to locate events, both spatially and temporally, with specific functionals of the gravitational and other fields. This important idea is of course motivated by the analysis of the 'hole argument' and spacetime diffeomorphisms (see Sections 2.2.2 and 2.3.2). Thus the idea is that for the example of a scalar field $\phi$, the value $\phi(X)$ of $\phi$ at a particular spacetime point $X$ has no physical meaning because of the action of the spacetime diffeomorphism group; but, the value of $\phi$ where something 'is' *does* have a physical meaning in the sense that '$\phi$(thing)' is diffeomorphism invariant. In practice, this strategy seems only to have been adopted by some of the approaches to the problem of time as it manifests itself in canonical quantum gravity.

(3) And indeed, one approach is to drop spacetime methods and instead adopt a canonical approach to general relativity, so that the basic ingredients are geometrical fields on a three-dimensional manifold. The problem then is to reconstruct some type of spatio-temporal picture within which the quantum calculations can be interpreted. This is the strategy adopted by the canonical quantum gravity programme (Section 2.4.5).

Studies of the problem of time in canonical quantum gravity raise the alluring question whether a meaningful quantum theory can be constructed in a way that contains no fundamental reference to time at all. That this is a far from trivial matter is shown by our earlier remarks about the crucial role of time in conventional quantum theory (see the references in note 37).

## 2.4 Research programmes in quantum gravity

As we have seen, we are far from having an 'axiomatic' framework for quantum gravity, or even a broad consensus about what to strive for beyond the minimal requirement that the theory should recover classical general relativity and normal quantum theory in the appropriate domains – usually taken to be all physical regimes well away from those characterized by the Planck length.

In this section, we shall focus on three specific research programmes. Our aim is not to review the technical status of these programmes, but rather to explore their treatments of spacetime. Of these three programmes, two are the main current focus for work in quantum gravity: superstring theory and canonical quantum gravity. These programmes complement each other nicely, and enable the special ideas of either of them to be viewed in a different perspective by invoking the other – a feature that is rather useful in a subject that lacks unequivocal experimental data! For example, they exemplify the choice of approaches we discussed in Section 2.3.1.3, about whether to quantize general relativity, or to have classical general relativity emerge from a quantum theory of something quite different. Superstring theory takes the latter approach; canonical quantum gravity the former.

The other programme we shall discuss is a spacetime-oriented quantization of general relativity, which we dub 'the particle-physics programme'. This programme is no longer regarded as capable of providing a full theory of quantum gravity; but it predated, and so influenced, both the other two programmes, and this means that discussing its own treatment of spacetime will form a helpful backdrop to discussing theirs.[40]

All three programmes postulate at the fundamental level a spacetime manifold. But it may differ in its dimension, metric structure, etc. from the four-dimensional Lorentzian manifold familiar from classical general relativity. And, in fact, these programmes suggest limitations to the applicability of the concept of a spacetime manifold itself. We shall explore this possibility further in Section 2.5.

We begin in Section 2.4.1 by listing four topics that will act as 'probes' in our survey of how concepts of space and time are treated in these programmes. This is followed by some historical orientation to these programmes (Section 2.4.2), and then we consider them *seriatim*, in the following order: the particle-physics programme (Section 2.4.3); superstring theory (Section 2.4.4); and canonical quantum gravity (Section 2.4.5).

### 2.4.1 Focussing the question: how is spacetime treated?

We will take the following four topics as 'probes' in our survey of how the concept of spacetime is treated in the various quantum gravity programmes. We present them as a sequence of questions; but of course they overlap with one another.

(1) *Use of standard quantum theory.* Are the technical formalism and conceptual framework of present-day quantum theory adequate for the programme's envisaged theory of quantum gravity? In particular, do any features of the programme suggest advantages, or indeed disadvantages, of the 'heterodox' interpretations of quantum theory, discussed in Sections 2.2.1.2–2.2.1.4?

(2) *Use of standard spacetime concepts.* How much of the familiar treatment of spatio-temporal concepts, adopted by general relativity and quantum theory, does the programme adopt? In particular, does it model spacetime as a four-dimensional differentiable manifold? If so, does it add to this manifold a quantized metric tensor? And if so, what exactly is the relation of this to a classical Lorentzian metric on the manifold?

(3) *The spacetime diffeomorphism group.* Assuming the programme models spacetime as a manifold, what role does it assign to the group of spacetime diffeomorphisms? Or if it decomposes spacetime into space and time, what role is given to spatial diffeomorphisms?

(4) *The problem of time.* How does the problem of time manifest itself in the programme, and how does the programme address it? In particular: how much of the familiar treatment of spacetime must be retained for the envisaged theory of quantum gravity to be constructed? Must both metric and manifold be fixed; or can we work with a fixed background manifold, but no background metric? Could we also do without the background manifold?

### 2.4.2 Some historical background to the three programmes

To introduce the survey of our three chosen research programmes, it is useful to sketch some of the historical development of quantum gravity research – a development from which these programmes have all sprung.

The early history of attempts to quantize general relativity goes back at least to the 1960s, and was marked by a deep division of opinion about whether quantum gravity should be tackled from a spacetime perspective – the so-called 'covariant' approach, whose leading champion was Bryce DeWitt – or from a 'canonical' approach, in which spacetime is decomposed into space plus time before the theory is quantized.

The early predominance of the canonical programme stemmed partly from the fact that the attitude in the 1960s towards quantum field theory was very different from that of today. With the exception of quantum electrodynamics, quantum field theory was poorly rated as a fundamental way of describing the interactions of elementary particles. Instead, this was the era of the S-matrix, the Chew axioms, Regge poles, and – towards the end of that period – the dual resonance model and the Veneziano amplitude that led eventually to superstring theory.

In so far as it was invoked at all in strong interaction physics, quantum field theory was mainly used as a phenomenological tool to explore the predictions of current algebra, which was thought to be more fundamental. When quantum field theory was studied seriously, it was largely in the context of an 'axiomatic' programme – such as the Wightman axioms for the $n$-point functions.

This neglect of quantum field theory influenced the way quantum gravity developed. In particular, with a few notable exceptions, physicists trained in particle physics and quantum field theory were not interested in quantum gravity, and the subject was mainly left to those whose primary training had been in general relativity. This imparted a special flavour to much of the work in that era. In particular, the geometrical aspects of the theory were often emphasized at the expense of quantum field theoretic issues – thereby giving rise to a tension that has affected the subject to this day.

However, a major change took place in the early 1970s when t'Hooft demonstrated the renormalizability of quantized Yang–Mills theory. Although not directly connected with gravity, these results had a strong effect on attitudes towards quantum field theory in general, and reawakened a wide interest in the subject. One spin-off was that many young workers in particle physics became intrigued by the challenge of applying the new methods to quantum gravity. This led to a revival of the covariant approach; more specifically, to the particle-physics programme (Section 2.4.3), and thereby eventually to supergravity and superstring theory (Section 2.4.4).

On the other hand, the canonical programme has also continued to flourish since the 1970s; indeed, until the relatively recent advent of 'superstring cosmology', canonical quantum gravity provided the only technical framework in which to discuss quantum cosmology. A major development in canonical quantum gravity was Ashtekar's discovery in 1986 of a new set of variables that dramatically simplifies the intractable Wheeler–DeWitt equation which lies at the heart of the programme's quantum formalism. This in turn led to Rovelli's and Smolin's discovery of yet other variables, labelled by loops in space.

### 2.4.3 The particle-physics programme

#### 2.4.3.1 *The basic ideas*

In this programme, the basic entity is the *graviton* – the quantum of the gravitational field. Such a particle is deemed to propagate in a background Minkowski spacetime, and – like all elementary particles – is associated with a specific representation of the Poincaré group which is labelled by its mass and its spin. The possible values of mass and spin are sharply limited by the physical functions which the graviton is to serve. In particular, replication of the inverse-square law behaviour of the static gravitational force requires the graviton to have mass zero, and the spin must be either 0 or $2\hbar$. However, zero spin is associated with a scalar field $\phi(X)$, whereas spin-2 comes from a symmetric Lorentz tensor field $h_{\mu\nu}(X)$; the obvious implication is that these spin values correspond to Newtonian gravity and general relativity respectively.

The key to relating spin-2 particles and general relativity is to fix the background topology and differential structure of spacetime $\mathcal{M}$ to be that of Minkowski spacetime, and then to write the Lorentzian metric $\gamma$ on $\mathcal{M}$ as

$$\gamma_{\alpha\beta}(X) = \eta_{\alpha\beta} + \kappa h_{\alpha\beta}(X). \tag{2.3}$$

Here, $h$ measures the departure of $\gamma$ from the flat spacetime metric $\eta$ and is regarded as the 'physical' gravitational field with the coupling constant $\kappa^2 = 8\pi G/c^2$, where $G$ is Newton's constant.

The use of the expansion in eqn. 2.3 strongly suggests a perturbative approach in which quantum gravity is seen as a theory of small quantum fluctuations around a background Minkowski spacetime. Indeed, when this expansion is substituted into the Einstein–Hilbert action for pure gravity, $S = \int d^4X |\gamma|^{1/2} R(\gamma)$ (where $R(\gamma)$ is the scalar curvature), it yields (i) a term that is bilinear in the fields $h$ and which – when quantized in a standard way – gives a theory of non-interacting, massless spin-2 gravitons; and (ii) a series of higher-order terms that describe the interactions of the gravitons with each other. Thus, a typical task would be to compute the probabilities for various numbers of gravitons to scatter with each other and with the quanta of whatever matter fields might be added to the system.

This approach to quantum gravity has some problematic conceptual features (see below). But, nonetheless, had it worked it would have been a major result, and would undoubtedly have triggered a substantial effort to construct a spacetime-focussed quantum gravity theory in a non-perturbative way. A good analogue is the great increase in studies of lattice gauge theory that followed the proof by t'Hooft that Yang–Mills theory is perturbatively renormalizable.

However, this is not what happened. Instead, a number of calculations were performed around 1973 that confirmed earlier suspicions that perturbative quantum gravity is non-renormalizable.[41] There have been four main reactions to this situation:

- Adopt the view in which general relativity is an 'effective field theory' and simply add as many empirically determined counterterms as are appropriate at the energy concerned. The ensuing structure will break down at the Planck scale, but a pragmatic particle physicist might argue that this is of no importance since the Planck energy is so much larger than anything that could be feasibly attainable in any foreseeable particle accelerator (see the discussion at the end of the second approach in Section 2.3.1.3).
- Continue to use standard perturbative quantum field theory, but change the classical theory of general relativity so that the quantum theory becomes renormalizable. Examples of such attempts include: (i) adding higher powers of the Riemann curvature $R^\alpha_{\beta\mu\nu}(\gamma)$ to the action; and (ii) supergravity (see Section 2.4.4).
- Keep classical general relativity as it is, but develop quantization methods that are intrinsically non-perturbative. Examples of this philosophy are 'Regge calculus'

(which involves simplicial approximations to spacetime) and techniques based on lattice gauge theory. Of particular importance in recent years is the Ashtekar programme for canonical quantization which is fundamentally non-perturbative (see Section 2.4.5.2).

- Adopt the view that the non-renormalizability of perturbative quantum gravity is a catastrophic failure that requires a very different type of approach. In terms of the classification in Section 2.3.1.3, this would mean adopting its second, or third or fourth types of approach: quantizing a classical theory that is quite different from general relativity (such as a string theory); or having general relativity emerge as a low-energy limit of a quantum theory that is not a quantization of any classical system; or having it and quantum theory both emerge from something completely different.

### 2.4.3.2 *Spacetime according to the particle-physics programme*

The response given by this programme to our four conceptual probes presented in Section 2.4.1 is as follows.

1. *Use of standard quantum theory*. The basic technical ideas of standard quantum theory are employed, suitably adapted to handle the gauge structure of the theory of massless spin-2 particles. Furthermore, the traditional, Copenhagen interpretation of the theory is applicable (even if not right!), in that the background Minkowski metric and spacetime manifold are available to serve as the classical framework, in which measurements of the quantum system, according to this interpretation, are to be made.

Of course, other interpretations of quantum theory discussed in Section 2.2.1, such as Everettian or pilot-wave interpretations, may well also be applicable to this programme. Our present point is simply that the particle-physics programme gives no special reasons in favour of such interpretations.

2. *Use of standard spacetime concepts*. The background manifold and metric are described in the language of standard differential geometry. Note that, from a physical perspective, the restriction to a specific background topology means a scheme of this type is not well adapted to addressing some of the most interesting questions in quantum gravity such as the role of black holes, quantum cosmology, the idea of possible spacetime 'phase changes', etc.

3. *The spacetime diffeomorphism group*. The action of the group of spacetime diffeomorphisms is usually studied infinitesimally. The transformation of the graviton field $h_{\mu\nu}(X)$ under a vector-field generator $\xi$ of such a diffeomorphism is simply $h_{\mu\nu}(X) \mapsto h_{\mu\nu}(X) + \partial_\mu \xi_\nu(X) + \partial_\nu \xi_\mu(X)$, which is very reminiscent of the gauge transformations of the electromagnetic vector potential, $A_\mu(X) \mapsto A_\mu(X) + \partial_\mu \phi(X)$.

Indeed, in this infinitesimal sense, the effect of spacetime diffeomorphisms is strictly analogous to the conventional gauge transformations of electromagnetism or Yang–Mills theory (in that spacetime points are fixed); and the same type of quantization procedure can be used. In particular, the invariance of the quantum theory under these transformations is reflected in a set of 'Ward identities' that must be satisfied by the vacuum expectation values of time-ordered products (the '$n$-point functions') of the operator field at different spacetime points.

4. *The problem of time*. The background metric $\eta$ provides a fixed causal structure with the associated family of Lorentzian inertial frames. Thus, at this level, there

is no problem of time. The causal structure also allows a notion of microcausality, thereby permitting a conventional type of relativistic quantum field theory to be applied to the field $h_{\alpha\beta}$.

However, many people object strongly to an expansion like eqn. 2.3 since it is unclear how this background causal structure is to be related to the physical one; or, indeed, what the latter really means. For example, does the 'physical' causal structure depend on the state of the quantum system? There have certainly been conjectures to the effect that a non-perturbative quantization of this system would lead to quantum fluctuations of the causal structure around a quantum-averaged background that is not the original Minkowskian metric. As emphasized earlier, it is not clear what happens to the microcausal commutativity condition in such circumstances; or, indeed, what is meant in general by 'causality' and 'time' in a system whose light-cones are themselves the subject of quantum fluctuations.

### 2.4.4 The superstrings programme

#### 2.4.4.1 *The introduction of supersymmetry*

When confronted with the non-renormalizability of covariant quantum gravity, the majority of particle physicists followed a line motivated by the successful transition from the old non-renormalizable theory of the weak interactions (the 'four-fermion' theory) to the new renormalizable unification of the weak and electromagnetic forces found by Salam, Glashow, and Weinberg. Thus the aim was to construct a well-defined theory of quantum gravity by adding carefully chosen matter fields to the classical theory of general relativity with the hope that the ultraviolet divergences would cancel, leaving a theory that is perturbatively well-behaved.

A key observation in this respect is that the divergence associated with a loop of gravitons might possibly be cancelled by introducing *fermions*, on the grounds that the numerical sign of a loop of virtual fermions is opposite to that of a loop of bosons. With this motivation, supergravity was born, the underlying supersymmetry invariance being associated with a spin-$\frac{3}{2}$ fermionic partner (the 'gravitino') for the bosonic spin-2 graviton. Moreover, since supersymmetry requires very special types of matter, such a scheme lends credence to the claim that a successful theory of quantum gravity must involve unifying the fundamental forces, i.e. the extra fields needed to cancel the graviton infinities might be precisely those associated with some grand unified scheme.

Recall from Section 2.3.1 that the mark of unification of two forces is the equality of their coupling constants; and that the energy-dependence of the coupling constants for the electromagnetic, weak and strong nuclear forces renders them at least approximately equal at around $10^{20}$ MeV. Here, it is of considerable interest to note that the introduction of supersymmetry greatly improves the prospects for exact equality: there is every expectation that these three fundamental forces do unify exactly in the supersymmetric version of the theory. The fact that $10^{20}$ MeV is not that far away from the Planck energy ($10^{22}$ MeV) adds further weight to the idea that a supersymmetric version of gravity may be needed to guarantee its inclusion in the pantheon of unified forces.

Early expectations for supergravity theory were high following successful low-order results, but it is now generally accepted that if higher-loop calculations could be performed (they are very complex) intractable divergences would appear once more. However, this line of thought continues and the torch is currently carried by the superstring programme, which in terms of number of papers produced per week is now by far the dominant research programme in quantum gravity.

The superstrings programme has had two phases. The first phase began in earnest in the mid-1980s (following seminal work in the mid-1970s) and used a perturbative approach; as we shall see, its treatment of spacetime can be presented readily enough in terms of our four probes listed in Section 2.4.1. The second phase began in the early 1990s, and 'still rages'. It has yielded rich insights into the underlying non-perturbative theory. But the dust has by no means settled. Even the overall structure of the underlying theory remains very unclear; and in particular, its treatment of spacetime is too uncertain for our four probes to be applied. Accordingly, our discussion of the second phase will forsake the probes, and just report how recent developments indicate some fundamental limitations in the manifold conception of spacetime.[42]

### 2.4.4.2 *Perturbative superstrings*

The perturbative superstrings programme involves quantizing a classical system; but the system concerned is not general relativity, but rather a system in which a one-dimensional closed string propagates in a spacetime $\mathcal{M}$ (whose dimension is in general not 4). More precisely, the propagation of the string is viewed as a map $X : \mathcal{W} \to \mathcal{M}$ from a two-dimensional 'world-sheet' $\mathcal{W}$ to spacetime $\mathcal{M}$ (the 'target spacetime'). The quantization procedure quantizes $X$, but not the metric $\gamma$ on $\mathcal{M}$, which remains classical. The appropriate classical theory for the simplest such system is described by the famous Polyakov action, which is invariant under conformal transformations on $\mathcal{W}$. To preserve this conformal invariance in the quantized theory, and to satisfy other desirable conditions, the following conditions are necessary:

(i) the theory is made supersymmetric;
(ii) the spacetime $\mathcal{M}$ has a certain critical dimension (the exact value depends on what other fields are added to the simple bosonic string); and
(iii) the classical spacetime metric $\gamma$ on $\mathcal{M}$ satisfies a set of field equations that are equivalent to the (supergravity version of) Einstein's field equations for general relativity plus small corrections of Planck size: this is the sense in which general relativity emerges from string theory as a low-energy limit.[43]

Superstring theory has the great advantage over the particle-physics programme that, for certain string theories, the individual terms in the appropriate perturbation expansion *are* finite. Furthermore, the particle content of theories of this type could be such as to relate the fundamental forces in a unified way. Thus these theories provide a concrete realization of the old hope that quantum gravity necessarily involves a unification with the other fundamental forces in nature.

### 2.4.4.3 *Spacetime according to the perturbative superstrings programme*

The response given by the perturbative superstrings programme to the four conceptual probes in Section 2.4.1 is broadly as follows.

1. *Use of standard quantum theory*. As in the case of the particle-physics programme, the basic technical ideas of standard quantum theory are employed, albeit with suitable adaptations to handle the gauge structure of the theory. Similarly, the Copenhagen interpretation of the theory is, arguably, applicable; namely, by using as the required classical framework or background structure, the solution of the low-energy field equations for the spacetime metric $\gamma$. On the other hand, one can also argue contrariwise, that it is unsatisfactory to have the interpretation of our fundamental quantum theory only apply in a special regime, viz. low energies.

And as for the particle-physics programme, other interpretations of quantum theory discussed in Section 2.2.1 might be applicable. But they have not been developed in connection with superstrings, and so far as we can see, perturbative superstring theory gives no special reason in favour of them.

2. *Use of standard spacetime concepts*. In perturbative superstring theories, the target spacetime $M$ is modelled using standard differential geometry, and there seems to be no room for any deviation from the classical view of spacetime. However, in so far as the dimension of $M$ is greater than four, some type of 'Kaluza–Klein' scenario is required in which the extra dimensions are sufficiently curled up to produce no perceivable effect in normal physics, whose arena is a four-dimensional spacetime.

3. *The spacetime diffeomorphism group*. Superstring theory shows clearly how general relativity can occur as a fragment of a much larger structure – thereby removing much of the fundamental significance formerly ascribed to the notions of space and time. True, the low-energy limit of these theories is a form of supergravity but, nevertheless, standard spacetime ideas do not play a central role. This is reflected by the graviton being only one of an infinite number of particles in the theory. In particular, the spacetime diffeomorphism group $D$ appears only as part of a much bigger structure, as in option (ii) of the discussion in Section 2.3.2. Consequently, its technical importance for the quantum scheme is largely subsumed by the bigger group.

4. *The problem of time*. The perturbative expansion in a superstring theory takes place around the background given by the solution to the low-energy field equations for the spacetime metric $\gamma$. In particular, this provides a background causal structure; and hence – in that sense – there is no problem of time.

On the other hand, the situation is similar in many respects to that of the particle-physics programme; in particular, there are the same worries about the meaning of 'causality' and 'time' in any precise sense. There is, however, an important difference between superstring theory and the simple perturbative quantization of general relativity – namely, there are realistic hopes that a proper non-perturbative version of the former is technically viable. If such a theory is found it will be possible to address the conceptual problems with time and causality in a direct way – something that is very difficult in the context of the (mathematically non-existent!) theory of perturbatively quantized general relativity.

#### 2.4.4.4 *The second phase of superstrings*

Since the early 1990s, a lot of work in the superstrings programme has focussed on exploring the underlying *non*-perturbative theory. These developments seem to have striking implications for our conception of space and time at the Planck scale. So now we turn to these, albeit with trepidation, since the dust has by no means yet settled in this area of research; and *what* the implications are is far from clear!

These developments are based on various types of 'duality' transformation or symmetry. For example, one of the simplest forms of duality ('$T$-duality') arises when the target space is a five-dimensional manifold of the form $\mathcal{M}_4 \times S^1$ ($S^1$ is the circle). It transpires that the physical predictions of the theory are invariant under replacement of the radius $R$ of the fifth dimension with $2\alpha'/R$ where $\alpha'$ is a fundamental constant in the theory. Thus we cannot differentiate physically between a very small, and a very large, radius for the additional dimension – indeed, there is a precise sense in which they are 'gauge' equivalent to each other. One of the most important implications of this invariance is that there exists a minimum length of $R_{min} = \sqrt{2}\alpha'$ – an idea that must surely have significant implications for our overall understanding of the conceptual implications of the theory.[44]

Another type of duality ('$S$-duality') involves the idea that, for certain theories, the physics in the large-coupling limit is given by the weak-coupling limit of a 'dual' theory whose fundamental entities can be identified with solitonic excitations of the original theory. It is believed that the several known consistent perturbative superstring theories are related in this way to each other[45] and also to the theories involving extended objects ('membranes') of dimension greater than one. Ideas of this type are certainly attractive, not least because they provide a real possibility of theoretically probing the physically interesting, high-energy regimes of such theories.

In short, these developments suggest rather strongly that the manifold conception of spacetime is not applicable at the Planck length; but is only an emergent notion, approximately valid at much larger length–scales. We shall take up this idea, in general terms, in Section 2.5. At a more technical level, the new ideas suggest that Lagrangian field-theoretic methods (which are used in the perturbative superstring theories) are reaching the limit of their domain of applicability, and should be replaced by (for example) a more algebraic approach to theory construction that places less reliance on an underlying classical system of fields (i.e. the third of the four types of approach we listed in Section 2.3.1.3).

### 2.4.5 The canonical quantum gravity programme

#### 2.4.5.1 *Quantum geometrodynamics*

The canonical approach to quantum gravity starts with a reference foliation of spacetime with respect to which the appropriate canonical variables are defined.[46] These are the 3-metric $g_{ab}(x)$ on a spatial slice $\Sigma$ of the foliation, and a canonical conjugate $p^{ab}(x)$ that – from a spacetime perspective – is related to the extrinsic curvature of $\Sigma$ as embedded in the four-dimensional spacetime.

A key property of general relativity – which reflects the role of the group $\mathcal{D}$ of spacetime diffeomorphisms – is that these variables are not independent, but instead satisfy certain constraints, usually written as

$$\mathcal{H}_a(x) = 0 \tag{2.4}$$

$$\mathcal{H}_\perp(x) = 0, \tag{2.5}$$

where $\mathcal{H}_a(x)$ and $\mathcal{H}_\perp(x)$ are complicated functions of the $g$ and $p$, and their derivatives.

The constraint functions $\mathcal{H}_a$ and $\mathcal{H}_\perp$ play a fundamental role in the theory since their Poisson bracket algebra (known as the 'Dirac algebra') is that of the group $\mathcal{D}$ of spacetime diffeomorphisms projected along, and normal to, the spacelike hypersurfaces of the reference foliation. Thus, the basic question of understanding the role of spacetime diffeomorphisms is coded in the structure of these constraints.

In addition to the constraint equations (eqns. 2.4–2.5), there is also a collection of dynamical equations that specify how the canonical fields $g_{ab}(x)$ and $p^{cd}(x)$ evolve with respect to the time variable associated with the given foliation. However, it transpires that these equations are redundant since it can be shown that if $\gamma$ is a spacetime metric on $\mathcal{M}$ that satisfies the constraint equations (eqns. 2.4–2.5) on any spacelike hypersurface, then *necessarily* the projected canonical variables $g_{ab}(x)$ and $p^{cd}(x)$ will satisfy the dynamical equations. In this sense, the entire theory is already coded into just the four constraint equations (eqns. 2.4–2.5); so, in practice, attention is almost invariably focussed on them alone. Furthermore, among these equations, (eqn. 2.5) is the crucial one, essentially because – when viewed from a spacetime perspective – $\mathcal{H}_\perp$ is associated with the canonical generators of displacements in timelike directions.

This system can be quantized in a variety of ways. One possibility is to impose a gauge for the invariance associated with the Dirac algebra; solve the constraint equations (eqns. 2.4–2.5) classically; and then quantize the resulting 'true' canonical system in a standard way. However, the final equations are intractable in anything other than a perturbative sense, where they promptly succumb to virulent ultraviolet divergences.

Most approaches to canonical quantum gravity do not proceed in this way. Instead, the full set of fields $(g_{ab}(x), p^{cd}(x))$ is quantized via the 'canonical commutation relations'

$$[\hat{g}_{ab}(x), \hat{g}_{cd}(x')] = 0 \tag{2.6}$$

$$[\hat{p}^{ab}(x), \hat{p}^{cd}(x')] = 0 \tag{2.7}$$

$$[\hat{g}_{ab}(x), \hat{p}^{cd}(x')] = i\hbar \delta^c{}_{(a}\delta^d_{b)}\delta^{(3)}(x, x') \tag{2.8}$$

of operators defined on the 3-manifold $\Sigma$. Following Dirac, the constraints are interpreted as constraints on the allowed state vectors $\Psi$, so that $\hat{\mathcal{H}}_a(x)\Psi = 0 = \hat{\mathcal{H}}_\perp(x)\Psi$ for all $x \in \Sigma$. In particular, on choosing the states as functions of the 3-geometry $g$ – and with operator representatives $(\hat{g}_{ab}(x)\Psi)[g] := g_{ab}(x)\Psi[g]$ and $(\hat{p}^{cd}(x)\Psi)[g] := -i\hbar\, \delta\Psi[g]/\delta g_{ab}(x)$ – the constraints $\hat{\mathcal{H}}_a\Psi = 0$ imply that $\Psi[g]$ is constant under changes of $g$ induced by infinitesimal diffeomorphisms of the spatial 3-manifold $\Sigma$; and the crucial constraint $\hat{\mathcal{H}}_\perp(x)\Psi = 0$ becomes the famous Wheeler–DeWitt equation.

But the Wheeler–DeWitt equation is horribly ill-defined in any exact mathematical sense and, unfortunately, perturbative approaches to its definition and solution are as virulently badly behaved as are its particle-physics-based cousins: in both cases the problem is trying to define products of operator fields defined at the same point. Indeed, until the rise of the Ashtekar programme (see below), most of the work developing and using the Wheeler–DeWitt equation in anything other than a purely heuristic sense relied on truncating the gravitational field to just a few degrees of freedom, so that it becomes a partial differential equation in a finite number of variables, which one can at least contemplate attempting to solve exactly.[47]

In this context, we should mention again the Euclidean programme in quantum gravity. Here, the central role is played by functional integrals over all the Riemannian – rather than Lorentzian – metrics on a four-dimensional manifold $\mathcal{M}$. (The motivation for Riemannian metrics is partly an analogy with the successful use of imaginary time in Yang–Mills theory.) It can be shown that the functional $\Psi[g]$ of $g$ defined by certain such functional integrals satisfies (at least, in a heuristic way) the Wheeler–DeWitt equation; in particular, this is the basis of the famous Hartle–Hawking 'no boundary' proposal for the 'wave function of the universe' in quantum cosmology. So in this sense, the Euclidean programme amounts to a way of constructing states for quantum geometrodynamics. For this reason, and also because we discuss this programme (and its use in quantum cosmology) in detail elsewhere (Butterfield and Isham 1999), we set it aside here.

### 2.4.5.2 *The Ashtekar programme and loop variables*

As emphasized above, the Wheeler–DeWitt equation is ill-defined in any exact mathematical sense. However, a major advance took place in 1986 when Ashtekar (1986) found a set of canonical variables which produce a dramatic simplification of the structure of the central constraint functions $\mathcal{H}_a(x)$ and $\mathcal{H}_\perp(x)$. Since then there has been a very active programme to exploit these new variables, both in classical general relativity and in quantum gravity.

From a technical perspective, one of the great dangers in canonical quantum gravity is the generation of anomalous quantum excitations of non-physical modes of the gravitational field. However, even to talk of such things requires the operators to be defined rigorously – a task that is highly non-trivial, since this is the point at which the infamous ultraviolet divergences are likely to appear. One of the main reasons why these new developments are potentially so important is the hope it offers of being able to define these operators properly, and hence address such crucial issues as the existence of anomalous excitations.

The developments during the past decade have been very impressive and, in particular, there is now real evidence in support of the old idea that non-perturbative methods must play a key role in constructing a theory of quantum gravity. If successful in its current form, this programme will yield a theory of quantum gravity in which unification of the forces is *not* a necessary ingredient. This demonstrates the importance of distinguishing between a quantum theory of gravity itself, and a 'theory of everything' which of necessity includes gravity.

In spite of their great structural significance, the use of the Ashtekar variables has had little impact so far on the conceptual problems in canonical quantum gravity, and so we shall not discuss the technical foundations of this programme here. However, it is important to note that one of the new variables is a spin-connection, which suggested the use of a gravitational analogue of the gauge-invariant loop variables introduced by Wilson in Yang–Mills theory. Seminal work in this area by Rovelli and Smolin (1990) has produced many fascinating ideas, including a demonstration that the area and volume of space are quantized – something that is evidently of philosophical interest and which has no analogue in quantum geometrodynamics. We shall return to these ideas briefly in Section 2.5.

2.4.5.3 *Spacetime according to the canonical quantum gravity programme*
The response given by the canonical quantum gravity programme to the four probes in Section 2.4.1 is broadly as follows.

1. *Use of standard quantum theory.* The basic technical ideas of standard quantum theory are employed, suitably adapted to handle the non-linear constraints satisfied by the canonical variables. On the other hand, the traditional, Copenhagen interpretation of quantum theory is certainly *not* applicable unless a background spatial metric is assumed, and a resolution is found of the problem of time, at least at some semiclassical level. However, most attempts to implement the canonical scheme abhor the introduction of any type of background metric, and hence major conceptual problems can be expected to arise if this programme is ever fully realized.

Of course, one radical strategy for coping with this situation is 'to make a virtue of necessity', as discussed in Section 2.2.1.3; i.e. to adopt the pilot-wave interpretation of quantum theory, and thereby introduce a background metric in a strong sense, involving a preferred foliation of spacetime. Note, however, that with its emphasis on configuration space, this interpretation would not seem appropriate for the Ashtekar and loops programmes where, unlike quantum geometrodynamics, the states are *not* functions on the configuration space of all 3-metrics.

Although we are setting aside quantum cosmology, we should add that since most work in quantum cosmology has been done within the canonical quantum gravity programme, there is another sense in which the Copenhagen interpretation is certainly not applicable to canonical quantum gravity; whereas (as discussed in Section 2.2.1.2), rivals such as Everettian interpretations might be.

2. *Use of standard spacetime concepts.* In this regard, canonical quantum gravity in effect 'lies between' the conservatism of the particle-physics programme, and the radicalism of superstrings. For like the former, it uses a background dimensional manifold (but it uses no background metric). More precisely, the canonical theory of classical relativity assumes *ab initio* that the spacetime manifold $\mathcal{M}$ is diffeomorphic to $\Sigma \times \mathbb{R}$ where $\Sigma$ is some 3-manifold; and this 3-manifold becomes part of the fixed background in the quantum theory – so that, as for the particle-physics programme, there is no immediate possibility of discussing quantum changes in the spatial topology.

3. *The spacetime diffeomorphism group.* In the canonical quantum gravity programme, the classical Poisson-bracket algebra of the constraint functions (i.e. the Dirac algebra) can be interpreted as the algebra of spacetime diffeomorphisms projected along, and normal to, spacelike hypersurfaces. In the quantum theory, it is usually assumed that this Poisson-bracket algebra is to be replaced with the analogous commutator algebra of the corresponding quantum operators.

The Dirac algebra contains the group of spatial diffeomorphisms, $\mathrm{Diff}(\Sigma)$, as a subgroup, but it is not itself a genuine group. Invariance under $\mathrm{Diff}(\Sigma)$ means that the functionals of the canonical variables that correspond to physical variables are naturally construed as being non-local with respect to $\Sigma$. Recent work with the loop-variable approach to canonical quantum gravity has been particularly productive in regard to the implications of invariance under spatial diffeomorphisms.

The role of the full Dirac algebra is more subtle and varies according to the precise canonical scheme that is followed. There are still contentious issues in this area – particularly in regard to exactly what counts as an 'observable' in the canonical scheme.

4. *The problem of time*. One of the main aspirations of the canonical approach to quantum gravity has always been to build a formalism with no background spatial, or spacetime, metric (this is particularly important, of course, in the context of quantum cosmology). In the absence of any such background structure, the problem of time becomes a major issue.

There are various obvious manifestations of this. One is that the Wheeler–DeWitt equation makes no apparent reference to time, and yet this is to be regarded as the crucial 'dynamical' equation of the theory! Another manifestation concerns the starting canonical commutation relations (eqns. 2.6–2.8). The vanishing of a commutator like eqn. 2.6 would normally reflect the fact that the points $x$ and $x'$ are 'spatially separated'. But what does this mean in a theory with no background causal structure?

The situation is usually understood to imply that, as mentioned in Section 2.3.3.2 (strategies 2 and 3), 'time' has to be reintroduced as the values of special *physical* entities in the theory – either gravitational or material – with which the values of other physical quantities are to be correlated. Thus, rather than talking about clocks *measuring* time – which suggests there is some external temporal reference system – we think of time as being *defined* by a clock, which in this case means part of the overall system that is being quantized. Thus, physical time is introduced as a reading on a 'physical' clock.

Unfortunately, it is a major unsolved problem whether: (i) this can be done at all in an exact way; and (ii) if so, how the results of two different such choices compare with each other, and how this is related to spatio-temporal concepts. In fact, there are good reasons for thinking that it is not possible to find any 'exact' internal time, and that the standard notion of time only applies in some semiclassical limit of the theory. In this way time would be an emergent or phenomenological concept, rather like temperature or pressure in statistical physics. We discuss this line of thought further in Section 2.5.4 of Butterfield and Isham (1999); here, we just emphasize that it is specifically about time, not spacetime – and in that sense, not this chapter's concern. Section 2.5 *will* discuss the idea that spacetime, though not distinctively time, is phenomenological.

## 2.5 Towards quantum spacetime?

### 2.5.1 Introduction: quantization and emergence

In this section, we turn to discuss some treatments of spacetime that are in various ways more radical than those given by the programmes in Section 2.4. We shall adopt the classification in Section 2.3.1.3 of four types of approach to quantum gravity.

Recall that they were:

(1) quantizing general relativity;
(2) quantizing a different classical theory, while still having general relativity emerge as a low-energy (large-distance) limit;
(3) having general relativity emerge as a low-energy limit of a quantum theory that is not a quantization of a classical theory; and finally, and most radically,
(4) having both general relativity and quantum theory emerge from a theory very different from both.

Thus in Section 2.5.2, we will discuss type (2); in Section 2.5.3, type (3); and in Section 2.5.4, type (4). But it will help set these discussions in context, to take up two topics as preliminaries: (i) the relation of the programmes in Section 2.4 to this classification; and (ii) the notion of emergence, and its relation to quantization, in general.

### 2.5.1.1 *Some suggestions from the three programmes*

It is easy to place the three programmes in Section 2.4 within the above classification. We have seen that the particle-physics and canonical programmes are examples of type (1), while the superstrings programme is an example of type (2) (at least in its perturbative version). But we should add two remarks to 'sketch in the landscape' of this classification.

(i) All three programmes are *similar* in that the main way they go beyond what we called the common treatment of spacetime of our 'ingredient theories' (viz. as a four-dimensional manifold with a (classical) Lorentzian metric), is by quantizing a quantity that is a standard type of physical variable within the context of classical physics defined using the familiar tools of differential geometry. For the particle-physics and canonical programmes (Sections 2.4.3, 2.4.5), this is the spatio-temporal metric $\gamma$, and the 3-metric $g$ respectively. In the perturbative superstrings programme, the variable concerned is the function $X$ that maps a loop into the target spacetime. However, in this section, we will discuss treatments that in some way or other 'go beyond' quantizing standard classical objects.

(ii) On the other hand, we should stress that the two main *current* programmes discussed in Section 2.4 make various radical suggestions about spacetime: suggestions which are not reflected by their being classified as types (1) and (2) in the schema above. We already saw some of these suggestions in Section 2.4. In particular, we saw that the superstrings programme requires spacetime to have a dimension which is in general not four; and – more strikingly – recent work on non-perturbative approaches suggests that more than one manifold may contribute at the Planck scale, or that models based on 'non-commutative geometry' may be appropriate. And for the canonical quantum gravity programme, we mentioned the discrete spectra of the spatial area and spatial volume quantities: results that arguably suggest some type of underlying discrete structure of space itself.

But these programmes also make other such suggestions. For example, we will see in Section 2.5.2 a general way in which quantizing a metric (as in these programmes) suggests a quantization of logically weaker structure such as differential or topological structure; these are called 'trickle-down effects'.

To sum up: Both these programmes threaten the ingredient theories' common treatment of spacetime, quite apart from their quantizing the metric; indeed they even threaten the manifold conception of space or spacetime.

### 2.5.1.2 *Emergence and quantization*

At this point it is worth developing the contrast (introduced in Section 2.1.1) between two general strategies which one can adopt when attempting to go beyond the common treatment of spacetime. The distinction can be made in terms of any part of the common treatment, not just metrical structure on which the programmes in Section 2.4 focus: for example, topological structure.

To 'go beyond' such a structure, one strategy is to argue that it is emergent (in physics' jargon: 'phenomenological'). 'Emergence' is vague, and indeed contentious; for along with related notions like reduction, it is involved in disputes, central in philosophy of science, about relations between theories and even sciences. But here, we only need the general idea of one theory $T_1$ being emergent from another $T_2$ if in a certain part of $T_2$'s domain of application (in physics' jargon, a 'regime': usually specified by certain ranges of values of certain of $T_2$'s quantities), the results of $T_2$ are well approximated by those of $T_1$ – where 'results' can include theoretical propositions as well as observational ones, and even 'larger structures' such as derivations and explanations.

This relation of emergence can of course be iterated, yielding the idea of a 'tower' of theories, each emerging from the one above it. Figure 2.1 portrays this idea, for the case of interest to us, viz. where the 'bottom theory' is classical general relativity. Figure 2.1 also uses the physics jargon of a theory being 'phenomenological'; and for the sake of definiteness, it assumes an uppermost 'ultimate' theory – an assumption to which, as is clear from Section 2.1.3, we are not committed. We should

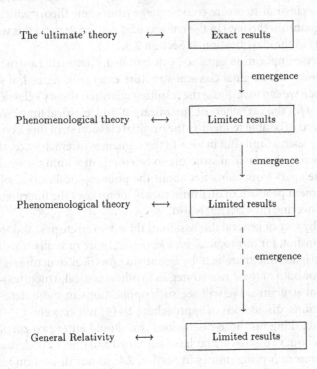

Fig. 2.1. A hierarchy of phenomenological theories.

add that of course many different towers will in general branch off from a given theory.[48]

So much for the general idea of emergence. The other strategy for 'going beyond a classical structure' is to try to quantize it, in some sense; and then to recover it as some sort of classical limit of the ensuing quantum theory. We say 'quantize in some sense', because although 'quantization' is certainly less vague than 'emergence', it is far from being precise. There is an open-ended family of 'quantization procedures' that are only provably equivalent on some simple cases (such as certain finite-dimensional, unconstrained Hamiltonian systems); and there is no procedure known whereby an arbitrary classical system can be quantized in an unequivocal way. Similarly, we say 'some sort of classical limit', because (as we said in Section 2.3.1.3) 'classical limit' also has various senses: it can refer to special states whose evolution over time follows classical laws, or to certain quantum quantities taking values in a range where classical theory is successful.

Though these two strategies are vague, the general idea of them is enough to make it clear that they are independent. That is: neither implies the other, though of course they *can* be combined (as the classification of approaches (1)–(4) makes clear).

Thus one can maintain that some classical structure is emergent, without quantizing it (i.e. exhibiting it as a classical limit of a quantization). Indeed, there are at least two ways one can do this. The first is familiar from our discussion of superstrings (Section 2.4.4): a classical structure (there, the metric geometry of general relativity) can emerge from a quantum theory (superstrings) which is not a quantized version of a classical theory of the structure. The second way is independent of quantum theory: viz. a classical structure could emerge from some theory which has nothing to do with quantum theory. (In the context of quantum gravity, this way is taken by approach (4) in the classification of Section 2.3.1.3.)

The converse implication can also be questioned. Quantizing a structure does not *ipso facto* render the original classical structure emergent. Agreed, if we quantize a structure, then we can investigate the resulting quantum theory's classical limits: and even allowing for the vagueness of 'quantization' and 'classical limit', we are more or less guaranteed to be able to identify the original classical structure as such a limit, or as a feature of such a limit. But in view of the vagueness of 'emergence', this might not count as showing the classical structure to be emergent within the quantum theory. Indeed, quite apart from subtleties about the philosophical notion of 'emergence', the measurement problem of quantum theory looms over the interpretation of such limits as 'recovering the classical world'.

So much by way of general discussion of these two strategies, and their independence. The upshot, for our topic of how to depart more radically from the common treatment of spacetime than 'just' by quantizing metrical structure, is that one can apply either or both of these two strategies to other classical structures – for example to topological structure. We will see such applications in more detail in the next three subsections' discussions of approaches (2)–(4) respectively.

Before embarking on those discussions, we should enter two *caveats*. First, the ideas we are about to discuss are far less established and understood, than are the trio of *bona fide* research programmes in Section 2.4; so our discussion needs must be much more tentative. Second, we will downplay the first strategy, i.e. emergence, on the grounds that our complementary paper Butterfield and Isham (1999) discusses it – both in general philosophical terms, and (for classical spacetime structure), in terms of classical limits of a theory of quantum gravity. Also *very* little is known about the prospects for this strategy used on its own, i.e. uncombined with the second strategy of quantization – as we will see in Sections 2.5.3 and 2.5.4.

### 2.5.2 Quantization 'below' the metric

We turn to discuss approach (2), where one quantizes a theory other than classical general relativity, but obtains it as a low-energy limit of the quantized theory. Needless to say, only a tiny fraction of the vast range of classical theories has been quantized, so that in full generality, little is known about this approach; and much of what is known concerns the superstrings programme.

So in this subsection, we will confine ourselves to general comments about the treatment of spacetime to be expected once one adopts this approach. Furthermore, we will only consider applying this approach to quantizing classical spacetime structures (hence this subsection's title). Given this restriction, our discussion is given a natural structure by the fact that the common treatment of spacetime – viz. as a pair $(\mathcal{M}, \gamma)$, with $\mathcal{M}$ a four-dimensional differentiable manifold, and $\gamma$ a Lorentzian metric – appears at one end of a hierarchical chain of structure; so that we can picture the introduction of a quantum effect 'below the metric $\gamma$' in terms of an earlier point in the chain. Of course, a given mathematical structure can often be placed in more than one such chain, and the question then arises of which chain to use.

The common treatment of spacetime as a pair $(\mathcal{M}, \gamma)$ fits naturally into the chain

$$\text{set of spacetime points} \rightarrow \text{topology} \rightarrow \text{differential structure}$$
$$\rightarrow (\mathcal{M}, \gamma). \tag{2.9}$$

Indeed, this chain is implicit in much of our previous discussion. The bottom level is a set $\mathcal{M}$, whose elements are to be identified with spacetime points; but this set is formless, its only general mathematical property being its cardinal number. In particular, there are no relations between the elements of $\mathcal{M}$, and no special way of labelling any such element. The next step is to impose a topology on $\mathcal{M}$ so that each point acquires a family of neighbourhoods. Then one can talk about relationships between points, albeit in a rather non-physical way. This defect is overcome by adding the key ingredient of the conventional treatment of spacetime: the topology of $\mathcal{M}$ must be compatible with that of a differentiable manifold, so that a point in $\mathcal{M}$ can be labelled smoothly (at least, locally) by giving the values of four real numbers. In the final step a Lorentzian metric $\gamma$ is placed on $\mathcal{M}$, thereby introducing the ideas of the lengths of a path joining two spacetime points, parallel transport with respect to a connection, causal relations between pairs of points, etc. Note that an analogous discussion applies to the usual modelling of space and time individually by a three-dimensional, and one-dimensional manifold respectively.

Note that a variety of intermediate stages can be inserted: for example, the link 'differential structure $\rightarrow (\mathcal{M}, \gamma)$' could be factored as

$$\text{differential structure} \rightarrow \text{causal structure} \rightarrow (\mathcal{M}, \gamma). \tag{2.10}$$

A quite different scheme arises by exploiting the fact that a differentiable manifold $\mathcal{M}$ is uniquely determined by the algebraic structure of its commutative ring of differentiable functions, $\mathcal{F}(\mathcal{M})$ (cf. our discussion in Section 2.2.2 of the spectral theorem for commutative $C^*$ algebras). And a ring is itself a complicated algebraic structure that can be analysed into a hierarchy of substructures in several ways. Thus, one alternative chain to eqn. 2.9 is

$$\text{set} \rightarrow \text{abelian group} \rightarrow \text{vector space} \rightarrow \mathcal{F}(\mathcal{M}) \rightarrow (\mathcal{M}, \gamma). \qquad (2.11)$$

Given these chains of structure leading to $(\mathcal{M}, \gamma)$, and others like them, it is clear what are the options facing approach (2). One must decide:

> (i) which of the chains to $(\mathcal{M}, \gamma)$ to use; and
> (ii) at what level in the chosen chain to try to quantize.

For example, if one uses the first chain, one faces such questions as: should we accept a fixed set of spacetime (or, for a canonical approach: spatial) points, but let the topology and/or differential structure be subject to quantum effects? Or should we say that the notion of a spacetime point itself is not meaningful at a fundamental level: i.e. it is a concept that should not appear in the theory, even in a mathematical sense?

We end by making two general comments about this situation, corresponding to the two decisions, (i) and (ii).

In regard to (i), we stress that the details of one's programme will depend strongly on the initial decision about which chain to use. Thus if one decides to apply quantization to the second chain (eqn. 2.11), one is led naturally to consideration of the algebraic approach to classical general relativity pioneered by Geroch (1972) ('Einstein algebras') and non-commutative analogues thereof (Parfionov and Zapatrin 1995). And, of course, the idea of a non-commutative version of the algebra $\mathcal{F}(\mathcal{M})$ was one of the motivating factors behind Connes' seminal ideas on non-commutative geometry (Connes 1994).

Concerning (ii), one should be aware that, once a chain has been chosen, quantization at one level could 'trickle down' to produce quantum effects at a 'lower', more general, level in the chain (i.e. to the left in our diagrams). For example, quantization of the metric could trickle down to the differential structure or topology.[49]

Such 'trickle-down' effects were envisaged by Wheeler (1968) in his original ideas about quantum topology in the context of canonical quantization. His idea was that large quantum fluctuations in a quantized 3-metric $\hat{g}_{ab}(x)$ would generate changes in the spatial topology; for the effects of quantum gravity would become more pronounced at decreasing distances, resulting eventually in a 'foam-like' structure at around the Planck length.

We stress that such effects depend on an appropriate mechanism. Thus, against Wheeler's intuition that quantum gravity effects become stronger at decreasing distances, one might hold that, in fact, quantum gravity is 'asymptotically-free' – so that the effects become *smaller* as the scale reduces. Under these circumstances, there would be no metric-driven topology changes.[50]

Another, more radical, example of trickle-down effects arises in connection with Penrose's thesis that a projective view of spacetime structure is more appropriate in quantum gravity: in particular, a spacetime point should be identified with the collection of all null rays that pass through it. Quantizing the spacetime metric will then induce quantum fluctuations in the null rays, which will therefore no longer intersect in a single point. In this way, quantum fluctuations at the top of the first chain (eqn. 2.9) trickle right down to the bottom of the chain, so that the very notion of a 'spacetime point' acquires quantum overtones.

### 2.5.3 Spacetime from a non-quantization

We turn to approach (3) in our classification, according to which general relativity emerges as a low-energy limit of a quantum theory that is *not* given as the quantization of a classical theory, but rather 'intrinsically' in some way.

By and large, programmes following this approach will reject the conception of spacetime as a manifold ('from the outset', rather than 'sneaking up' on this conclusion in the way the programmes in Section 2.4 do). Agreed, the envisaged quantum theory might in principle include the postulation of a spacetime manifold (though it is not a quantization of a classical theory defined on that manifold): a manifold which then turns out to be the manifold on which the low-energy limit, general relativity, is defined. But by and large, this circumstance would be odd: why should the quantum theory postulate just what the emergent approximation needs? Accordingly, we will here consider programmes (or, more precisely, mainly one fragment of one possible programme!) which do indeed reject the spacetime manifold at the fundamental level.

An example of the way the envisaged quantum theory might have general relativity as a low-energy limit is as follows. Return to the tower of theories of Section 2.5.2, and imagine that one result of emergence at some level in the tower is: (i) the idea of a 'local region' – not regarded as a subset of something called 'spacetime', but rather as an emergent concept in its own right; together with (ii) an algebra of such regions that specifies their theoretical use, and that can be identified mathematically as the algebra of a certain open covering of a genuine continuum manifold $\mathcal{M}$. Hence – as long as one keeps to the phenomena appropriate to this level – it is *as if* physics is based on the spacetime manifold $\mathcal{M}$, even though this plays no fundamental role in the 'ultimate' theory with which we started. Specific ideas of this type have arisen in the context of attempts to quantize the point-set topology of a set (Sorkin 1991, Isham 1990). We note in passing that the mathematics of locales, and more generally topos theory (in particular, the idea of a 'Grothendieck topos') provides a natural framework in which to develop the idea that regions are more important than points.

The possible significance of regions, rather than points, arises also in recent ideas about the nature of quantum physics in a bounded region. These go back to an old remark of Bekenstein (1974) to the effect that any attempt to place a quantity of energy $E$ in a spatial region with boundary area $A$ (and such that $E > \sqrt{A}$) will cause a black hole to form, and this puts a natural upper bound on the value of the energy in the region. The implication is that in any theory of quantum gravity whose semiclassical states contain something like black-hole backgrounds, the quantum physics of a bounded region will involve only a *finite*-dimensional Hilbert space.

This intriguing possibility is closely related to the so-called 'holographic' hypothesis of t'Hooft (1993) and Susskind (1995) to the effect that physical states in a bounded region are described by a quantum field theory on the *surface* of the region, with a Hilbert space of states that has a finite dimension.

Ideas of this type could have major implications for quantum gravity. In particular – and in terms of the tower in Fig. 2.1 – the implication is that at one level of phenomenological theory the idea of local spacetime regions makes sense, and in those regions the quantum theory of gravity is finite-dimensional. However, in the (possibly different) tower of phenomenological approximations that includes weak-field perturbative approaches to quantum gravity, the effective theory uses an infinite-dimensional Hilbert space to describe the states of weakly excited gravitons. And, at this level, spacetime is modelled by a continuum manifold, with a full complement of spacetime points.

We should emphasize that an approach like this does not necessarily exclude the proposals of the more familiar research programmes in quantum gravity: such proposals would however become phenomenological, i.e. part of the emergence of general relativity and quantum theory. If our present understanding of quantum gravity is any guide, this effective quantization of the gravitational field will involve a non-local – possibly string-like – structure. This raises the intriguing question of whether superstring theory and the loop-variable approach to canonical quantum gravity can both be regarded as different modes (or phases) of a more basic, common structure (Smolin 1998).

### 2.5.4 Spacetime emergent from a non-quantum theory

Finally, let us raise the question of the justification in quantum gravity of the use of standard quantum theory itself. So, in terms of the portrayal of the emergence of general relativity shown in Fig. 2.1, the idea now is that there would also be a tower of theories leading down to the emergence of standard quantum theory. Of course, in accordance with the point that towers can branch off from each other, this tower may well not be the same one as that leading to general relativity.

One particularly relevant issue in regard to quantum theory – and the only one that we shall discuss in any detail – is the question of what justifies its use of *continuum* concepts: specifically, its use of real and complex numbers. This question is very pertinent if one is already worried about the use of continuum ideas in the manifold model of space or time.

The formalism of quantum theory immediately suggests two answers: one concerning eigenvalues, and the other probabilities. Thus, one might answer by saying that real numbers represent the possible results of measurements (so that if eigenvalues of operators are to represent results, we want the operators to be self-adjoint). But why should measurement results be represented by real numbers? One natural (if not compelling) answer is that apparently all measurement results can in principle be reduced to the positions of a pointer in space – and space is modelled using real numbers. At the very least, this is certainly true of the elementary wave mechanics of a point particle moving in Euclidean space; and this example, particularly the

Hilbert space generalization of its specific mathematical structure – has become one of the paradigms for quantum theory in general.

So according to this line of thought, the use of real numbers (and similarly, complex numbers) in quantum theory in effect involves a prior assumption that space should be modelled as a continuum. If so, then the suggestion that standard spacetime concepts break down at the scale of the Planck length and time, and must be replaced by some discrete structure which only 'looks like' a differentiable manifold at larger scales, means that we cannot expect to construct a theory of this discrete structure using standard quantum theory – with its real and complex numbers. Of course, this argument is not water-tight, but it does illustrate how potentially unwarranted assumptions can enter speculative theoretical physics, and thereby undermine the enterprise.

The second possible answer to the question 'what justifies quantum theory's use of real and complex numbers?' is that probabilities are real numbers between 0 and 1 (so that if probabilities are to be given by the squared norm of state vectors, the vector space must have $\mathbb{R}$ or $\mathbb{C}$ as its ground-field). But why should probabilities be represented by real numbers? Of course, if probability is construed as 'relative frequencies' of sequences of measurements, then the real numbers do arise naturally as the ideal limits of collections of rational numbers. However, the idea of measurement 'at the Planck length' is distinctly problematic (cf. our remarks in Section 2.1.3.3 about 'in principle inaccessibility'), and if the concept of probability is relevant at all in such regimes, one may feel that a different interpretation of this concept is more appropriate: for example, the propensity interpretation.

But there is no *a priori* reason why a 'propensity' (whatever that may mean!) should be modelled by a real number lying between 0 and 1. Agreed, it may well be appropriate sometimes to say that one propensity is 'larger' than another; but there may also be propensities that cannot be compared at all (a not unreasonable suggestion in the context of non-commuting operators in a quantum theory), and this suggests that a minimal model for such probabilities would be a partially ordered set with some type of additional algebraic structure (so that 'sums' of probabilities can be defined for disjoint propositions).

For these reasons, a good case can be made that a complete theory of quantum gravity may require a revision of quantum theory itself in a way that removes the *a priori* use of continuum numbers in its mathematical formalism.

Finally, we note that (from time to time) a few hardy souls have suggested that a full theory of quantum gravity may require changing the foundations of mathematics itself. A typical argument is that standard mathematics is based on set theory, and certain aspects of the latter (for example, the notion of the continuum) are grounded ultimately in our spatial perceptions. However, our perceptions probe only the world of classical physics – and hence we feed into the mathematical structures currently used in *all* domains of physics, ideas that are essentially classical in nature. The ensuing category error can be remedied only by thinking quantum theoretically from the very outset – in other words, we must look for 'quantum analogues' of the categories of standard mathematics.

How this might be done is by no means obvious.[51] One approach is to claim that, since classical logic and set theory are so closely linked (a proposition $P$

determines – and is determined by – the class of all entities for which $P$ can be rightly asserted), one should start instead with the formal structure of *quantum* logic and try to derive an analogous 'non-Boolean set theory'. Such ideas are related to the exciting subject of topos theory, which can be viewed as a far-reaching generalization of standard set theory. This is why, as mentioned in Section 2.5.3, topos theory is a natural arena within which to develop speculative schemes in which 'regions' of spacetime (or space, or time) are more important than 'points' (which may not exist at all).[52]

### 2.5.5 Envoi

Clearly, this section has opened up a Pandora's box of possibilities for the overall shape of a theory of quantum gravity: possibilities that it is well-nigh impossible to adjudicate between – not least because it is very hard even to make individual possibilities precise and detailed. So we will make no pretence of judging them here.

Instead, by way of a modest and (mercifully brief!) conclusion, we want to stress some of this chapter's lines of thought, that led to this Pandora's box. We began with the question how spacetime might be treated in a theory of quantum gravity. We described how the search for such a theory was beset by various conceptual difficulties, including difficulties about the ingredient theories, and about spacetime concepts – and also beset by a dire lack of data (a predicament reviewed in Sections 2.1–2.3). On the other hand, we could not 'duck out' of searching for some such theory (Sections 2.3.1.1–2.3.1.2). We reviewed in Section 2.4 three main programmes that all proceed by quantizing a classical theory which postulates a spacetime manifold. In various ways, these programmes suggest that there are fundamental limitations to the applicability of the manifold conception of spacetime; (of course, they also have various more specific problems, both physical and conceptual). Thus our attention was turned to the more radical programmes of this section . . . where we admit to having to suspend judgement.

To sum up: Quantum gravity is most unusual in comparison with other branches of physics, and indeed with most other branches of human enquiry – or with other 'games people play'. It is an exciting unpredictable game, with very few rules – and yet, as the sports commentators say, 'there is everything to play for!'.

### Notes

Chris Isham would like to thank the Mrs. L. D. Rope Third Charitable Settlement for financial support. We thank Joy Christian, Oliver Pooley, Roberto Torretti, and the Editors for comments on a previous version.

1. Though this construal of 'quantum gravity' is broad, we take it to exclude studies of a quantum field propagating in a spacetime manifold equipped with a fixed background Lorentzian metric. That is to say, 'quantum gravity' must involve in some way a quantum interaction of the gravitational field with itself. Quantum field theory in a fixed background is a rich subject in its own right, and it is a useful way of probing certain aspects of quantum gravity. But it does not raise the philosophical problems that we will pursue in this paper.
2. We choose this topic partly in the light of this volume's emphasis on spacetime, especially in Chapters 5–7 and 12–14.

3. See Butterfield and Isham (1999), which discusses philosophical aspects of emergence, which here we only treat very briefly, in Section 2.5. But in another regard, it is more limited: it discusses the emergence of spatio-temporal concepts for only one of the three programmes in Section 2.4.

4. We should add that though this chapter focuses on the treatment of spacetime, quantum gravity also 'puts pressure' on the formalism of, and usual interpretative ideas about, quantum theory; as we shall briefly discuss in Section 2.2. Furthermore, quantum gravity even puts pressure on standard mathematics itself, in that constructing a theory of quantum gravity might require some non-standard mathematical ideas: for example, spacetime might be modelled mathematically by something that is not a *set*. We will briefly discuss this line of thought in Section 2.5.

5. We note however that there have been recent suggestions about possible tests for quantum gravity: for example, Ellis, Mavromatos, and Nanopoulos (1999).

6. Here $G$ is Newton's constant, $\hbar$ is Planck's constant (divided by $2\pi$), and $c$ is the speed of light.

7. There is another connection with the conceptual problems about quantum theory. Namely, the lack of data in quantum gravity research is analogous to that which – until relatively recently – faced research in conceptual problems about quantum theory. In view of these connections, it is curious that there has been so little interaction between the two research communities.

8. As mentioned, lack of space prevents a full defence of this point. Suffice it here to add two comments. (1) The point does not assume that new doctrine which does *not* change the subject matter must be cumulative, i.e. must not contradict old doctrine. Suppose doctrine can be withdrawn or adjusted, without the subject matter changing: the point still holds good. (2) The point is independent of physicalism. For it turns on physics aspiring to give a complete description of its subject matter. But this implies nothing about whether the subject matter of physics exhausts all (empirical) subject matters, i.e. about whether physicalism is true.

9. After all, even if one confines oneself to the topic of space and time, and to authors who seem to deserve the name 'transcendental idealists', there are several positions to evaluate; for example, Kant's view of space and time as transcendentally ideal, though empirically real – the *a priori* conditions for all experience – and Kuhn's view (on one reading) that space and time are not transcendentally ideal but are instead part of the world in itself.

10. In fact, the theoretical physicist (C.J.I.) in this collaboration does *not* hold this view.

11. The suggestion also allows what we hoped for in Section 2.1.2: namely, the existence of quantum gravity effects at much more accessible length–scales. It only contends that aspects we cannot thus probe are not empirical.

12. For further discussion, see Chapters 8, 12, 13, and 14.

13. For more such pressure, see Chapter 8.

14. Kantian themes about the *a priori* nature of space and time arise here, just as in Section 2.1.3; for Kantian aspects of Bohr, see Kaiser (1992).

15. About the second, suffice it to say that such views propose to revise the logic of discourse about quantum systems (so we do not intend the type to include purely technical investigations of non-Boolean structures). But it is unclear how such proposals solve the interpretative problems of quantum theory, such as the measurement problem or non-locality; and indeed, these proposals now seem to have few advocates – at least compared with 25 years ago. In any case, their advocates hardly connect them with quantum gravity.

16. We do so, albeit briefly, in Butterfield and Isham (1999); see also Butterfield (1995) and (1996).

17. This asserts that a system has a real number $r$ as its value for a quantity $Q$ if and only if the quantum state is an eigenstate of $Q$ with eigenvalue $r$.

18. Agreed, this second difference is a matter of degree. Furthermore, Everettians' imprecision about the preferred quantity, and the dynamics of its values, is partly just an accident, due to the facts that: (i) their view first developed within traditional quantum measurement theory, which invokes imprecise notions like 'apparatus' and 'pointer-position'; and (ii) they are willing to secure only the appearance of a definite macroscopic realm, not a truly definite one, and are therefore able to leave future psychophysics to specify the preferred quantities. But this matter of degree is no problem for us: our taxonomy of four approaches is not intended to be rigid.

19. For further discussion, see Chapter 12 and Valentini (2000).

20. For further discussion and references, see Chapter 14.

21. We should stress that here we take the word 'object' in the 'post-Fregean' sense of anything that could be the referent of a singular term. In particular, it does not necessarily mean a 'thing out there' as a physicist might construe the phrase.

22. Agreed, there are many other conceptual aspects of general relativity that bear on quantum gravity. We mentioned the philosophy of geometry at the end of Section 2.1.2. Another obvious example is the global structure of time, which bears on quantum cosmology. Here, one faces such issues as: In what sense could a quantum event 'precede' the big-bang? This is related to the 'problem of time' in quantum gravity; which will be discussed later (Section 2.3.3); see also Chapters 5–7, and Section 5 of our essay (Butterfield and Isham 1999). For a philosophical discussion of several other conceptual aspects of general relativity bearing on quantum gravity, see Earman (1995).

23. In any case, the rise of field theory undermines the contrast one naively learns in everyday life, between empty space and material bodies.

24. This of course reflects the rise, from the mid-nineteenth century onwards, of the field-theoretic conception of matter, whether classical or quantum. From another perspective, it reflects the dominant position of set theory in the foundations of mathematics.

25. For further discussion, see Chapters 9 and 14.

26. This view of the argument seems to have been Einstein's own view, from the time of his discovery of general relativity in 1915 onwards. For discussion and references (including replies on behalf of points being physically real) see Earman and Norton (1987) and Earman (1989).

27. The four 'fundamental' forces recognized by present-day physicists are the electromagnetic force; the 'weak' nuclear force, that is responsible for radioactive decay; the 'strong' nuclear force, that binds together the constituents of nuclei; and the gravitational force.

28. This does not exclude a Kaluza–Klein-type higher dimension at Planckian scales. Indeed, superstring theory suggests strongly that something like this does occur.

29. Our complementary essay, Butterfield and Isham (1999), discusses this in some detail, especially as regards the Euclidean programme – which is mentioned in Section 2.4 below only as a species of canonical quantum gravity.

30. In 1971, one of us (C.J.I.) took part in a public debate with John Stachel, who challenged the former on this very issue. As a keen young quantum field theorist, C.J.I. replied that he was delighted to quantize everything in sight. These days, not least because of the moderating influence of his philosopher friends, he is more cautious!

31. More precisely, the spacetime metric must be static or stationary.

32. Essentially this approach was also used in developing important elementary-particle physics theories, where there was no pre-existing classical theory; for example, the Salam–Glashow–Weinberg electro-weak theory, and the quantum chromodynamics description of the strong nuclear force.

33. For further discussion, see Chapters 9–11.

34. The diffeomorphisms concerned are those of compact support, i.e. they are equal to the unit map outside some closed and bounded region of the spacetime (for some purposes, more subtle 'fall-off' rates at infinity may be appropriate). Thus, for example, a Poincaré-group transformation of Minkowski spacetime is not included. This restriction is imposed because the role of transformations with a non-trivial action in the asymptotic regions of spacetime is quite different from those that act trivially.

35. Another possibility – not exemplified in the programmes surveyed in this chapter – is that the diffeomorphism group $\mathcal{D}$ could be related to a bigger group $G$ in a projective way, i.e. there is some normal subgroup $K$ of $G$ so that $G/K \simeq \mathcal{D}$.

36. Here we should distinguish the invariance of each individual expression in a theory's formalism from the invariance of all the physically measurable values of quantities; the latter being of course a weaker property. In fact, most of the programmes whose basic framework treats $\mathcal{D}$ as an exact covariance group *in practice* 'fix a gauge' for their formalism, and so work with non-covariant/non-diffeomorphism-invariant expressions, and then show physically measurable values to be gauge-invariant. So in practice, these programmes enjoy only the weaker property.

37. For much fuller expositions of the problem from the point of view of canonical quantum gravity, see Chapters 3 and 10; see also Kuchař (1992) and Isham (1993).
38. This is why the meaning assigned to the time-energy uncertainty relation $\delta t \, \delta E \geq \frac{1}{2}\hbar$ is quite different from that associated with, for example, the position and the momentum of a particle.
39. Note that this general statement of the problem of time would have an analogue in a stochastic version of *classical* general relativity, in which the metric tensor is regarded as a random variable. Of course, the other difficult problem of understanding what is meant by 'superpositions' of spacetime geometries would be absent in this case.
40. Clearly, one could also discuss other programmes so as to provide still more of a backdrop to the two main ones. One obvious choice is the Euclidean programme, which can be viewed as a spacetime-oriented species of the traditional geometrodynamic version of canonical quantum gravity. But we ignore it here, since (i) it is especially connected with quantum cosmology, which we have set aside; and (ii) we discuss it in Butterfield and Isham (1999).
41. Full references can be found in reviews written around that time; for example, in the proceedings of the first two Oxford conferences on quantum gravity (Isham, Penrose, and Sciama 1975, 1981).
42. For more detailed discussion of superstrings, see Chapters 9–11.
43. The more realistic superstring theories involve an additional massless 'dilaton' scalar field $\phi$, and a massless vector particle described by a three-component field strength $H_{\mu\nu\rho}$. The presence of these extra fundamental fields has a major effect on the classical solutions of the field equations; in particular, there have been many studies recently of black-hole and cosmological solutions.
44. The phenomenon can be generalized to more than one extra dimension and with a topology that is more complex than just a product of circles.
45. A third type of duality known as 'mirror symmetry' also plays an important role here.
46. A fairly comprehensive bibliography of papers on canonical general relativity can be found in the review papers Isham (1991) and Kuchař (1993).
47. So one should not attach too much weight to the results of such simple approximations. Admittedly, models of this type can be valuable tools for exploring the many conceptual problems that arise in quantum cosmology.
48. For more discussion of emergence, especially in relation to reduction, see Section 2 of our complementary essay (Butterfield and Isham 1999).
49. This was mentioned in Section 2.5.1.1, as a way in which the programmes in Section 2.4 put pressure on the manifold conception of spacetime; a way additional to those already discussed.
50. The idea that gravity might be asymptotically-free was studied some years ago by Fradkin and Tseytlin (1981) in the context of $R + R^2$ theories of gravity.
51. A recent example of this type of thinking can be found in Finkelstein (1996).
52. Topos theory has a related deep connection with non-standard logical structures: something we have exploited in recent work using presheaf logic to analyse the Kochen–Specker theorem in standard quantum theory (Butterfield and Isham 1998, 1999a).

# 3    Naive quantum gravity

Steven Weinstein

## 3.1 Introduction

The world of classical, relativistic physics is a world in which the interactions between material bodies are mediated by fields. The 'black body catastrophe' provided the first indication that these fields (in particular the electromagnetic field) should be 'quantized'.[1] Modern field theory contains quantum field–theoretic descriptions of three of the four known interactions (forces) – all except gravity. It is characteristic of the theories of these three forces that the values of the fields carrying the forces are subject to the Heisenberg uncertainty relations, such that not all the field strengths at any given point can be specified with arbitrary precision.

Gravity, however, has resisted quantization. There exist several current research programmes in this area, including superstring theory and canonical quantum gravity.[2] One often comes across the claim that the gravitational field must be quantized, and that quantization will give rise to a similar local uncertainty in the gravitational field. Here we will examine this claim, and see how the very things that make general relativity such an unusual 'field' theory not only make the quantization of the theory so technically difficult, but make the very idea of a 'fluctuating gravitational field' so problematic.

## 3.2 What is a field?

Maxwell's theory of electromagnetism describes the interaction of electrically charged matter (consisting of 'charges') and the electromagnetic field.[3] Charges act as 'sources' for the field, and the field in turn exerts a force on the charges, causing them to accelerate. The field is specified against a background of space and time, assigning values for the various components $E_i$, $B_j$, etc., of the electric and magnetic fields to each point in space at a given time.[4] The acceleration of a charged object at a given point is then given by the Lorentz law

$$\ddot{\vec{x}} = \frac{\vec{F}}{m} = \frac{q_E(\vec{E} + \dot{\vec{x}} \times \vec{B})}{m}, \tag{3.1}$$

where $m$ is the mass of the object, $q_E$ is its electrical charge, and $\dot{\vec{x}}$ the velocity of the object (in appropriate units). In short, the acceleration of a given object in a given field is directly proportional to the charge, and inversely proportional to the mass.

Maxwell theory is the paradigmatic field theory, yet there are three other 'interactions' known in nature, associated with different sorts of charge.[5] As noted in the Introduction to this chapter, the theories of the strong and weak nuclear interactions are also field theories, specifying the fields associated with their respective charges, and the resulting forces on the charges.

The remaining interaction is gravity. As in the theories of other interactions, objects carry a charge, the charge acts as a source for something like a field, and the theory quantifies how the properties of this field affect the behaviour of the object (and vice versa). What is uniquely characteristic of gravity is that the gravitational charge $q_G$ of an object is identical to its mass (in Newtonian theory) or mass-energy (in general relativity). This has far-reaching consequences. One, it means that gravity is universal, since all objects have a mass (respectively, mass-energy). Two, it means that all objects behave the same in a gravitational field (because the ratio of the charge to the mass $q_G/m = m/m = 1$). This equivalence of gravitational charge and inertial mass is what we shall refer to as the 'principle of equivalence' or 'equivalence principle'.[6]

If gravity were universal, yet objects reacted differently to gravitational effects, then there would be no particular reason to associate the gravitational field with spacetime geometry. It is the fact that objects behave the same in a gravitational field that leads to describing gravity as a property of spacetime itself.[7] The reason for this is that 'behave the same' means 'follow the same spacetime trajectory'. Einstein noticed that if these trajectories could be construed as characteristic features of a curved spacetime geometry, then gravity could be represented geometrically. They can – the special trajectories are 'geodesics'.[8]

An alternative way to conceive of gravity would of course be to follow the lead of other theories, and regard the gravitational field as simply a distribution of properties (the field strengths) in *flat* spacetime.[9] What ultimately makes this unattractive is that the distinctive properties of this spacetime would be completely unobservable, because all matter and fields gravitate. In particular, light rays would not lie on the 'light cone' in a flat spacetime, once one incorporated the influence of gravity. It was ultimately the unobservability of the inertial structure of Minkowski spacetime that led Einstein to eliminate it from his theory of gravitation and embrace the geometric approach.

Nonetheless, we shall see that this attribution of gravity to the curvature of spacetime leads to great conceptual and technical difficulties, essentially because it makes it difficult, if not impossible, to treat gravity within the conceptual and mathematical framework of other field theories. Thus, it is worth asking whether it is at all *possible* to construe gravitation as a universal interaction that nonetheless propagates in flat, Minkowski spacetime. The idea might be to still construe the field geometrically (retaining part of Einstein's insight into the significance of the equivalence principle), but to construe the geometrical aspect as 'bumps' on a special, flat background.

The short answer is, 'No', for three reasons. First, the 'invisibility' of the flat spacetime means that there is no privileged way to decompose a given curved spacetime into a flat background and a curved perturbation about that background. Though this non-uniqueness is not particularly problematical for the classical theory, it is quite problematical for the quantum theory, because different ways of decomposing the geometry (and thus retrieving a flat background geometry) yield different quantum theories.[10] Second, not all topologies admit a flat metric, and therefore spacetimes formulated on such topologies do not admit a decomposition into flat metric and curved perturbation.[11] Third, it is not clear *a priori* that, in seeking to make a decomposition into background and perturbations about the background, that the background should be *flat*. For example, why not use a background of constant curvature?

The upshot is that, for general spacetimes, the gravitational field can only be *locally* decomposed into a flat Minkowski background and a curved foreground, and even then there is no unique way to do it. Thus, we are stuck with a theory in which the gravitational field seems irrevocably tied to a fully geometric description, which in particular means that the field, such as it is, defines its own background – it is both 'stage' and 'actor'.

### 3.3 The uncertainty of quantization

Quantum theory applies to all sorts of systems. In a quantum theory, the determinate properties of classical mechanics are replaced by indeterminate properties, represented by self-adjoint operators on a Hilbert space. For example, objects such as low-energy particles have indeterminate position $\vec{x} = (x_i, x_j, x_k)$ and momentum $\vec{p} = (p^i, p^j, p^k)$. These quantities (the components of the vectors) are represented by self-adjoint operators $\hat{x}_i$ and $\hat{p}_j$ satisfying commutation relations

$$[\hat{x}_i, \hat{p}_j] = i\hbar\delta_{ij}. \tag{3.2}$$

As conventionally understood, the commutation relations imply that the position and momentum of the particle cannot be specified with arbitrary accuracy at a given time.

Quantum fields, such as the quantum electromagnetic field, are similarly represented. The relevant observable properties of the electromagnetic field are the various components of the electric and magnetic field at each point in space (at some time – we are working in the canonical framework), and these are formally represented by the six operators $\hat{E}_i(\vec{x})$ and $\hat{B}_i(\vec{x})$ ($i = 1, 2, 3$).[12] For a scalar field, we have simply $\hat{\phi}(x)$ and its conjugate $\hat{\pi}(x)$. These operators all satisfy canonical commutation relations.[13] For example,

$$[\hat{\phi}(\vec{x}), \hat{\pi}(\vec{x}')] = i\hbar\delta^3(\vec{x} - \vec{x}'). \tag{3.3}$$

One might well think that the gravitational field should also be quantized, and that analogous commutation relations should hold for the operators representing *its* properties. This line of thinking is implicit in the writings of many physicists. But in fact it is not at all obvious what it even *means* for the gravitational 'field' to be subject to uncertainty relations. The two obstacles are:

(1) The uncertainty relations apply to physical, observable quantities, such as the position and the momentum of a particle, or the values of the magnetic and electric fields at each point. Such observable quantities correspond to the canonical degrees of freedom of the theory. But no one has succeeded in isolating such quantities for the gravitational field.[14]

(2) We use classical matter and fields to physically identify points of spacetime. If all fields except for the gravitational field are treated quantum-mechanically, we can still use the gravitational field. But what does it even mean to talk about the values of the gravitational field *at a point* (or commutation relations *between* points) if the field itself is subject to quantum 'fluctuations'?[15]

Regarding the first obstacle, even though it is relatively well known that gravity has not been reduced to a true canonical system, the relevance of this to the lack of local observables seems to be quite underappreciated. Perhaps the reason for this is that, insofar as one understands the gravitational field to be represented by the Riemann tensor $R^{\alpha}{}_{\beta\gamma\delta}$ (itself composed of first- and second-derivatives of the metric $g_{\alpha\beta}$), and insofar as this tensor has a value at every point, it is thought that the gravitational field is well-defined at every point. In the second half of the next section we will discuss the utility of this characterization of the field.

The second point is more straightforward. The significance of the 'diffeomorphism invariance' of general relativity is that one needs some sort of classical structure like the metric or other physically meaningful tensorial objects (such as the Maxwell tensor $F_{\alpha\beta}$ corresponding to the electromagnetic field) in order to give physical meaning to 'spacetime points'. Thus, if we quantize the metric and other fields, it is difficult to see how to talk meaningfully about the relation between the quantum fluctuations of a field at, and between, points.[16] We shall explore this idea further in the subsequent section.

## 3.4 Quantifying the effects of gravity: local field strength

### 3.4.1 Absolute acceleration

Traditionally, the physical significance of the values of a field at a given point is to determine the motion of a charge at that point. More specifically, the strength and direction of the field at the point determines the acceleration of the charge. So for instance in Maxwell theory, the acceleration of an electric charge at a given point is directly proportional to the strength and direction of the electric and magnetic fields at that point (see eqn. 3.1 above).

Note that the energy density of the field is calculated from these field strengths. For Maxwell theory, the energy $H$ in an infinitesimal spatial volume $dV$ is $dH = \left( |\vec{E}|^2 + |\vec{B}|^2 \right) dV$. This is significant in that the fact that the function giving the total energy $H = \int_{\Sigma} dH$ over a region of space $\Sigma$ is the Hamiltonian, and the Hamiltonian is the 'generator' of time-evolution in the canonical formalism. Thus, the fact that the energy of the field is well defined corresponds to the fact that the time-evolution is well defined. As we shall see, in general relativity the energy is not well defined in general, and the time-evolution is ambiguous.[17]

Implicit in the definition of field strength here is the use of inertial frames as canonical reference frames. The acceleration of the charged particle is defined with respect to inertial frames – acceleration is the deviation from inertial motion. But as we saw above, the presence of gravity means that there *are* no inertial frames. On the face of it, this presents a problem for the definition of field strength in Maxwell theory in the presence of gravity. But one can recover a useful analogue of the previous definition by utilizing the nearest analogue to inertial motion in curved spacetime, which is motion along a geodesic. The field strengths in curved spacetime then just give the acceleration with respect to a given geodesic, i.e. the deviation from geodesic motion. (Technically, they assign a 'four-acceleration' to a charge at each point, the value of which determines the extent to which the charge will deviate from geodesic motion.)

It now follows that, according to this definition of field strength, the gravitational field strength at a point is always zero, no matter what the value of the Riemann tensor is at that point! If a freely falling observer (i.e. one following a geodesic) releases a gravitational 'test charge' (any massive object, i.e. any object at all), then the test charge will not accelerate relative to the observer.[18] Rather, it will remain stationary with respect to the observer. In short, if one conceives of field strength as deviation from geodesic motion, then the gravitational field strength must be zero everywhere. Similarly, the energy density must be zero everywhere, since the magnitude of the velocity of a test particle never changes.

This claim, that the gravitational field strength is zero at each point, must be taken with a grain of salt. The argument is really that *if* one carries over to gravity the traditional notion of field strength, then one finds that the gravitational field strength is zero. Though it will turn out that there is no fully adequate local characterization of the gravitational field, we can do a bit better, and it is instructive to see how.

### 3.4.2 Relative acceleration and the Riemann tensor

Of course, we can and do observe the effects of gravitation. But as we have seen, what we observe is neither the acceleration of test objects relative to inertial observers (for there are no inertial observers) nor, with respect to their nearest gravitational analogues, geodesics in curved spacetime. Typically, what we observe are *tidal* effects, which involve the way in which bits of matter (or observers) distributed in space accelerate toward or away from each other. This relative acceleration is encoded in the Riemann curvature tensor $R^{\alpha}{}_{\beta\gamma\delta}$.

As with any tensor, the Riemann tensor is defined at every spacetime point, and thus it might seem that it offers a way of characterizing the local properties of the gravitational field. Given an observer at a point, we can find the relative acceleration $a^{\alpha}$ of nearby matter (which will follow nearby geodesics) by the geodesic deviation equation

$$a^{\alpha} = R^{\alpha}{}_{\beta\gamma\delta}u^{\beta}v^{\gamma}u^{\delta}, \tag{3.4}$$

where $u^{\alpha}$ is the tangent vector to the observer's worldline (representing his or her velocity) and $v^{\alpha}$ is a 'geodesic selector', the purpose of which is to select a particular neighbouring geodesic to compare with the geodesic traced out by the observer (i.e. his or her worldline). The quantity $a^\alpha$ then represents the relative acceleration of the two geodesics.

The fact that the Riemann curvature tensor seems to encode the effects of gravity in the neighbourhood of any given point might suggest that it, like the Maxwell tensor in electromagnetism, fully characterizes the gravitational field. If this were the case, then one might expect that knowing the Riemann tensor at a given time would determine the Riemann tensor at future times, just as knowing the Maxwell tensor at a given time (the $\vec{E}$ and $\vec{B}$ field at a given time) determines the Maxwell tensor in the future.[19]

Looked at in a certain light, this construal of the Riemann tensor has a certain plausibility. After all, just as one can form the Maxwell tensor $F_{\alpha\beta}$ from the derivatives of a vector potential $A_\alpha$ via the equation $F_{\alpha\beta} = \nabla_{[\alpha}A_{\beta]}$, one can form the Riemann tensor $R^\alpha{}_{\beta\gamma\delta}$ from derivatives of the metric $g_{\alpha\beta}$. In the electromagnetic case, the quantities of physical significance are captured in the Maxwell tensor, and transformations $A_\alpha \longrightarrow A'_\alpha$ of the vector potential that leave the Maxwell tensor unchanged are thus regarded as non-physical 'gauge' transformations. One might be tempted to guess, then, that since the Riemann tensor may be formed from derivatives of the metric, transformations of the metric which leave the Riemann tensor invariant are physically meaningless 'gauge' transformations. However, this is not the case.

To understand this, let us consider a situation in which one is given a manifold (thus a topology) and the Riemann tensor on the manifold. Suppose we have the manifold $S^1 \times \mathbb{R}$, and we are told that the Riemann tensor vanishes everywhere. This means that the metric is flat, and therefore that we are considering a cylinder. Does this information determine the metric on the cylinder? No, it does not. If it *did*, it would tell us the circumference of the cylinder, hence its radius. But since all cylinders have the same (vanishing) curvature, the curvature underdetermines the metric.

To what extent does the curvature underdetermine the metric? In cases of high symmetry, e.g. the hypersurfaces of constant curvature in typical idealized cosmological models, it underdetermines it by quite a bit. In the general case, where the Riemann tensor varies from point to point, one can often determine the metric up to a conformal factor. But this is insufficient to extract unambiguous physical information from the Riemann tensor.

For example, suppose one is given the Riemann tensor at a point, and one wants to know the way in which particles in the neighbourhood of the point will accelerate toward or away from a given observer at the point. An observer is characterized by a worldline in spacetime, and an observer at a given point is characterized by the tangent vector $u^\alpha$ to the worldline at that point. Therefore, one could construct a tensor

$$Z^\alpha{}_\gamma = R^\alpha{}_{\beta\gamma\delta}u^\beta u^\delta \tag{3.5}$$

which represents the acceleration of nearby matter relative to an observer moving along a worldline with tangent vector $u^\alpha$. However, there is something wrong with this picture, and it has to do with how we choose the tangent vector. A tangent vector is constrained to be of unit length, but we cannot tell how long a vector is without the metric. Therefore, to each of the conformally related metrics $g_{\alpha\beta}$ associated with a given Riemann tensor, there is associated a different set of candidate tangent vectors $u^\alpha$. In the absence of a specific metric, one cannot even form the tensor $Z^\alpha{}_\gamma$, because one has no way of normalizing candidate tangent vectors $u^\alpha$. In short, the fact that the Riemann tensor by itself contains no physical information suggests that it is a mistake to regard it as fully characterizing the gravitational field, in any conventional sense.

### 3.5 Causal structure

In the previous section, we examined one of the difficulties in applying the uncertainty principle to the gravitational field, the difficulty that the values of the gravitational field at a point are not even well defined in general relativity. In this section and the next, we will address another difficulty, which concerns the status of commutation relations in a theory in which the spacetime geometry itself is quantized.

In a conventional classical field theory in a flat background spacetime, the causal structure tells us the 'domain of dependence' of the field values at a point. In other words, we know that the values of a field at spacetime point $x$ are related to the values of the field at points in its forward and backward lightcones. In the corresponding quantum field theory, this is reflected in the fact that the (covariant) field operators $\hat{\phi}_i(x)$ at spacelike separated points $x$ and $y$ commute:

$$[\hat{\phi}_i(x), \hat{\phi}_j(y)] = 0. \tag{3.6}$$

The intuitive physical picture behind this is that measurements of the field at point $x$ do not reveal anything about the field at point $y$, because they are not in causal contact with each other.

Of course, when one incorporates the gravitational effects of a classical field, one formulates it in a curved spacetime, where the curvature respects the stress–energy properties of the matter in accord with Einstein's equation. If one wants to treat the fields quantum-mechanically, there are two choices: one can attempt to leave the spacetime classical and use that structure in the quantization ('semiclassical' gravity), or one can attempt to quantize gravity. The difficulty with the former is that one wants a determinate spacetime structure despite the indeterminate (because quantum) stress–energy of the field. One can pursue this by using the 'expectation value' (the average value) of the fields in a given state to determine the spacetime curvature – this is the approach taken by those working in the field of 'quantum field theory in curved spacetime'.[20] In such a theory, one can make sense of commutation relations like eqn. 3.6, because one can determine whether or not two points are spacelike separated. But such a theory can make no claim to being fundamental (Duff 1981, Kibble 1981).

Suppose, then, that we opt for the second alternative and allow some of the components of the gravitational field to 'fluctuate', so that, for example, the curvature at each point is subject to quantum fluctuations. In that case, we would expect that the metric itself is subject to quantum fluctuations (since the curvature is built from derivatives of the metric). But if the metric is indefinite, then it is by no means clear that it will be meaningful to talk about whether $x$ and $y$ are spacelike separated, unless the metric fluctuations somehow leave the causal (i.e. conformal) structure alone.

Assuming that the metric fluctuations do affect the causal structure, one would expect that the commutation relations themselves should reflect this by also undergoing quantum fluctuations of some sort. However, it is not at all clear what this means, or how it might be represented. And in particular, it should be noted that such a commutator would have no apparent counterpart in the classical theory. In allowing metric fluctuations to affect causal structure, one is clearly at some remove from ordinary field–theoretic quantization schemes.

### 3.6 What's the point?

We began by taking a close look at how one might characterize the gravitational field at a given point, and we then went on to examine the consequences of turning whatever local quantities we might find into operators, in a quantum theory of gravity. Both of these are problems peculiar to gravitational physics, in that they arise as a result of the principle of equivalence, the equivalence of gravitational and inertial mass that practically compels us to regard classical gravity as a theory of spacetime geometry.

The final point, however, has less to do with gravity per se than the fact that any theoretical framework incorporating gravity must seemingly be diffeomorphism-invariant. Up to this point we have adopted the polite fiction that, for example, the Riemann tensor at a point $x$ is a physically meaningful quantity. In practice, however, we need to know how to locate $x$ in order to extract such information. Classically, this is not a problem, as long as reference objects or observers are part of the model. Thus, we can make sense of the value of the Riemann tensor at $x$, if $x$ means something like 'in the southeast corner of the lab at 5 o'clock' or 'where Jim will be standing in 10 minutes'. But it is entirely unclear how to carry this sort of thing over to the case in which all matter (including the lab and Jim) and all fields (including gravity) are quantized. If we treat the lab quantum mechanically, then the location of the southeast corner of the lab at 5 o'clock will not designate a particular point at all.

This is also true in ordinary quantum field theory. In order to give any physical content to a field operator defined at a spacetime point $x$, we need a physical object that we can identify with the point $x$. In practice, this means that we need objects which are very *massive*, so that, barring macroscopic Schrödinger-cat states, they track a definite spacetime trajectory. However, this will not do for a quantum theory of gravity, for although increasing the mass of an object localizes it, it also amounts to increasing its gravitational 'charge'. This means that the more accurate (with respect to a classical background) one's reference system might be expected to be, the more it actually interacts with the quantum-gravitational background one is trying to measure. Ultimately, of course, this is why classical gravitational observables are diffeomorphism-invariant – one cannot isolate a system gravitationally, and all matter, including the reference objects, must be included in the description. But this raises profound difficulties at the quantum level.[21]

### 3.7 Conclusion

We began by looking at the idea of a gravitational field subject to quantum fluctuations at each point of spacetime – a naive yet popular conception of what a quantum theory of gravity might entail. Upon examination, it turned out that the only way to quantify the effects of gravity at a point makes use of relational properties, which fail to capture all observable gravitational phenomena. Furthermore, because any fluctuations in the field would mean fluctuations in the spacetime structure itself, one is left with no way of individuating the points that lends itself to the structure of quantum theory.

In the real world of quantum gravity research, one finds these problems cropping up, albeit in sometimes oblique ways. In canonical quantum gravity, the most obvious counterpart of the first problem is the extreme difficulty of finding any observables (Torre 1994).[22] Should one find them, one would expect that they would not be local observables, but some sort of non-local or perhaps global (i.e. over all of space) observables. It is worth noting in this connection that one of the great ironies of quantum gravity is that it is a theory which is generally supposed to be applicable only at an incredibly small scale (the Planck length is $10^{-33}$ cm), yet any candidate gravitational observables would have to be highly non-local.

Another counterpart to the first problem is the notorious 'problem of time' in quantum gravity. As we saw, the lack of any complete specification of local field strength for gravity implies that there is no adequate definition of local energy density. For the important case of spatially closed spacetimes, this raises great difficulty for a global characterization of energy. In conventional physics, the function that characterizes the energy (the 'Hamiltonian') is the function that mathematically generates time translation, and the ill-definedness of energy in general relativity corresponds to our inability to isolate a Hamiltonian for the theory. In this light, it is not surprising that time-evolution is inherently ambiguous, and that consequently there are great difficulties in even formally constructing a quantum theory.[23]

The counterpart of the second problem, identifying the causal structure, is skirted in canonical quantum gravity by positing a split of spacetime into space and time at the outset. This is not without consequences, however. Among the most serious is the fact that the diffeomorphism group (the invariance group of the full theory of general relativity) is represented in a distorted way in the canonical theory, so that it is unclear that one is actually quantizing general relativity at all. Furthermore, it is characteristic of the classical theory that hypersurfaces which begin as spacelike can evolve into null surfaces, thus killing the evolution. One should expect an analogue of this problem in the canonical quantum theory, though how this would arise depends on how the problem of time is resolved. All this suggests that a theory that truly unifies quantum theory and gravity will be one in which the idea of local fluctuations in a field plays no role, and so a theory which is radically different from any quantum field theory with which we are familiar at present.[24]

## Notes

1. See the first chapter of Bohm's textbook (Bohm 1951) for a concise history of the origins of quantum theory.
2. See Ròvelli (1998) for a recent review.
3. Here we understand 'field' to mean an assignment of properties (the 'values' of the field) to each point in space or spacetime. 'Spacetime' will be represented, for our purposes, by a four-dimensional differentiable manifold equipped with a Lorentz metric satisfying Einstein's equations. 'Space' then refers to a spacelike hypersurface in some spacetime.
4. For simplicity of presentation, we use the canonical picture for electromagnetism and thus make an arbitrary split of spacetime into space and time.
5. Quantum-mechanically, these fields are usually identified with various symmetry groups: $U(1)$ for the electromagnetic interaction (Maxwell theory), $SU(2) \times U(1)$ for the electroweak interaction (combined theory of the electromagnetic and weak interaction), and $SU(3)$ for the strong interaction.
6. There are many different versions of the equivalence principle in the literature – the version here is what Ciufolini and Wheeler (1995) call the 'weak equivalence principle'. The review article by Norton (1993) contains an excellent taxonomy of the various senses of 'equivalence principle'.
7. We are actually considering an idealized limit in which we ignore the contribution of the object itself to the gravitational field. It is only in this limit that it makes sense to talk about two different massive objects moving in the same field, for the objects themselves change the field in proportion to their energy and momentum.
8. The ambiguities that arise in the geometry when the equivalence principle is not respected are discussed in Weinstein (1996).
9. An interesting philosophical analysis of this line of thinking may be found in Reichenbach (1958).
10. However, the decomposition of a curved spacetime into a flat part and a curved perturbation is useful in many classical (i.e. non-quantum) applications. See Chapter 11 of Thorne (1994) for a popular exposition.
11. For example, $S^4$ does not admit a flat metric. See the classic paper by Geroch and Horowitz (1979) for further discussion of this and related topological issues.
12. Technically, these objects correspond to operator-valued distributions, which must be 'smeared' with test functions in order to yield well-defined operators. Chapter 3 of Fulling (1989) contains a lucid discussion.
13. The canonical commutation relations for the electromagnetic field are rather messy, due to the presence of the constraints $\vec{\nabla} \cdot \vec{B} = 0$ and $\vec{\nabla} \cdot \vec{E} = 0$ (in the vacuum case). The commutation relations are $[\hat{E}_i(\vec{x}), \hat{B}_j(\vec{x}')] = i\hbar(\delta_{ij} - \partial_j\partial_i/\nabla^2)\,\delta^3(\vec{x} - \vec{x}')$.
14. Technically, the point here is that we lack an explicit characterization of the reduced phase space of general relativity.
15. This point is taken up at greater length in Weinstein (2001).
16. We are speaking loosely here. There are only four canonical degrees of freedom per spacetime point, whereas the metric has ten components. Thus, if we are attempting to quantize general relativity as we would quantize an ordinary field theory, only four of the ten components should be subject to 'quantum fluctuations'.
17. This ambiguity is behind the notion of 'many-fingered time' that one finds in texts such as Misner, Thorne, and Wheeler (1973).
18. To be precise, the test charge will not accelerate relative to the observer as long as its centre-of-mass and the observer's centre-of-mass coincide at the time of release. If the observer holds the test object out to one side and lets it go, then the difference in the gravitational field at the points where the centres-of-mass of the two objects are located will result in a relative acceleration if the Riemann tensor is non-zero at those points. (This is known as a 'tidal effect'.)

19. Here we are supposing that the Cauchy problem is well-posed, i.e. that the spacetime is spatially closed or that appropriate boundary conditions have been specified.
20. See Fulling (1989) or Wald (1994) for a modern introduction.
21. See Weinstein (2001) for a more extensive discussion of the difficulties of diffeomorphism-invariant quantum theory.
22. Rovelli and Smolin (1995) claim that the area and volume operators in loop quantum gravity are observables, but this relies on a procedure in which matter, treated classically, is used to 'gauge fix' the theory.
23. Excellent reviews of the problem of time are Isham (1993) and Kuchař (1992). See also Weinstein (1998, 1998a).
24. For examples of some of the more radical speculations, see Smolin (1995, 1995a), Susskind (1995), 't Hooft (1997).

# 4     Quantum spacetime: What do we know?

Carlo Rovelli

## 4.1 The incomplete revolution

Quantum mechanics (QM) and general relativity (GR) have profoundly modified our understanding of the physical world. However, they have left us with a general picture of the physical world which is unclear, incomplete, and fragmented. Combining what we have learned about our world from the two theories and finding a new synthesis is a major challenge, perhaps *the* major challenge, in today's fundamental physics.

The two theories have opened a major scientific revolution, but this revolution is not completed. Most of the physics of this century has been a series of triumphant explorations of the new worlds opened by QM and GR. QM leads to nuclear physics, solid state physics, and particle physics; GR leads to relativistic astrophysics, cosmology, and is today leading us towards gravitational astronomy. The urgency of applying the two theories to increasingly larger domains, and the momentous developments and the dominant pragmatic attitude of the middle of the twentieth century have obscured the fact that a consistent picture of the physical world, more or less stable for three centuries, has been lost with the advent of QM and GR. This pragmatic attitude cannot be satisfactory, or productive, in the long run. The basic Cartesian–Newtonian notions such as matter, space, time, and causality, have been deeply modified, and the new notions do not stay together. At the basis of our understanding of the world reigns a surprising confusion. From QM and GR we know that . we live in a spacetime with quantum properties: a *quantum spacetime*. But what is a quantum spacetime?

During the past decade, the attention of the theoretical physicists has been increasingly focussed on this major problem. Whatever the outcome of the enterprise, we are witnessing a large scale intellectual effort directed at accomplishing a major aim: completing the twentieth century scientific revolution, and finding a new synthesis.

In this effort, physics is once more facing *conceptual* problems: What is matter? What is causality? What is the role of the observer in physics? What is time? What is the meaning of 'being somewhere'? What is the meaning of 'now'? What is the meaning of 'moving'? Is motion to be defined with respect to objects or with respect to space? These foundational questions, or sophisticated versions of these questions, were central in the thinking and results of Einstein, Heisenberg, Bohr, Dirac, and their colleagues. But these are also precisely the same questions that Descartes, Galileo, Huygens, Newton, and their contemporaries debated with passion – the questions that led them to create modern science. For the physicists of the middle of the twentieth century, these questions were irrelevant: one does not need to worry about first principles in order to apply the Schrödinger equation to the helium atom, or to understand how a neutron star stays together. But today, if we want to find a novel picture of the world, if we want to understand what quantum spacetime is, we have to return, once again, to those foundational issues. We have to find a new answer to these questions – different from Newton's – which takes into account what we have learned about the world with QM and GR.

Of course, we have little, if any, direct empirical access to the regimes in which we expect genuine quantum gravitational phenomena to appear. Anything could happen at those fantastically small distance scales, far removed from our experience. Nevertheless, we do have information about quantum gravity, and we do have indications of how to search for it. In fact, we are precisely in one of the very typical situations in which good fundamental theoretical physics has worked at its best in the past: we have learned two new extremely general 'facts' about our world, QM and GR, and we have 'just' to figure out what they imply, when taken together. The most striking advances in theoretical physics happened in situations analogous to this one.

Here, I present some reflections on these issues.[1] What have we learned about the world from QM and, especially, GR? What do we know about space, time, and matter? What can we expect from a quantum theory of spacetime? To what extent does taking QM and GR into account force us to modify the notion of time? What can we already say about quantum spacetime?

I also present a few reflections on issues raised by the relationship between philosophy of science and research in quantum gravity. I am not a philosopher, and I can touch on philosophical issues only at the risk of being naive. I nevertheless take this risk here, encouraged by Craig Callender and Nick Huggett, within the extremely stimulating concept of this book. I present some methodological considerations: – How shall we search? How can present successful theories can lead us towards a theory that does not yet exist? – as well as some general considerations. I also discuss the relationship between physical theories that supersede each other and the attitude we may have with respect to the truth-content of a physical theory – with respect to the reality of the theoretical objects the theory postulates in particular, and to its factual statements about the world in general.

I am convinced of the reciprocal usefulness of a dialogue between physics and philosophy (Rovelli 1997). This dialogue has played a major role during other periods in which science faced foundational problems. In my opinion, most physicists underestimate the effect of their own epistemological prejudices on their research, and many philosophers underestimate the influence – either positive or negative – they have on fundamental research. On the one hand, a more acute philosphical awarness would greatly help physicists engaged in fundamental research: Newton, Heisenberg, and Einstein could not have done what they did if they were not nurtured by (good or bad) philosophy. On the other hand, I wish that contemporary philosophers concerned with science would be more interested in the foundational problems that science is facing today. It is here, I believe, that stimulating and vital issues lie.

## 4.2 The problem

What is the task of a quantum theory of gravity, and how should we search for such a theory? The task of the search is clear and well defined. It is determined by recalling the three major steps that led to the present problematical situation.

### 4.2.1 First step. A new actor on the stage: the field

The first step is in the works of Faraday, Maxwell, and Einstein. Faraday and Maxwell introduced a new fundamental notion in physics, the field. Faraday's (1991) book includes a fascinating chapter with the discussion of whether a field (in Faraday's terminology, the 'lines of force') is 'real'. As far as I understand this subtle chapter (understanding Faraday is tricky: it took the genius of Maxwell), in modern terms what Faraday is asking is whether there are independent degrees of freedom in the electric and magnetic fields. A degree of freedom is a quantity that I need to specify (more precisely: whose value and whose time derivative I need to specify) in order to be able to predict univocally the future evolution of the system. Thus Faraday is asking: if we have a system of interacting charges, and we know their positions and velocities, is this knowledge sufficient to predict the future motions of the charges? Or rather, in order to predict the future, do we also have to specify the instantaneous configuration of the field (the field's degrees of freedom)? The answer is in Maxwell equations: the field has independent degrees of freedom. We cannot predict the future evolution of the system from its present state unless we know the instantaneous field configuration. Learning to use these degrees of freedom led to radio, television, and cellular telephone.

To which physical entity do the degrees of freedom of the electromagnetic field refer? This was one of the most debated issues in physics towards the end of the nineteenth century. Electromagnetic waves have aspects in common with water waves, or with sound waves, which describe vibrations of some material medium, so the natural interpretation of the electromagnetic field was that it too describes the vibrations of some material medium – for which the name 'ether' was chosen. A strong argument supports this idea: the wave equations for water or sound waves fail to be Galilean-invariant. They do so because they describe propagation over a medium (water, air) whose state of motion breaks Galilean invariance and defines a preferred reference frame. Maxwell equations break Galilean invariance as well, and it was thus natural to hypothesize a material medium determining the preferred reference frame. But a convincing dynamical theory of the ether compatible with the various experiments (for instance on the constancy of the speed of light) could not be found.

Rather, physics took a different course. Einstein *believed* Maxwell theory as a fundamental theory and *believed* the Galilean insight that velocity is relative and inertial systems are equivalent. Merging the two, he found special relativity. A main result of special relativity is that the field cannot be regarded as describing vibrations of underlying matter. The idea of the ether is abandoned, and *the field has to be taken seriously as an elementary constituent of reality*. This is a major change from the ontology of Cartesian–Newtonian physics. In the best description we can give of the physical world, there is a new actor: the field, for instance, the electromagnetic field described by the Maxwell potential $A_\mu(x)$, $\mu = 0, 1, 2, 3$. The entity described by $A_\mu(x)$ (more precisely, by a gauge-equivalent class of $A_\mu(x)$'s) is one of the elementary constituents of the physical world, according to the best conceptual scheme physics has found, so far, for grasping our world.

### 4.2.2 Second step. Dynamical entities have quantum properties

The second step (out of chronological order) is the replacement of the mechanics of Newton, Lagrange, and Hamilton with QM. Like classical mechanics, QM provides a very general framework. By formulating a specific dynamical theory within this framework, one obtains a number of important physical consequences, substantially different from those implied by the Newtonian scheme. Evolution is determined only probabilistically; some physical quantities can take only certain discrete values (they are 'quantized'); if a system can be in a state $A$, where a physical quantity $q$ has value $a$, as well as in state $B$, where $q$ has value $b$, then the system can also be in states (denoted $\Psi = c_a A + c_b B$) where $q$ has value $a$ with probability $|c_a|^2/(|c_a|^2 + |c_b|^2)$, or, alternatively, $b$ with probability $|c_b|^2/(|c_a|^2 + |c_b|^2)$ (superposition principle); conjugate variables cannot be assumed to have value at the same time (uncertainty principle); and what we can say about the properties that the system will have the day after tomorrow is not determined just by what we can say about the system today, but also on what we will be able to say about the system tomorrow. (Bohr would have simply said that observations affect the system. Formulations such as Bohm's or consistent histories force us to use intricate wordings to describe the same physical fact.)

The formalism of QM exists in a number of more or less equivalent versions: Hilbert spaces and self-adjoint observables, Feynman's sum over histories, the algebraic formulation, and others. Often, we are able to translate from one formulation to another. However, often we cannot do easily in one formulation what we can do in another.

QM is not the theory of micro-objects. It is our best form of mechanics. If quantum mechanics failed for macro-objects, we would have detected the boundary of its domain of validity in mesoscopic physics. We have not.[2] The classical regime raises some problems (why are effects of macroscopic superposition difficult to detect?). Solving these problems requires good understanding of physical decoherence and perhaps more. But there is no reason to doubt that QM represents a deeper – not a shallower – level of understanding of nature than classical mechanics. Trying to resolve the difficulties in our grasping of our quantum world by resorting to old classical intuition is just lack of courage. We have learned that the world has quantum properties. This discovery will stay with us, like the discovery that velocity is only relational, or that the Earth is not the centre of the universe.

The empirical success of QM is immense, and its physical obscurity is undeniable. Physicists do not yet agree on what QM precisely says about the world (the difficulty, of course, refers to the physical meaning of notions such as 'measurement', 'history', 'hidden variable' ... ). It is a bit like the Lorentz transformations before Einstein: correct, but what do they mean?

In my opinion, what QM means is that the contingent (variable) properties of any physical system, or the state of the system, are relational notions which only make sense when referred to a second physical system. I have argued for this thesis (Rovelli 1996, 1998a), but I will not enter into this discussion here, because the issue of the interpretation of QM has no direct connection with quantum gravity. Quantum gravity and the interpretation of QM are two major but (virtually) completely unrelated problems.

QM was first developed for systems with a finite number of degrees of freedom. As discussed in the previous section, Faraday, Maxwell, and Einstein introduced the field, which has an infinite number of degrees of freedom. Dirac put the two ideas together. He *believed* quantum mechanics and he *believed* Maxwell's field theory much beyond their established domain of validity (respectively: the dynamics of finite dimensional systems, and the classical regime) and constructed quantum field theory (QFT), in its first two incarnations, the quantum theory of the electromagnetic field and the relativistic quantum theory of the electron. In this exercise, Dirac derived the existence of the photon just from Maxwell theory and the basics of QM. Furthermore, by just *believing* special relativity and *believing* quantum theory, namely assuming their validity far beyond their empirically explored domains of validity, he predicted the existence of antimatter.

The two embrionic QFTs of Dirac were combined in the 1950s by Feynman and his colleagues, giving rise to quantum electrodynamics, the first non-trivial interacting QFT. A remarkable picture of the world was born: quantum fields over Minkowski space. Equivalently, à la Feynman: the world as a quantum superposition of histories of real and virtual interacting particles. QFT had its ups and downs, then triumphed with the standard model: a consistent QFT for all interactions (except gravity), which, in principle, can be used to predict anything we can measure (except gravitational phenomena), and which, in the past fifteen years has received nothing but empirical verifications.

### 4.2.3 Third step. The stage becomes an actor

Descartes, in *The Principles*, gave a fully relational definition of localization (space) and motion (on the relational/substantivalist issue see for instance Barbour 1989, Earman 1989, Rovelli 1991). According to Descartes, there is no 'empty space'; there are only objects, and it makes sense to say that an object A is contiguous to an object B.

The 'location' of an object A is the set of the objects to which A is contiguous. 'Motion' is change in location; that is, when we say that A moves we mean that A goes from the contiguity of an object B to the contiguity of an object C.[3] A consequence of this relationalism is that there is no meaning in saying 'A moves', except if we specify with respect to which other objects (B, C . . . ) it is moving. Thus, there is no 'absolute' motion. This is the same definition of space, location, and motion, that we find in Aristotle.[4]

Relationalism, namely the idea that motion can be defined only in relation to other objects – should not be confused with Galilean relativity. Galilean relativity is the statement that 'rectilinear uniform motion' is indistinguishable from stasis: that velocity (but just velocity!) is relative to other bodies. Relationalism holds that *any* motion (however zigzagging) is indistinguishable from stasis. The very formulation of Galilean relativity requires a non-relational definition of motion ('rectilinear and uniform' with respect to what?).

Newton took a very different course. He devotes much energy to criticize Descartes' relationalism, and to introduce a different view. According to him, *space* exists, and it exists even if there are no bodies in it. Location of an object is the part of space that the object occupies, and motion is change of location.[5] Thus, we can say whether an object moves or not, independently of surrounding objects. Newton argues that the notion of absolute motion is necessary for constructing mechanics, for instance with his famous discussion of the experiment of the rotating bucket in the *Principia*.

This point has often raised confusion because one of the corollaries of Newtonian mechanics is that there is no detectable preferred reference frame. Therefore, in Newtonian mechanics the notion of *absolute velocity* is actually meaningless. The important point, however, is that in Newtonian mechanics velocity is relative, but any other feature of motion is not relative: it is absolute. In particular, acceleration is absolute. It is acceleration that Newton needs to construct his mechanics; it is acceleration that the bucket experiment is supposed to prove to be absolute, against Descartes. In a sense, Newton overdid things somewhat, introducing the notion of absolute position and velocity (perhaps even just for explanatory purposes), and many people later criticized Newton for his unnecessary use of absolute position, but this is irrelevant for the present discussion. The important point here is that Newtonian mechanics requires absolute acceleration, against Aristotle and against Descartes. Precisely the same is true for special relativistic mechanics.

Similarly, Newton introduced absolute time. Newtonian space and time or, in modern terms spacetime, are like a *stage* over which the action of physics takes place, the various dynamical entities being the actors.

The key feature of this stage – Newtonian spacetime – is its metrical structure. Curves have length, surfaces have area, regions of spacetime have volume. Spacetime points are at fixed *distances* one from another. Revealing, or measuring, this distance, is very simple. It is sufficient to take a rod and put it between two points. Any two points which are one rod apart are at the same distance. Using modern terminology, physical space is a linear three-dimensional space, with a preferred metric. On this space there exist preferred coordinates $x^i$, $i = 1, 2, 3$, in terms of which the metric is just $\delta_{ij}$. Time is described by a single variable $t$. The metric $\delta_{ij}$ determines lengths, areas and volumes and defines what we mean by straight lines in space. If a particle deviates with respect to this straight line, it is, according to Newton, accelerating. It is not accelerating with respect to this or that dynamical object: it is accelerating in absolute terms.

Special relativity changes this picture only marginally, loosing up the strict distinction between the 'space' and the 'time' components of spacetime. In Newtonian spacetime, space is given by fixed three-dimensional planes. In special relativistic spacetime, which three-dimensional plane you call space depends on your state of motion. Spacetime is now a four-dimensional manifold $M$ with a flat Lorentzian metric $\eta_{\mu\nu}$. Again, there are preferred coordinates $x^{\mu}$, $\mu = 0, 1, 2, 3$, in terms of which $\eta_{\mu\nu} = \text{diag}[1, -1, -1, -1]$. This tensor, $\eta_{\mu\nu}$, enters all physical equations, representing the influence of the stage and of its metrical properties on the motion of anything. Absolute acceleration is deviation of the world line of a particle from the straight lines defined by $\eta_{\mu\nu}$. The only essential novelty with special relativity is that the 'dynamical objects' or 'bodies' moving over spacetime now also include the fields. For example, a violent burst of electromagnetic waves coming from a distant supernova has *travelled across space* and reached our instruments. For the rest, the Newtonian construct of a fixed background stage over which physics happens is not altered by special relativity.

The profound change comes with general relativity (GR). The central discovery of GR, can be enunciated in three points. One of these is conceptually simple, the other two are tremendous. First, the gravitational force is mediated by a field, very much like the electromagnetic field: the gravitational field. Second, Newton's *spacetime* – the background stage that Newton introduced against most of the earlier European tradition – and the *gravitational field*, are the *same thing*. Third, the dynamics of the gravitational field, of the other fields such as the electromagnetic field, and any other dynamical object, is fully relational, in the Aristotelian–Cartesian sense. Let me illustrate these three points.

First, the gravitational field is represented by a field on spacetime, $g_{\mu\nu}(x)$, just like the electromagnetic field $A_{\mu}(x)$. They are both very concrete entities: a strong electromagnetic wave can hit you and knock you down; and so can a strong gravitational wave. The gravitational field has independent degrees of freedom, and is governed by dynamical equations, the Einstein equations.

Second, the spacetime metric $\eta_{\mu\nu}$ disappears from all equations of physics (recall it was ubiquitous). In its place – we are instructed by GR – we must insert the gravitational field $g_{\mu\nu}(x)$. This is a spectacular step: Newton's background spacetime was nothing but the gravitational field! The stage is promoted to one of the actors. Thus, in all physical equations one now sees the direct influence of the gravitational field. How can the gravitational field determine the metrical properties of things, which are revealed, say, by rods and clocks? Simply, the interatomic separation of the rods' atoms, and the frequency of the clock's pendulum are determined by explicit couplings of the rod's and clock's variables with the gravitational field $g_{\mu\nu}(x)$, which enters the equations of motion of these variables. Thus, any measurement of length, area or volume is, in reality, a measurement of features of the gravitational field.

But what is really formidable in GR, the truly momentous novelty, is the third point: the Einstein equations, as well as *all other equations of physics* appropriately modified according to GR instructions, are fully relational in the Aristotelian–Cartesian sense. This point is independent of the previous one. Let me give first a conceptual, and then a technical account of it.

The point is that the only physically meaningful definition of location within GR is relational. GR describes the world as a set of interacting fields including $g_{\mu\nu}(x)$, and possibly other objects, and motion can be defined only by positions and displacements of these dynamical objects relative to each other (for more details on this, see Rovelli 1991, and especially 1997).

To describe the motion of a dynamical object, Newton had to assume that acceleration is absolute, namely that it is not relative to this or that other dynamical object. Rather, it is relative to a background space. Faraday, Maxwell, and Einstein extended the notion of 'dynamical object': the stuff of the world is fields, not just bodies. Finally, GR tells us that the background space is itself one of these fields. Thus, the circle is closed, and we are back to relationalism: Newton's motion with respect to space is indeed motion with respect to a dynamical object – the gravitational field.

All this is coded in the active diffeomorphism invariance (diff invariance) of GR.[6] Because active diff invariance is a gauge, the physical content of GR is expressed only by those quantities, derived from the basic dynamical variables, which are fully independent from the points of the manifold.

In introducing the background stage, Newton introduced two structures: a spacetime manifold, and its non-dynamical metric structure. GR gets rid of the non-dynamical metric, by replacing it with the gravitational field. More importantly, it gets rid of the manifold, by means of active diff invariance. In GR, the objects of which the world is made do not live over a stage and do not live on spacetime; they live, so to say, over each other's shoulders.

Of course, nothing prevents us, if we wish so to do, from singling out the gravitational field as 'the more equal among equals', and declaring that location is absolute in GR, because it can be defined with respect to it. But this can be done within any relationalism: we can always single out a set of objects, and declare them as not moving by definition.[7] The problem with this attitude is that it fully misses the great Einsteinian insight: that Newtonian spacetime is just one field among the others. More seriously, this attitude sends us into a nightmare when we have to deal with the motion of the gravitational field itself (which certainly 'moves': we are spending millions of dollars for constructing gravity wave detectors to detect its tiny vibrations). There is no absolute referent of motion in GR: the dynamical fields 'move' with respect to each other.

Notice that the third step was not easy for Einstein, and came later than the previous two. Having well understood the first two, but still missing the third, Einstein actively searched for non-generally covariant equations of motion for the gravitational field between 1912 and 1915. With his famous 'hole argument' he had convinced himself that generally covariant equations of motion (and therefore, in this context, active diffeomorphism invariance) would imply a truly dramatic revolution with respect to the Newtonian notions of space and time (with regard to the hole argument, see Earman and Norton 1987, Rovelli 1991, Belot 1998a). In

1912, Einstein was unable to take this profoundly revolutionary step (Norton 1984, Stachel 1989), but in 1915 he took it, and found what Landau calls 'the most beautiful of the physical theories'.

### 4.2.4 Bringing the three steps together

In the light of the three steps described above, the task of quantum gravity is clear and well defined. We have learned from GR that spacetime is a dynamical field among others, obeying dynamical equations, and having independent degrees of freedom: a gravitational wave is extremely similar to an electromagnetic wave. We have learned from QM that every dynamical object has quantum properties, which can be captured by appropriately formulating its dynamical theory within the general scheme of QM.

*Therefore*, spacetime itself must exhibit quantum properties. Its properties – including the metrical properties it defines – must be represented in quantum mechanical terms. Notice that the strength of this 'therefore' derives from the confidence we have in the two theories, QM and GR.

Now, there is nothing in the basis of QM which contradicts the physical ideas of GR. Similarly, there is nothing in the basis of GR that contradicts the physical ideas of QM. Therefore, there is no *a priori* impediment in searching for a quantum theory of the gravitational fields, that is, a quantum theory of spacetime. The problem is (with some qualification) rather well posed: is there a quantum theory (say, in one formulation, a Hilbert space $H$, and a set of self-adjoint operators) whose classical limit is GR?

On the other hand, all previous applications of QM to *field* theory, namely conventional QFTs, rely heavily on the existence of the 'stage', the fixed, non-dynamical, background metric structure. The Minkowski metric $\eta_{\mu\nu}$ is essential for the construction of a conventional QFT (it enters everywhere; for instance, in the canonical commutation relations, in the propagator, in the Gaussian measure . . . ). We certainly cannot simply replace $\eta_{\mu\nu}$ with a quantum field, because these equations become nonsense.

Therefore, to search for a quantum theory of gravity, we have two possible directions. One possibility is to 'disvalue' the GR conceptual revolution, reintroduce a background spacetime with a non-dynamical metric $\eta_{\mu\nu}$, expand the gravitational field $g_{\mu\nu}$ as $g_{\mu\nu} = \eta_{\mu\nu} + fluctuations$, quantize only the fluctuations, and hope to recover full GR somewhere down the road. This is the path followed, for instance, by perturbative string theory.

The second direction means being faithful to what we have learned about the world so far, namely to QM and GR's insights. We must then search for a QFT that, genuinely, does not require a background space. But the past three decades have been characterized by the great success of conventional QFT, which neglects GR and is based on the existence of a background spacetime, and we live in the aftermath of this success. It is not easy to get away from the conceptual and technical habits of conventional QFT. Still, this is necessary if we want to build a QFT which fully incorporates active diff invariance, and in which localization is fully relational. In my opinion, this is the right way to go.

### 4.3 Quantum spacetime

#### 4.3.1 Space

Spacetime, or the gravitational field, is a dynamical entity (GR). All dynamical entities have quantum properties (QM). Therefore, spacetime is a quantum object. It must be described (picking one formulation of QM, but keeping in mind that others may be equivalent, or more effective) in terms of states $\Psi$ in a Hilbert space. Localization is relational. Therefore, these states cannot represent quantum excitations localized in some space, but must define space themselves. They must be quantum excitations 'of' space, not 'in' space. Physical quantities in GR that capture the true degrees of freedom of the theory are invariant under active diff, therefore the self-adjoint operators that correspond to physical (predictable) observables in quantum gravity must be associated with diff invariant quantities.

Examples of diff invariant geometric quantities are physical lengths, areas, volumes, or time intervals of regions determined by dynamical physical objects. These must be represented by operators. Indeed, a measurement of length, area or volume is a measurement of features of the gravitational field. If the gravitational field is a quantum field, then length, area and volume are quantum observables. If the corresponding operator has a discrete spectrum, they will be quantized; namely, they can take certain discrete values only. In this sense we should expect a discrete geometry. This discreteness of the geometry, implied by the conjunction of GR and QM, is very different from the naive idea that the world is made by discrete bits of something. It is like the discreteness of the quanta of the excitations of a harmonic oscillator. A generic state of spacetime will be a continuous quantum superposition of states whose geometry has discrete features, not a collection of elementary discrete objects.

A concrete attempt to construct such a theory is loop quantum gravity. (For an introduction to the theory, an overview of its structure and results, and full references, see Rovelli 1998b.) Here, I present only a few remarks on the theory. Loop quantum gravity is a rather straightforward application of quantum mechanics to Hamiltonian general relativity. It is a QFT in the sense that it is a quantum version of a field theory, or a quantum theory for an infinite number of degrees of freedom, but it is profoundly different from conventional, non-general-relativistic QFT theory. In conventional QFT, states are quantum excitations of a field over Minkowski (or over a curved) spacetime. In loop quantum gravity, the quantum states turn out to be represented by (suitable linear combinations of) spin networks (Rovelli and Smolin 1995a, Baez 1996, Smolin 1997). A spin network is an abstract graph with links labelled by half-integers (Fig. 4.1).

Intuitively, we can view each node of the graph as an elementary 'quantum chunk of space'; the links represent (transverse) surfaces separating the quanta of space, and the half-integers associated with the links determine the (quantized) area of these surfaces. The spin network represent relational quantum states: they are not located in a space. Localization must be defined in relation to them. For instance, if we have, say, a matter quantum excitation, this will be located on the spin network; while the spin network itself is not located anywhere.

The operators corresponding to area and volume have been constructed in the theory, simply by starting from the classical expression for the area in terms of the

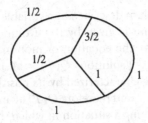

Fig. 4.1. A simple spin network.

metric, then replacing the metric with the gravitational field (this is the input of GR) and then replacing the gravitational field with the corresponding quantum field operator (this is the input of QM). The construction of these operators requires appropriate generally covariant regularization techniques, but no renormalization: no infinities appear. The spectrum of these operators has been computed and turns out to be discrete (Rovelli and Smolin 1995, Ashtekar and Lewandowski 1997, 1997a). Thus, loop quantum gravity provides a family of precise quantitative predictions: the quantized values of area and volume. For instance, the (main sequence of the) spectrum of the area is

$$A = 8\pi\hbar G \sum_{i=1,n} \sqrt{j_i(j_i + 1)},$$

where $(j_i) = (j_1 \ldots j_n)$ is any finite sequence of half-integers. This formula gives the area of a surface pinched by $n$ links of a spin network state. The half-integers $j_1 \ldots j_n$ are associated with the $n$ links that pinch the surface. This illustrates how the links of the spin network states can be viewed as transverse 'quanta of area'. The picture of macroscopic physical space that emerges is then that of a tangle of one-dimensional intersecting quantum excitations, called the weave (Ashtekar, Rovelli, and Smolin 1992). Continuous space is formed by the weave in the same manner in which the continuous two-dimensional surface of a T-shirt is formed by woven threads.

### 4.3.2 Time

The aspect of GR's relationalism that concerns space was largely anticipated by earlier European thinking. Much less so (as far as I am aware) was the aspect of this relationalism that concerns time. GR's treatment of time is surprising, difficult to fully appreciate, and hard to digest, for the time of our perceptions is very different from the time that theoretical physics finds in the world as soon as one exits the minuscule range of physical regimes we are accustomed to. We seem to have a very special difficulty in being open-minded about this particular notion.

Already, special relativity teaches us something about time which many of us have difficulty accepting. According to special relativity, there is absolutely no meaning in saying 'right now on Andromeda': there is no physical meaning in the idea of 'the state of the world right now', because which set of events we consider as 'now' is perspectival, so the 'now' on Andromeda for me might correspond to 'a century ago' on Andromeda for you. Thus, there is no single well-defined universal time in which the history of the universe 'happens'. The modification of the concept of time introduced by GR is much deeper. Let me illustrate this modification.

Consider a simple pendulum described by a variable $Q$. In Newtonian mechanics, the motion of the pendulum is given by the evolution of $Q$ in time, namely by $Q(T)$, which is governed by the equation of motion, say $\ddot{Q} = -\omega Q$, which has (the two-parameter family of) solutions $Q(T) = A \sin(\omega T + \phi)$. The state of the pendulum at time $T$ can be characterized by its position and velocity. From these two, we can compute $A$ and $\phi$ and therefore $Q(T)$ at any $T$. From the physical point of view, we are really describing a situation in which there are *two* physical objects: a pendulum, whose position is $Q$, and a clock, indicating $T$. If we want to take data, we have repeatedly to observe $Q$ and $T$. Their *relation* will be given by the equation above. The relation can be represented (for given $A$ and $\phi$) by a line in the $(Q, T)$ plane.

In Newtonian terms, time flows in its absolute way, the clock is just a device to keep track of it, and the dynamical system is formed by the pendulum alone. But we can view the same physical situation from a different perspective. We can say that we have a physical system formed by the clock and the pendulum together, and view the dynamical system as expressing the relative motion of one with respect to the other. This is precisely the perspective of GR: to express the relative motion of the variables, with respect to each other, in a 'democratic' fashion.

To do that, we can introduce an 'arbitrary parameter time' $\tau$ as a co-ordinate on the line in the $(Q, T)$ plane. (But bear in mind that the physically relevant information is in the line, not in its co-ordinatization!) Then the line is represented by two functions, $Q(\tau)$ and $T(\tau)$, but a reparametrization of $\tau$ in the two functions is a gauge, namely it does not modify the physics described. Indeed, $\tau$ does not correspond to anything observable, and the equations of motion satisfied by $Q(\tau)$ and $T(\tau)$ (easy to write, but I will not write them down here) will be invariant under arbitrary reparametrizations of $\tau$. Only $\tau$-independent quantities have physical meaning.

This is precisely what happens in GR, where the 'arbitrary parameters', analogous to the $\tau$ of the example, are the co-ordinates $x^\mu$. Namely, the spatial co-ordinate $\vec{x}$ and the temporal co-ordinate $t$. These have no physical meaning whatsoever in GR: the connection between the theory and the measurable physical quantities that the theory predict is only via quantities independent from $\vec{x}$ and $t$. Thus, $\vec{x}$ and $t$ in GR have a very different physical meaning than their homonyms in non-general-relativistic physics. The latter correspond to readings on rods and clocks; the former correspond to nothing at all. Recall that Einstein described his great intellectual struggle to find GR as 'understanding the meaning of the co-ordinates'.

In the example, the invariance of the equations of motion for $Q(\tau)$ and $T(\tau)$ under reparametrization of $\tau$, implies that if we develop the Hamiltonian formalism in $\tau$ we obtain a constrained system with a (weakly) vanishing Hamiltonian. This is because the Hamiltonian generates evolutions in $\tau$, evolution in $\tau$ is a gauge, and the generators of gauge transformations are constraints. In canonical GR we have precisely the same situation: the Hamiltonian vanishes, and the constraints generate evolution in $t$, which is unobservable – it is gauge. GR does not describe evolution in time: it describes the relative evolution of many variables with respect to each other. All these variables are democratically equal: there is not a preferred one that 'is the true time'. This is the temporal aspect of GR's relationalism.

A large part of the machinery of theoretical physics relies on the notion of time (on the different meanings of time in different physical theories, see Rovelli 1995). A theory of quantum gravity should do without it. Fortunately, many essential tools that are usually introduced using the notion of time can equally well be defined without mentioning time at all. This, by the way, shows that time plays a much weaker role in the structure of theoretical physics than what is mostly assumed. Two crucial examples are 'phase space' and 'state'.

The phase space is usually introduced in textbooks as the space of the states of the systems 'at a given time'. In a general relativistic context, this definition is useless. However, it has been known since Lagrange that there is an alternative, equivalent, definition of phase space as the space of solutions of the equations of motion, and this definition does not require that we know what we mean by time. Thus, in the example above the phase space can be co-ordinatized by $A$ and $\phi$, which co-ordinatize the space of the solutions of the equations of motion.

A time-independent notion of 'state' is then provided by a point of this phase space, namely by a particular solution of the equations of motion. For instance, for an oscillator a 'state', in this atemporal sense, is characterized by an amplitude $A$ and a phase $\phi$. Notice that given the (time-independent) state ($A$ and $\phi$), we can compute any observable: in particular, the value $Q_T$ of $Q$ at any desired $T$. Notice also that $Q_T$ is independent from $\tau$. This point often raises confusion: one may think that if we restrict to $\tau$-independent quantities then we cannot describe evolution. This is wrong: the true evolution is the relation between $Q$ and $T$, which is $\tau$-independent. This relation is expressed in particular by the value (let us denote it $Q_T$) of $Q$ at a given $T$. $Q_T$ is given, obviously, by

$$Q_T(A, \phi) = A \sin(\omega T + \phi).$$

This can be seen as a one-parameter (the parameter is $T$) family of observables on the gauge-invariant phase space co-ordinatized by $A$ and $\phi$. Notice that this is a perfectly $\tau$-independent expression. In fact, an explicit computation shows that the Poisson bracket between $Q_T$ and the Hamiltonian constraint that generates evolution in $\tau$ vanishes.

This time-independent notion of states is well known in its quantum mechanical version: it is the Heisenberg state (as opposed to Schrödinger state). Similarly, the operator corresponding to the observable $Q_T$ is the Heisenberg operator that gives the value of $Q$ at $T$. The Heisenberg and Schrödinger pictures are equivalent if there is a normal time evolution in the theory, but in the absence of a normal notion of time evolution, the Heisenberg picture remains viable, and the Schrödinger picture becomes meaningless.[8] In quantum gravity, only the Heisenberg picture makes sense (Rovelli 1991c, 1991e).

In classical GR, a point in the physical phase space, or a state, is a solution of the Einstein equations, up to active diffeomorphisms, so a state represents a 'history' of spacetime. The quantities that can be univocally predicted are those that are independent from the co-ordinates, namely those that are invariant under diffeomorphisms. These quantities have vanishing Poisson brackets with all the constraints, and given a state, the value of each of these quantities is determined. In quantum gravity, a quantum state represents a 'history' of quantum spacetime, and the observables are represented by operators that commute with *all* the quantum constraints. If we know the quantum state of spacetime, we can then compute the expectation value of any diffeomorphism invariant quantity, by taking the mean value of the corresponding operator. The observable quantities in quantum gravity are precisely the same as in classical GR.

Some of these quantities may express the value of certain variables 'when and where' certain other quantities have certain given values. They are the analogue of the reparametrization invariant observable $Q_T$ in the example above. These quantities describe evolution in a way which is fully invariant under the parameter time, unphysical gauge evolution (Rovelli 1991b, 1991c). The corresponding quantum operators are Heisenberg operators. There is no Schrödinger picture, because there is no unitary time evolution. There is no need to expect or to search for unitary time evolution in quantum gravity, because there is no time in which we could have unitary evolution. It is just a tenacious prejudice that unitary evolution is required for the consistency of the probabilistic interpretation. This idea is wrong.

What I have described is the general form that one may expect a quantum theory of GR to have. I have used the Hilbert space version of QM; but this structure can be translated in other formulations of QM. Of course, physics then works with dirty hands: gauge-dependent quantities, approximations, expansions, unphysical structures, and so on. A fully satisfactory construction of the above kind does not yet exist, but a concrete attempt to construct the physical states and the physical observables in loop quantum gravity is given by the spin foam models approach, which is the formulation one obtains by starting from loop quantum gravity and constructing a Feynman sum over histories (Reisenberger and Rovelli 1997, Baez 1998, Barrett and Crane 1998). (See Chapter 8 for more details on ideas underlying these developments.)

In quantum gravity, I see no reason to expect a fundamental notion of time to play any role. But the *nostalgia for time* is hard to resist for technical as well as emotional reasons. Many approaches to quantum gravity go out of their way to reinsert in the theory what GR teaches us to abandon: a preferred time. The time 'along which' things happen is a notion which makes sense only for describing a limited regime of reality. This notion is meaningless already in the (gauge-invariant) general relativistic classical dynamics of the gravitational field. At the fundamental level, we should, simply, forget time.

### 4.3.3 Glimpses

I close this section by briefly mentioning two more speculative ideas. One regards the emergence of time; the second the connection between relationalism in GR and relationalism in QM.

First, in the previous section, I argued that we should search for a quantum theory of gravity in which there is no independent time variable 'along' which dynamics 'happens'. A problem left open by this position is to understand the emergence of time in our world, with those features which are familiar to us. An idea discussed (Rovelli 1993a, 1993b, Connes and Rovelli 1994) is that the notion of time is not dynamical but rather thermodynamical. We can never give a complete account of the state of a system in a field theory (we cannot access the infinite amount of data needed to characterize completely a state). Therefore we have at best a statistical description of the state. But given a statistical state of a generally covariant system, a notion of a flow (more precisely a one-parameter group of automorphisms of the algebra of the observables) follows immediately. (In the quantum context, this corresponds to the Tomita flow of the state.) The relationship between this flow and the state is the relationship between the time flow generated by the Hamiltonian and a Gibbs state: the two essentially determine each other. In the absence of a preferred time, however, any statistical state selects its own notion of statistical time. This statistical time has a striking number of properties that allow us to identify it with the time of non-general relativistic physics. In particular, a Schrödinger equation with respect to this statistical time holds, in an appropriate sense. In addition, the time flows generated by different states are equivalent up to inner automorphisms of the observable algebra and therefore define a common 'outer' flow: a one-paramater group of outer automorphisms. This determines a state independent notion of time flow, which shows that a general covariant QFT has an intrinsic 'dynamics', even in the absence of a Hamiltonian and of a time variable. The suggestion is therefore that the temporal aspects of our world have statistical and thermodynamical origin, rather than dynamical. 'Time' is ignorance: a reflex of our incomplete knowleddge of the state of the world.

Second, what is QM really telling us about our world? In Rovelli (1996, 1998a), I have argued that what QM is telling us is that the contingent properties of any system – the state of any system – must be seen as relative to a second physical system, the 'observing system'. That is, quantum states and values that observables take are relational notions, in the same sense in which velocity is relational in classical mechanics (it is a relationship between two systems, not a property of a single system). I find the consonance between this relationalism in QM and the relationalism in GR quite striking, and it is tempting to speculate that they are related. Any quantum interaction (or quantum measurement) involving a system $A$ and a system $B$ requires $A$ and $B$ to be spatio-temporally contiguous. Vice versa, spatio-temporal contiguity, which ground the notions of space and time (derived and dynamical, not primary, in GR) can only be verified quantum mechanically (just because any interaction is quantum mechanical in nature). Thus, the net of quantum mechanical elementary interactions and the spacetime fabric are actually the same thing. Can we build a consistent picture in which we take this fact into account? To do that, we must identify two notions: the notion of a spatio-temporal (or spatial?) region, and the notion of a quantum system. For intriguing ideas in this direction, see Crane (1991) and Chapter 8 of this book.

## 4.4 Considerations on method and content

### 4.4.1 Method

Part of recent reflection on science has emphasized the 'non-cumulative' aspect of the development of scientific knowledge. According to this view, the evolution of scientific theories is marked by large or small breaking points, in which, to put it very crudely, the empirical facts are just reorganized within new theories. These would be to some extent 'incommensurable' with respect to their antecedents. Such ideas have influenced physicists.

The reader may have remarked that the discussion of quantum gravity I have given above assumes a different reading of the evolution of scientific knowledge. I have based the above discussion of quantum gravity on the idea that the central physical ideas of QM and GR represent our best guide for accessing the extreme and unexplored territories of the quantum-gravitational regime. In my opinion, the emphasis on the incommensurability between theories has probably clarified an important aspect of science, but risks obscuring something of the internal logic according to which, historically, physics finds knowledge. There is a subtle, but definite, cumulative aspect in the progress of physics, which goes far beyond the growth of validity and precision of the empirical content of theories. In moving from a theory to the theory that supersedes it, we do not save just the verified empirical content of the old theory, but more. This 'more' is a central concern for good physics. It is the source, I think, of the spectacular and undeniable predictive power of theoretical physics.

Let me illustrate the point I am trying to make with a historical case. There was a conflict between Maxwell's equations and Galileo's transformations, and there were two obvious ways out: disvalue Maxwell theory, degrading it to a phenomenological theory of some yet-to-be-discovered ether's dynamics; or disvalue Galilean invariance, accepting the idea that inertial systems are not equivalent in electromagnetic phenomena. Both approaches were pursued at the end of the century. Both are sound applications of the idea that a scientific revolution may very well change in depth what old theories teach us about the world. Which of the two ways did Einstein take?

None of them. For Einstein, Maxwell theory was a source of great awe. Einstein rhapsodizes about his admiration for Maxwell theory. For him, Maxwell had opened a new window on the world. Given the astonishing success of Maxwell theory, empirical (electromagnetic waves), technological (radio) as well as conceptual (understanding what light is), Einstein's admiration is comprehensible. But Einstein also had a tremendous respect for Galileo's insight; young Einstein was amazed by a book with Huygens' derivation of collision theory virtually out of Galilean invariance alone. Einstein understood that Galileo's great intuition – that the notion of velocity is only relative – *could not be wrong*. I am convinced that in Einstein's faith in the core of the great Galilean discovery there is very much to learn, for philosophers of science, as well as for contemporary theoretical physicists. So, Einstein believed the two theories, Maxwell and Galileo: he assumed that they would hold far beyond the regime in which they had been tested. Moreover, he assumed that Galileo had grasped something about the physical world, which was, simply, *correct*. And so had Maxwell. Of course, details had to be adjusted. The core of Galileo's insight was that all inertial systems are equivalent and that velocity is relative, not the details of the Galilean transformations. Einstein knew the Lorentz transformations (found, of course, by Lorentz, not by Einstein), and was able to see that they do not contradict Galileo's insight. If there was contradiction in putting the two together, the problem was ours: we were surreptitiously sneaking some incorrect assumption into our deductions. He found the incorrect assumption, which, of course, was that simultaneity could be well defined. It was Einstein's faith in the *essential physical correctness* of the old theories that guided him to his spectacular discovery.

There are innumerable similar examples in the history of physics that could equally well illustrate this point. Einstein found GR 'out of pure thought', having Newton theory on the one hand and special relativity – the understanding that any interaction is mediated by a field – on the other; Dirac found quantum field theory from Maxwell's equations and quantum mechanics; Newton combined Galileo's insight that acceleration governs dynamics with Kepler's insight that the source of the force that governs the motion of the planets is the sun . . . The list could be long. In all these cases, confidence in the insight that came with some theory, or 'taking a theory seriously', led to major advances that largely extended the original theory itself. Of course, it is far from me to suggest that there is anything simple, or automatic, in figuring out where the true insights are and in finding the way of making them work together. But what I am saying is that figuring out where the true insights are and finding the way of making them work together is the work of fundamental physics. This work is grounded on *confidence* in the old theories, not on random searches for new ones.

One of the central concerns of modern philosophy of science is to face the apparent paradox that scientific theories change, but are nevertheless credible. Modern philosophy of science is to some extent an after-shock reaction to the fall of Newtonian mechanics, a tormented recognition that an extremely successful scientific theory can nevertheless be untrue. But it is a narrow-minded notion of truth that is called into question by a successful physical theory being superseded by a more successful one.

A physical theory, in my view, is a conceptual structure that we use in order to organize, read and understand the world, and to make predictions about it. A successful physical theory is a theory that does so effectively and consistently. In the light of our experience, there is no reason not to expect that a more effective conceptual structure might always exist. Therefore an effective theory may always show its limits and be replaced by a better one. On the other hand, however, a novel conceptualization cannot but rely on what the previous one has already achieved.

When we move to a new city, we are at first confused about its geography. Then we find a few reference points, and we make a rough mental map of the city in terms of these points. Perhaps we see that there is part of the city on the hills and part on the plane. As time goes on, the map gets better, but there are moments in which we suddenly realize that we had it wrong. Perhaps there were indeed two areas with hills, and we were previously confusing the two. Or we had mistaken a big red building for the City Hall, when it was only a residential construction. So we adjust the mental map. Sometime later, we learn names and features of neighbourhoods and streets, and the hills, as references, fade away, because the neighbourhoods structure of knowledge is more effective than the hill/plane one . . . . The structure changes, but knowledge increases. And the big red building, now we know it, is not the City Hall – and we know it forever.

There are discoveries that are forever: that the Earth is not the centre of the universe, that simultaneity is relative, that we do not get rain by dancing. These are steps that humanity takes, and does not take back. Some of these discoveries amount simply to cleaning from our thinking wrong, encrusted, or provisional credences. But also discovering classical mechanics, or discovering electromagnetism, or quantum mechanics, are discoveries forever. Not because the details of these theories cannot change, but because we have discovered that a large portion of the world admits to understanding in certain terms, and this is a *fact* that we will have to face forever.

One of the theses of this chapter, is that general relativity is the expression of one of these insights, which will stay with us 'forever'. The insight is that the physical world does not have a stage, that localization and motion are relational only, that diff invariance (or something physically analogous) is required for any fundamental description of our world.

How can a theory be effective even outside the domain in which it was found? How could Maxwell predict radio waves, Dirac predict antimatter, and GR predict black holes? How can theoretical thinking be so magically powerful? Of course, we may think that these successes are chance, and only appear striking because of a deformed historical perspective: hundreds of theories are proposed, most of them die, and the ones that survive are the ones remembered – just as there is always somebody who wins the lottery, but this is not a sign that humans can magically predict the outcome of the lottery. My opinion is that such an interpretation of the development of science is unjust and, worse, misleading. It may explain something, but there is more in science. Tens of thousands play the lottery, but there were only two relativistic theories of gravity in 1916, when Einstein predicted that the light would be deflected by the sun precisely by an angle of $1.75''$. Familiarity with the history of physics, I feel confident to claim, rules out the lottery picture.

I think that the answer is simpler. Somebody predicts that the sun will rise tomorrow, and the sun rises. It is not a matter of chance (there aren't hundreds of people making random predictions of every sort of strange object appearing at the horizon). The prediction that tomorrow the sun will rise, is sound. However, it is not certain either: a neutron star could rush in, close to the speed of light, and sweep the sun away. More philosophically, who grants me the right of induction? Why should I be confident that the sun will rise, just because it has risen so many times in the past? I do not know the answer to *this* question. But what I know is that the predictive power of a theory beyond its own domain is *precisely of the same sort*. Simply, we learn something about nature (whatever this means), and what we learn is effective in guiding us to predict nature's behaviour. Thus, the spectacular predictive power of theoretical physics is nothing less and nothing more than common induction, and it is as comprehensible (or as incomprehensible) as my ability to predict that the sun will rise tomorrow. Simply, nature around us happens to be full of regularities *that we understand*, whether or not we understand why regularities exist at all. These regularities give us strong confidence – although not certainty – that the sun will rise tomorrow, and that the basic facts about the world found with QM and GR will be confirmed, not violated, in quantum-gravitational regimes that we have not empirically probed.

This view is not dominant nowadays in theoretical physics. Other attitudes dominate. The 'pragmatic' scientist ignores conceptual questions and physical insights, and only cares about developing a theory. This attitude was successful in the 1960s in getting to the standard model. The 'pessimistic' scientist has little faith in the possibilities of theoretical physics, because he or she worries that all possibilities are open, and anything might happen between here and the Planck length. The 'wild' scientist observes that great scientists had the courage to break with old and respected ideas and assumptions, and to explore new and strange hypotheses. From this observation, the 'wild' scientist concludes that to do great science one has to explore strange hypotheses and *violate respected ideas*: the wilder the hypothesis, the better. I think such wildness in physics is sterile; the greatest revolutionaries in science were extremely, almost obsessively, conservative. The greatest revolutionary, Copernicus, certainly was – and so was Planck. Copernicus was pushed to the great jump from his pedantic labour on the minute technicalities of the Ptolemaic system (fixing the equant). Kepler was forced to abandon circles by his extremely technical work on the details of Mars' orbit. He was using ellipses as approximations to the epicycle-deferent system, when he began to realize that the approximation was fitting the data better than the (supposedly) exact curve. And Einstein and Dirac were also extremely conservative: their vertiginous steps ahead were not pulled out of blue sky, from violating respected ideas but, on the contrary, they came from respect towards physical insights. In physics, novelty has always emerged from new data and from a humble, devoted interrogation of old theories: from turning these theories around and around, immerging into them, making them clash, merge, and talk until, through them, the missing gear could be seen. In my opinion, precious research energies are today lost in these attitudes. I worry that a philosophy of science that downplays the component of factual knowledge in physical theories might have part of the responsibility.

### 4.4.2 On content and truth in physical theories

If a physical theory is a conceptual structure that we use to organize, read and understand the world, then scientific thinking is not much different from common-sense thinking. In fact, it is only a better instance of the same activity: thinking about the world. Science is the enterprise of continuously exploring the possible ways of thinking about the world, and constantly selecting the ones that work best.

If so, there cannot be any qualitative difference between the theoretical notions introduced in science and the terms in our everyday language. A fundamental intuition of classical empiricism is that nothing grants us the 'reality' of the referents of the notions we use to organize our perceptions. Some modern philosophy of science has emphasized the application of this intuition to the concepts introduced by science. Thus, we are warned to doubt the 'reality' of theoretical objects (electrons, fields, black holes . . . ). I find these warnings incomprehensible. Not because they are ill-founded, but because they are not applied consistently. The fathers of empiricism consistently applied this intuition to *any* physical object. Who grants me the reality of a chair? Why should a chair be more than a theoretical concept organizing certain regularities in my perceptions? I will not venture here to dispute or agree with this doctrine. What I find incomprehensible is the position of those who grant the solid status of reality to a chair, but not to an electron; the arguments against the reality of the electron apply to the chair as well, and the arguments in favour of the reality of the chair apply to the electron as well. A chair, as well as an electron, is a concept that we use to organize, read, and understand the world. They are equally real. They are equally volatile and uncertain.

Perhaps, this curious schizophrenic attitude of antirealism about electrons and iron realism about chairs is the result of a complex historical evolution. First there was the rebellion against 'metaphysics', and, with it, the granting of confidence to science alone. From this point of view, metaphysical questioning of the reality of chairs is sterile – true knowledge is in science. Thus, it is to scientific knowledge that we apply empiricist rigour. But understanding science in empiricist terms required making sense of the raw empirical data on which science is based. With time, the idea of raw empirical data increasingly showed its limits, and the common-sense view of the world was reconsidered as a player in our picture of knowledge. This common-sense view gives us a language and a ground from which to start – the old antimetaphysical prejudice still preventing us, however, from applying empiricist rigor to this common sense view of the world as well. But if one is not interested in questioning the reality of chairs, for the very same reason, why should one be interested in questioning the 'reality of the electrons'?

Again, I think this point is important for science itself. The factual content of a theory is our best tool. Faith in this factual content does not prevent us from being ready to question the theory itself, if sufficiently compelled to do so by novel empirical evidence or by putting the theory in relation to other things *we know* about the world. Scientific antirealism, in my opinion, is not only a short sighted application of a deep classical empiricist insight; it is also a negative influence on the development of science. H. Stein (1999) has recently beautifully illustrated a case in which a great scientist, Poincaré, was blocked from getting to a major discovery (special relativity) by a philosophy that restrained him from 'taking seriously' his own findings.

Science teaches us that our naive view of the world is imprecise, inappropriate, and biased. It constructs better views of the world. Electrons, if anything at all, are 'more real' than chairs, not 'less real', in the sense that they ground a more powerful way of conceptualizing the world. On the other hand, the process of scientific discovery, and the experience of this century in particular, has made us painfully aware of the provisional character of *any* form of knowledge. Our mental and mathematical pictures of the world are only mental and mathematical pictures. This is true for abstract scientific theories as well as from the image we have of our dining room. Nevertheless, the pictures are powerful and effective and we can't do any better than that.

So, is there anything we can say with confidence about the 'real world'? A large part of recent reflection on science has taught us that raw data do not exist, and that any information about the world is already deeply filtered and interpreted by theory. Further than that, we could even think, as in the dream of Berkeley, that there is no 'reality' outside. Recent European reflection (and part of the American as well) has emphasized the fact that truth is always internal to the theory, that we can never exit language, and that we can never exit the circle of discourse within which we are speaking. It might very well be so. But, if the only notion of truth is internal to the theory, then this *internal truth* is what we mean by truth. We cannot exit from our own conceptual scheme. We cannot put ourself outside our discourse: outside our theory. There may be no notion of truth outside our own discourse. But it is precisely 'from within the language' that we can assert the reality of the world. And we certainly do so. Indeed, it is more than that: it is structural to our language to be a language *about* the world, and to our thinking to be a thinking *of* the world. Therefore, precisely because there is no notion of truth except the one in our own discourse, there is no sense in denying the reality of the world. The world is real, solid, and understandable by science. The best we can say about the physical world, and about what is there in the world, is what good physics says about it.

At the same time, our perceiving, understanding, and conceptualizing the world is in continuous evolution, and science is the form of this evolution. At every stage, the best we can say about the reality of the world is precisely what we are saying. The fact we will understand it better later on does not make our present understanding less valuable, or less credible. A map is not false because there is a better map, even if the better one looks quite different. Searching for a fixed point on which to rest is, in my opinion, naive, useless, and counterproductive for the development of science. It is only by believing our insights and, at the same time, questioning our mental habits, that we can go ahead. This process of cautious faith and self-confident doubt is the core of scientific thinking. Exploring the possible ways of thinking of the world, being ready to subvert (if required) our ancient prejudices, is among the greatest and the most beautiful of human adventures. In my view, quantum gravity – in its effort to conceptualize quantum spacetime, and to modify in depth the notion of time – is a step of this adventure.

### Notes

1. For recent general overviews of current approaches to quantum gravity, see Butterfield and Isham (1999) and Rovelli (1998).
2. Following Roger Penrose's opposite suggestions of a failure of conventional QM induced by gravity (Penrose 1994a), Antony Zeilinger is preparing an experiment to test such a possible failure of QM (Zeilinger 1997). It would be very exciting if Roger turned out to be right, but I am afraid that QM, as usual, will win.
3. 'We can say that movement is the transference of one part of matter or of one body, from the vicinity of those bodies immediately contiguous to it, and considered at rest, into the vicinity of some others' (Descartes, *Principia Philosophiae*, Sec. II-25, p. 51).
4. Aristotle insists on this point, using the example of the river that moves with respect to the ground, in which there is a boat that moves with respect to the water, on which there is a man that walks with respect to the boat . . . . Aristotle's relationalism is tempered by the fact that there is, after all, a preferred set of objects that we can use as universal reference: the Earth at the centre of the universe, the celestial spheres, the fixed stars. Thus, we can say, if we desire so, that something is moving 'in absolute terms', if it moves with respect to the Earth. Of course, there are *two* preferred frames in ancient cosmology: the one of the Earth and the one of the fixed stars; the two rotate with respect to each other. It is interesting to notice that the thinkers of the Middle Ages did not miss this point, and discussed whether we can say that the stars rotate around the Earth, rather than being the Earth that rotates under the *fixed* stars. Buridan concluded that, on grounds of reason, in no way is one view more defensible than the other. For Descartes, who writes, of course, after the great Copernican divide, the Earth is not anymore the centre of the Universe and cannot offer a naturally preferred definition of stillness. According to certain (possibly unfairly) unsympathetic commentators, Descartes, fearing the Church and scared by what happened to Galileo's stubborn defence of the idea that 'the Earth moves', resorted to relationalism, in *The Principles*, precisely to be able to hold Copernicanism without having to commit himself to the absolute motion of the Earth!

5. 'So, it is necessary that the definition of places, and hence local motion, be referred to some motionless thing such as extension alone or *space*, in so far as space is seen truly distinct from moving bodies' (Newton, *De gravitatione et Aequipondio Fluidorum*, 89–156). Compare with the quotation from Descartes in footnote 3.

6. Active diff invariance should not be confused with passive diff invariance, or invariance under change of co-ordinates. GR can be formulated in a co-ordinate-free manner, where there are no co-ordinates, and no changes of co-ordinates. In this formulation, the field equations are *still* invariant under active diffs. Passive diff invariance is a property of a formulation of a dynamical theory, while active diff invariance is a property of the dynamical theory itself. A field theory is formulated in manner invariant under passive diffs (or change of co-ordinates), if we can change the co-ordinates of the manifold, re-express all the geometric quantities (dynamical *and non-dynamical*) in the new co-ordinates, and the form of the equations of motion does not change. A theory is invariant under active diffs, when a smooth displacement of the dynamical fields (*the dynamical fields alone*) over the manifold, sends solutions of the equations of motion into solutions of the equations of motion. Distinguishing a truly dynamical field, namely a field with independent degrees of freedom, from a non-dynamical field disguised as dynamical (such as a metric field $g$ with the equations of motion Riemann$[g] = 0$) might require a detailed analysis (of, for instance, the Hamiltonian) of the theory.

7. Notice that Newton, in the passage quoted in footnote 5 argues that motion must be defined with respect to motionless space 'in so far as space is seen truly distinct from moving bodies'. That is: motion should be defined with respect to something that has no dynamics.

8. In the first edition of his celebrated book on quantum mechanics, Dirac (1930) used Heisenberg states (he calls them relativistic). In later editions, he switched to Schrödinger states, explaining in a preface that it was easier to calculate with these, but it was nevertheless a pity to give up the Heisenberg states, which are more fundamental. In what was perhaps his last public seminar, in Sicily, Dirac used just a single transparency, with just one sentence: 'The Heisenberg picture is the right one'.

# Part II

# Strings

# Reflections on the fate of spacetime

Edward Witten

## 5.1 Introduction

Our basic ideas about physics went through several upheavals early this century. Quantum mechanics taught us that the classical notions of the position and velocity of a particle were only approximations of the truth. With general relativity, spacetime became a dynamical variable, curving in response to mass and energy. Contemporary developments in theoretical physics suggest that another revolution may be in progress, through which a new source of 'fuzziness' may enter physics, and spacetime itself may be reinterpreted as an approximate, derived concept (see Fig. 5.1). In this article I survey some of these developments.

Let us begin our excursion by reviewing a few facts about ordinary quantum field theory. Much of what we know about field theory comes from perturbation theory; perturbation theory can be described by means of Feynman diagrams, or graphs, which are used to calculate scattering amplitudes. Textbooks give efficient algorithms for evaluating the amplitude derived from a diagram. But let us think about a Feynman diagram intuitively, as Feynman did, as representing a history of a spacetime process in which particles interact by the branching and rejoining of their worldlines. For instance, Fig. 5.2 shows two incident particles, coming in at $a$ and $b$, and two outgoing particles, at $c$ and $d$. These particles branch and rejoin at spacetime events labelled $x$, $y$, $z$, and $w$ in the figure.

According to Feynman, to calculate a scattering amplitude one sums over all possible arrangements of particles branching and rejoining. Moreover, for a particle travelling between two spacetime events $x$ and $y$, one must in quantum mechanics allow for all possible classical trajectories, as in Fig. 5.3. To evaluate the propagator of a particle from $x$ to $y$, one integrates over all possible paths between $x$ and $y$, using a weight factor derived from the classical action for the path.

So when one sees a Feynman diagram such as that of Fig. 5.2, one should contemplate a sum over all physical processes that the diagram could describe. One must

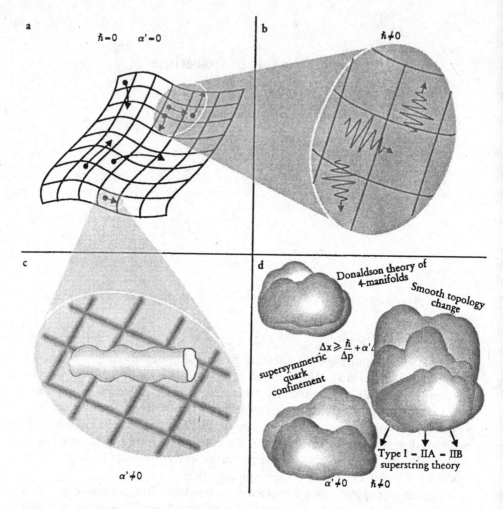

a   $\hbar = 0$   $\alpha' = 0$

b   $\hbar \neq 0$

c   $\alpha' \neq 0$

d   Donaldson theory of
4-manifolds

Smooth topology
change

$$\Delta x \gtrsim \frac{\hbar}{\Delta p} + \alpha' \Delta p$$

supersymmetric
quark
confinement

Type I – IIA – IIB
superstring theory

$\alpha' \neq 0$   $\hbar \neq 0$

Fig. 5.1. Four views of reality: (a) In classical physics, particles have definite locations and follow exact trajectories in a precise, curved spacetime. (b) Closer examination reveals the effects of quantum mechanics, $\hbar \neq 0$. Wavepackets propagate through spacetime, their positions and velocities uncertain according to Heisenberg. (c) In string theory, point particles are replaced by tiny loops having a 'string tension' $\alpha' \neq 0$. Even ignoring quantum mechanics ($\hbar = 0$), the concept of spacetime becomes 'fuzzy' at scales comparable to $\sqrt{\alpha'}$. (d) The full theory, employing both a string tension and quantum effects, is only beginning to take shape. Remarkable results are being uncovered that may overturn our conventional notions of spacetime.

integrate over all spacetime events at which interactions – branching and rejoining of particles – could have occurred, and integrate over the trajectories followed by the particles between the various vertices. And, of course, to actually predict the outcome of an experiment, one must (as in Fig. 5.4) sum over all possible Feynman diagrams – that is, all possible sequences of interactions by which a given initial state can evolve into a given final state.

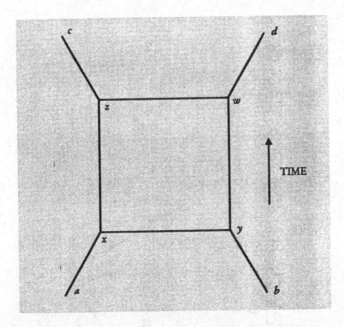

Fig. 5.2. A Feynman diagram with two incident particles at spacetime events $a$, $b$ and two outgoing particles at $c$, $d$. The particles interact by branching and rejoining at the spacetime events $x$, $y$, $z$, and $w$. Those vertices lead to fundamental problems in field theory.

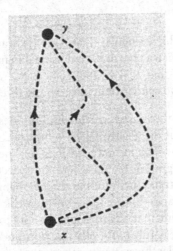

Fig. 5.3. Some classical trajectories for a particle propagating from $x$ to $y$; they all contribute to the Feynman propagator.

This beautiful recipe – formulated in the early days of quantum field theory – brought marvelous success and efficient, precision computations. Yet this recipe also exhibits certain of the present-day troubles in physics. One important property of a Feynman graph is that the graph itself, regarded as a

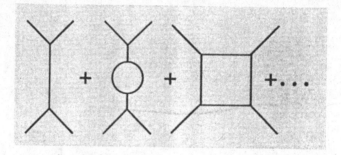

Fig. 5.4. Several Feynman diagrams contributing to the same physical process.

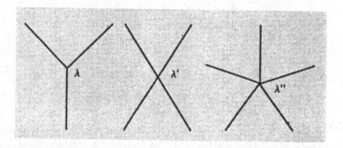

Fig. 5.5. Arbitrary factors associated with arbitrary branchings of particles in conventional field theory. In the 'standard model' of particle physics, this freedom leads to about seventeen parameters whose values are not understood theoretically.

one-dimensional manifold, is singular; that is, at the branching and joining points, the graph does not look like a true one-dimensional manifold. Everyone can agree, in Fig. 5.2 for instance, that $x, y, z$, and $w$ were the spacetime events at which interactions occurred. Two central difficulties spring directly from this:

*Infinities.* Quantum field theory is plagued with infinities, starting with the infinite electrostatic self-energy of the electron. The infinities come from the singularities of the Feynman diagrams. For instance, in Fig. 5.2, the potential infinities come from the part of the integration region where the spacetime events $x, y, z$, and $w$ all nearly coincide. Sometimes the infinities can be 'renormalized' away; that is the case for electrodynamics and – ultimately – for the weak and strong interactions in the Standard Model of elementary-particle physics. But for gravity, renormalization theory fails, because of the nature of the inherent non-linearities in general relativity. So we come to a key puzzle: the existence of gravity clashes with our description of the rest of physics by quantum fields.

*Too many theories.* There are many quantum field theories, depending on many free parameters, because one can introduce fairly arbitrary rules governing the branching and joining of particles. For instance, one could permit higher-order branchings of particles, as in Fig. 5.5. With every elementary branching process, one can (with certain restrictions) associate a 'coupling constant', an extra factor included in the evaluation of a Feynman diagram. In practice, the Standard Model describes the equations that underlie almost all the phenomena we know, in a framework that is compelling and highly predictive – but that also has (depending on precisely how one counts) roughly seventeen free parameters whose values are not understood theoretically. The seventeen parameters enter as special factors associated with the singularities of the Feynman diagrams. There must be some way to reduce this ambiguity!

### 5.2 String theory

We have one real candidate for changing the rules; this is string theory. In string theory the one-dimensional trajectory of a particle in spacetime is replaced by a two-dimensional orbit of a string (see Fig. 5.6). Such strings can be of any size, but under ordinary circumstances they are quite tiny, around $10^{-32}$ cm in diameter, a value determined by comparing the predictions of the theory for Newton's constant and the fine structure constant to experimental values. This is so small (about sixteen orders of magnitude less than the distances directly probed by high-energy experiments) that for many purposes the replacement of particles by strings is not very important; for other purposes, though, it changes everything. The situation is somewhat analogous to the introduction of Planck's constant $\hbar$ in passing from classical to quantum physics: for many purposes, $\hbar$ is so tiny as to be unimportant, but for many other purposes it is crucial. Likewise, in string theory one introduces a new fundamental constant $\alpha' \approx (10^{-32} \text{ cm})^2$ controlling the tension of the string. Many things then change.

One consequence of replacing worldlines of particles by worldtubes of strings is that Feynman diagrams get smoothed out. Worldlines join abruptly at interaction events, as in Fig. 5.7(a), but worldtubes join smoothly, as in Fig. 5.7(b).

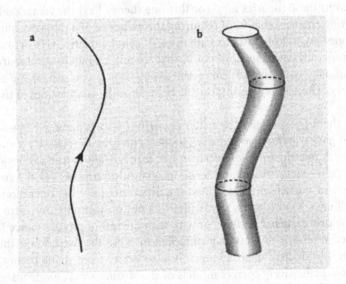

Fig. 5.6. Particles and strings. (a) A point particle traces out a one-dimensional worldline in spacetime. (b) The orbit of a closed string is a two-dimensional tube or 'worldsheet' in spacetime.

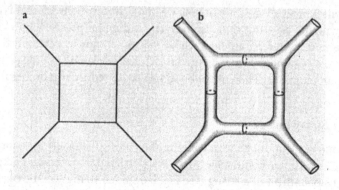

Fig. 5.7. String theory's smoothing effect is apparent when one compares a Feynman graph (a) with its stringy counterpart (b). The string diagram has no singular interaction points.

There is no longer an invariant notion of when and where interactions occur, so from the description above of the origin of the problems of field theory, we might optimistically hope to have finiteness, and only a few theories.

These hopes are realized. In fact, once one replaces worldlines by worldtubes, it is all but impossible to construct any consistent theories at all. That such theories do exist was established through a long and complex process stretching over roughly fifteen years, from the late 1960s to the early 1980s.[1] Moreover, there are only a few such theories; in fact, the very latest discoveries strongly suggest that they are all equivalent to each other so that apparently there is really only one such theory.

Moreover, these theories have (or this one theory has) the remarkable property of *predicting gravity* – that is, of requiring the existence of a massless spin-2 particle whose couplings at long distances are those of general relativity. (There are also calculable, generally covariant corrections that are unfortunately unmeasureably small under ordinary conditions.) This result is in striking contrast to the situation in field theory, where gravity is impossible because of the singularities of the Feynman graphs.

String theory (especially the heterotic string) also generates Yang–Mills gauge fields and gauge invariance in close parallel with gravity. Further, if one assumes that the weak interactions violate parity, one is practically forced to consider models with the right gauge groups and fermion quantum numbers for the conventional description of particle physics. Thus, the innocent-sounding operation of replacing worldlines by worldtubes forces upon us not only gravity but extra degrees of freedom appropriate for unifying gravity with the rest of physics. Since 1984, when generalized methods of 'anomaly cancellation' were discovered and the heterotic string was introduced, one has known how to derive from string theory uncannily simple and qualitatively correct models of the strong, weak, electromagnetic, and gravitational interactions.

Apart from gravity and gauge invariance, the most important general prediction of string theory is supersymmetry, a symmetry between bosons and fermions that string theory requires (at some energy scale). Searching for supersymmetry is one of the main goals of the next generation of particle accelerators. Its discovery would be quite a statement about nature and would undoubtedly provide a lot of clues about how theorists should proceed.

If this is the good news, what is the bad news? Perhaps what is most glaringly unsatisfactory is this: Crudely speaking there is wave-particle duality in physics, but in reality everything comes from the description by waves, which are then quantized to give particles. Thus a massless classical particle follows a light-like geodesic, while the wave description involves the Einstein, or Maxwell, or Yang–Mills equations, which are certainly much closer to the fundamental concepts of physics. Unfortunately, in string theory so far, one has generalized only the less fundamental point of view. As a result, we understand in a practical sense how to do many computations in string theory, but we do not yet understand the new underlying principles analogous to gauge invariance. The situation is illustrated in Fig. 5.8.

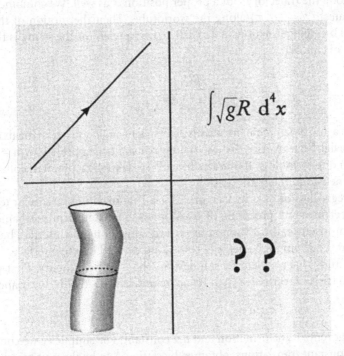

$$\int \sqrt{g}R \, d^4x$$

? ?

Fig. 5.8. The 'magic square' of string theory. The two rows represent ordinary physics and string theory, respectively, while the two columns represent particles and waves. In the upper left-hand corner, a line drawn at a 45 degree angle to the horizontal symbolizes a classical massless particle, propagating at the speed of light. In the lower left, we show the stringy analogue of this, the worldtube. In the upper right are crown jewels such as the Einstein–Hilbert action of general relativity. In the lower right should be the synthesis, related to the Einstein–Hilbert action as worldtubes are related to worldlines.

### 5.3 Some of the symptoms

Not knowing the concepts by which string theory will eventually be understood, here I can only describe some of the symptoms, some of the curious phenomena that occur in physics when $\alpha' \neq 0$. In so doing, I hope to give the reader a taste of the conceptual issues that theoretical physicists are grappling with.

But first we need some more background. A point particle moving in Minkowski space with proper time $\tau$ is described by giving its position $X^i(\tau)$ as a function of $\tau$ – here $X^i$ are the Minkowski co-ordinates. The action, or Lagrangian, for this particle is

$$I = \frac{1}{2} \int d\tau \sum_{ij} \eta_{ij} \frac{dX^i}{d\tau} \frac{dX^j}{d\tau}, \tag{5.1}$$

with $\eta_{ij}$ the metric of Minkowski space. If the particle is massless, the Lagrangian must be supplemented with a constraint saying that the velocity is light-like.

For a string, because the worldtube is two-dimensional, one has not just a proper time $\tau$ along the trajectory, but a proper position $\sigma$ as well. We combine them into co-ordinates $\sigma^\alpha = (\sigma, \tau)$ along the worldtube. Then the motion of the string is described by giving functions $X^i(\sigma^\alpha)$. The Lagrangian for the string is the obvious analogue of eqn. 5.1:

$$I = \frac{1}{2\alpha'} \int d^2\sigma \sum_{ij\alpha} \eta_{ij} \frac{dX^i}{d\sigma^\alpha} \frac{dX^j}{d\sigma^\alpha}. \tag{5.2}$$

This must again be supplemented with a constraint analogous to saying that a particle velocity is light-like. Notice that the stringy constant $\alpha'$ appears in eqn. 5.2 to make the action dimensionless. If one sets $\hbar = c = 1$, as particle physicists often do, then $\alpha'$ has dimensions of length squared.

Now, regardless of its origins, eqn. 5.2 is a Lagrangian quite similar to what one might meet in many problems of two-dimensional statistical mechanics or field theory. For instance, the $\sigma^\alpha$ might be co-ordinates along the interface between two media and the $X^i$ might be fields of some kind defined on the interface.

Let us study this problem by standard methods of field theory. First we look at the symmetries. Our problem had Poincaré invariance – that is, invariance under

$$X^i \rightarrow \Lambda^i_j X^j + a^i, \tag{5.3}$$

with $\Lambda$ a Lorentz transformation and $a$ a constant. For simplicity we consider here only the constant translations, obtained by setting $\Lambda$ to be the unit matrix:

$$X^i \rightarrow X^i + a^i. \tag{5.4}$$

In field theory or statistical mechanics, one of the first things that one calculates is the propagator or two-point correlation function $\langle X^i(\sigma)X^j(0)\rangle$. In the present problem, we have a conundrum because it is impossible for the two-point function to be invariant under transformation 5.4: under 5.4, $\langle X^i(\sigma)X^j(0)\rangle$ picks up a non-zero term $a^i a^j$. This number is a $c$-number, that is, an ordinary number and not an operator, and so is non-zero and cannot be cancelled for arbitrary $a^i$ by other contributions, as they are lower order in $a^i$.

Thus, there are two options. Either the two-point function in question is ill-defined, or Poincaré invariance is spontaneously broken in this theory and would not be observed as a symmetry of physical processes.

In fact, the first option prevails. By the standard recipe, the two-point function of this theory should be

$$\langle X^i(\sigma) X^j(0) \rangle = \eta^{ij} \int \frac{d^2k}{(2\pi)^2} \frac{e^{ik \cdot \sigma}}{k^2}. \tag{5.5}$$

The integral is infrared divergent. This divergence means that the 'elementary field' $X^i$ is ill-behaved quantum mechanically (but other fields are well-behaved and the theory exists).

This infrared divergence – which is central in string theory – was in fact first studied in the theory of two-dimensional $XY$ ferromagnets. In that context, the infrared divergence means that there is a low temperature phase with power law correlations but no long-range order. This is an example of a general theme: properties of spacetime in string theory (in this case, unbroken Poincaré invariance) reflect phenomena in two-dimensional statistical mechanics and field theory.

For instance, condensed matter theorists and field theorists are often interested in the anomalous dimensions of operators – how the renormalized operators scale with changes in the length or energy scale. In this case, by studying the anomalous dimension of a certain operator – namely, $(\partial X)^2 e^{ik \cdot X}$ – we could go on to explain why string theory predicts the existence of gravity. This tale has been told many times.[2] Here I prefer to convey the radical change that taking $\alpha' \neq 0$ brings in physics.

In analysing Poincaré invariance, we took the spacetime metric to be flat – we used the Minkowski metric $\eta_{ij}$ in eqn. 5.2. Nothing prevents us from replacing the flat metric by a general spacetime metric $g_{ij}(X)$, taking the worldtube Lagrangian to be

$$I = \frac{1}{2\alpha'} \int d^2\sigma \sum_{ij\alpha} g_{ij}(X) \frac{dX^i}{d\sigma^\alpha} \frac{dX^j}{d\sigma^\alpha}. \tag{5.6}$$

Simply by writing eqn. 5.6, we get, for each classical spacetime metric $g$, a two-dimensional quantum field theory, or at least the Lagrangian for one.

So spacetime with its metric determines a two-dimensional field theory. And that two-dimensional field theory is *all* one needs to compute stringy Feynman diagrams. The reason that theory suffices is that (as explained above) stringy Feynman diagrams are non-singular. Thus, in a field theory diagram, as in Fig. 5.7(a), even when one explains how free particles propagate (what factor is associated with the lines in the Feynman diagram) one must separately explain how particles interact (what vertices are permitted and what factors are associated with them). As the stringy Feynman diagram of Fig. 5.7(b) is non-singular, once one understands the propagation of the free string there is nothing else to say – there are no interaction points whose properties must be described.

Thus, once one replaces ordinary Feynman diagrams with stringy ones, one does not really need spacetime any more; one just needs a two-dimensional field theory describing the propagation of strings. And perhaps more fatefully still, one does not have spacetime any more, except to the extent that one can extract it from a two-dimensional field theory.

So we arrive at a quite beautiful paradigm. Whereas in ordinary physics one talks about spacetime and classical fields it may contain, in string theory one talks about an auxiliary two-dimensional field theory that encodes the information. The paradigm has a quite beautiful extension: a spacetime that obeys its classical field equations corresponds to a two-dimensional field theory that is conformally invariant (that is, invariant under changes in how one measures distances along the string). If one computes the conditions needed for conformal invariance of the quantum theory derived from eqn. 5.6, assuming the fields to be slowly varying on the stringy scale, one gets generally covariant equations that are simply the Einstein equations plus corrections of order $\alpha'$.

We are far from coming to grips fully with this paradigm, and one can scarcely now imagine how it will all turn out. But two remarks seem fairly safe. All the vicissitudes of two-dimensional field theory and statistical mechanics are reflected in 'spacetime', leading to many striking phenomena. And once $\alpha'$ is turned on, even in the classical world with $\hbar = 0$, 'spacetime' seems destined to turn out to be only an approximate, derived notion, much as classical concepts such as the position and velocity of a particle are understood as approximate concepts in the light of quantum mechanics.

### 5.4 Duality and the minimum length

A famous vicissitude of two-dimensional statistical mechanics is the duality of the Ising model. The Ising model is a simple model of a ferromagnet in two dimensions. As was discovered sixty years ago, the Ising model on a square lattice is equivalent to a 'dual' spin system on a dual lattice, as sketched in Fig. 5.9. If the original system is at temperature $T$, the dual system has temperature $1/T$. Thus, high and low temperatures are exchanged, and if there is precisely one phase transition, it must occur at the critical temperature, $T = 1$.

This duality has an analogue if the $Z_2$ symmetry of the Ising model (spin up and spin down) is replaced by $Z_n$ (spins pointing in any of $n$ directions equispaced around a circle). For large $n$, there is an interesting 'continuum limit', which leads to the following assertion: there is a smallest circle in string theory; a circle of radius $R$ is equivalent to a circle of radius $\alpha'/R$. By this we mean most simply the following. Imagine that the universe as a whole is not infinite in spatial extent, but that one of the three space dimensions is wrapped in a circle, making it a periodic variable with period $2\pi R$. Then there is a smallest possible value of $R$. When $R$ is large, things will look normal, but if one tries to shrink things down until the period is less than $2\pi\sqrt{\alpha'}$, space will re-expand in another 'direction' peculiar to string theory, and one will not really succeed in creating a circle with radius less than $\sqrt{\alpha'}$.

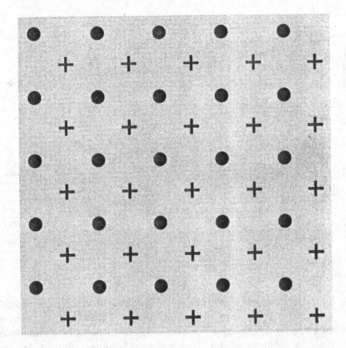

Fig. 5.9. A system of Ising spins on the lattice indicated by dots is equivalent to another spin system on the 'dual' lattice indicated by crosses. In string theory, analogous dualities of an underlying two-dimensional field theory result in dualities of spacetimes.

Technically, this arises as follows. A massless particle – or string – on a circle of radius $R$ has quantized momentum $p = n/R$, with integer $n$, and energy levels

$$E_n = \frac{|n|}{R}. \tag{5.7}$$

A string can also wrap $m$ times around the circle, with energy

$$\bar{E}_n = \frac{|m|R}{\alpha'}. \tag{5.8}$$

There is a duality symmetry – generalizing the duality of the Ising model – that exchanges the two spectra, exchanging also $R$ with $\alpha'/R$ (see Fig. 5.10).

As presented here, the argument might seem to apply only to circles wrapped around a periodic dimension of the universe. In fact, similar arguments can be made for any circle in spacetime.

The fact that one cannot compress a circle below a certain length–scale might be taken to suggest that the smaller distances just are not there. Let us try to disprove this. A traditional way to go to short distances is to go to large momenta. According to Heisenberg, at a momentum scale $p$, one can probe a distance $x \approx \hbar/p$. It would appear that by going to large $p$, one can probe small $x$ and verify that the small distances do exist. However (as described in Gross and Mende 1988), the Heisenberg microscope does not work in string theory if the energy is too large.

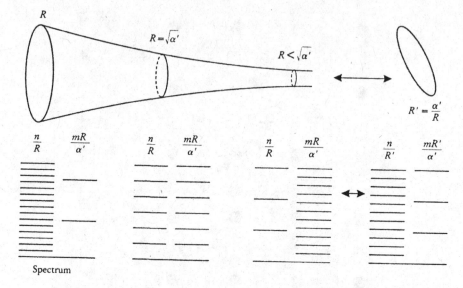

Fig. 5.10. Small circles don't exist. The spectrum of string states on a circle has two components, $n/R$ due to momentum quantization and $mR/\alpha'$ due to wrapping of the string around the circle $m$ times. When the circle radius shrinks to about size $\sqrt{\alpha'}$, the 'momentum' and 'wrapping' states become equivalent. As one tries to compress the circle further, the states become equivalent to those on a large 'dual' circle with 'momentum' and 'wrapping' states swapped.

Instead, the strings expand and – when one accelerates past the string scale – instead of probing short distances one just watches the propagation of large strings. It is roughly as if the uncertainty principle has two terms,

$$\Delta x \geq \frac{\hbar}{\Delta p} + \alpha' \frac{\Delta p}{\hbar}, \tag{5.9}$$

where the first term is the familiar quantum uncertainty and the second term reflects a new uncertainty or fuzziness due to string theory. With the two terms together, there is an absolute minimum uncertainty in length – of order $\sqrt{\alpha'} \approx 10^{-32}$ cm – in any experiment. But a proper theoretical framework for the extra term has not yet emerged.

A somewhat similar conclusion arises if one tries to compute the free energy at high temperature. In field theory, at high temperature $T$, one gets (in four dimensions) a free energy per unit volume $F \approx T^4/(\hbar c)^3$, as if each box of linear size $\hbar c/T$ contains one quantum of energy $T$. In string theory, the behaviour is similar until one reaches 'stringy' temperatures, after which the free energy seems to grow more slowly, roughly as if one cannot divide space into boxes less than $10^{-32}$ cm on a side, with each such box containing one string.

The duality symmetry described above also has a number of non-linear analogues, such as 'mirror symmetry', which is a relationship between two spacetimes that would be quite distinct in ordinary physics but turn out to be equivalent in string theory. The equivalence is possible because in string theory one does not really have a classical spacetime, but only the corresponding two-dimensional field theory;

136

two apparently different spacetimes $X$ and $Y$ might correspond to equivalent two-dimensional field theories.

A cousin of mirror symmetry is the phenomenon of topology change. Here one considers how space changes as a parameter – which might be the time – is varied. One starts with a spatial manifold $X$ so large that stringy effects are unimportant. As time goes on, $X$ shrinks and stringy effects become large; the classical ideas of spacetime break down. At still later times, the distances are large again and classical ideas are again valid, but one is on an entirely different spatial manifold $Y$! Quite precise computations of such processes have been developed.

### 5.5 Strings and quantum mechanics

In this article, I have generally suppressed the effects of quantum mechanics, or $\hbar$, and have attempted to explain how physics changes when one turns on $\alpha'$. My goal has been to explain that the phenomena and the change in viewpoint associated with $\alpha'$ – or string theory – are as striking as those associated with $\hbar$ – or quantum mechanics.

Of course, in the real world, $\hbar$ and (if string theory is correct) $\alpha'$ are both non-zero. What happens then? That is perhaps the main focus of current work in the field. We are far from getting to the bottom of things, but lately there have been enough surprising new ideas and discoveries to make up what some have characterized as 'the second superstring revolution'. (From that point of view, the 'first superstring revolution' was the period in the mid-1980s when the scope of superstring theory first came to be widely appreciated.) New dualities – generalizing the duality of Maxwell's equations between electric and magnetic fields – appear when $\hbar$ and $\alpha'$ are considered together. These new symmetries have enabled us to understand that – as I mentioned earlier – there is apparently only one string theory, the previously formulated theories being equivalent. Their richness is illustrated by the fact that (in their field theory limit) they have provided new insights about quark confinement, the geometry of four-dimensional spacetime and many other things.

Moreover, these new dualities mix $\hbar$ and $\alpha'$ in a way quite unlike anything previously encountered in physics. The existence of such symmetries that hold only for $\hbar \neq 0$ gives one the feeling that the natural formulation of the theory may eventually prove to be inherently quantum mechanical and thus, in a sense, may entail an explanation of quantum mechanics.

We shall have to leave further discussion of these matters for another occasion. Even so, I hope to have communicated a sense of some of the storm clouds in theoretical physics, and a feeling for the likely fate of the concept of spacetime.

### Notes

Reprinted with permission from Witten, E., 'Reflections on the Fate of Spacetime', *Physics Today*, 96(4), 1996, pp. 24–30. Copyright 1996, American Institute of Physics.

1. See Schwarz (1985).
2. For instance, see Volume 1, Chapter 1 of Green et al. (1987).

# 6     A philosopher looks at string theory

Robert Weingard

Before we, as philosophers, take a look at string theory I want to mention that more than one person has suggested to me that it is still too early for philosophical and foundational studies of string theory. Indeed, the suggestion emphasizes, since string theory is still in the process of development, and its physical and mathematical principles are not completely formulated, there is, in a sense, no theory for the philosopher to analyse. And I must admit that I think there is something to this suggestion. In a sense I hope I will make clear, there does not yet exist a precise mathematical formulation for string theory as there is Hilbert space for (elementary) quantum theory, and Riemann spacetime for general relativity. Because these latter formulations exist, we can ask precise questions and prove precise theorems about their interpretation. The Kochen–Specker theorem about non-contextualist hidden variable theories, the Fine–Brown proof of the insolvability of the quantum measurement problem, and the current determinism–hole argument debate are some examples. Without a clearly formulated mathematical structure, I don't think we can expect to get analogous distinctly stringy results.

This suggests a related worry. String theory, at least in the first quantized theory, is a relativistic quantum theory of strings (one-dimensional extended objects). One may well agree that of course all of the standard philosophical and foundational issues of quantum theory and relativity are still there, but be sceptical about whether string theory will either shed any light on these old problems, or give rise to stringy problems. Some philosophers have expressed analogous doubts about quantum field theory.

Again, there is something to this worry. But only something. I think it is true, as I have said, that string theory is not ready for certain kinds of foundational studies. And I don't think string theory (or quantum field theory) will shed light on those favourite topics of philosophers of physics, Bell's theorem and

138

realism-hidden variable theories in quantum mechanics. But there is more to physics than non-relativistic quantum mechanics.

For the past fifteen years, quantum gauge field theories have been our best theories of the physical universe. If we wanted to understand the basic ontology of the universe, at least according to the best source of information available, then we needed to understand these theories. And we needed to answer such questions as 'What is the ontological significance of a quantum field?', 'Do Fermi fields have a different ontological status than Bose fields?', 'What is the geometrical significance of the gauge connection and its associated fibre bundle?', among many others.

These questions still need answering. But now many physicists believe that string theory may (will?) replace quantum field theory as our best fundamental theory of physics. And in the process it is claimed that it will give us a unified theory of all the fundamental interactions, a consistent quantum theory of gravity (and thus of the metric field), and an explanation of the family structure of elementary particles, to name a few. If this is correct, then we will have to ask of string theory the kind of questions we should have been asking of quantum field theory.

Therefore I would like to talk about the following. First, I want to describe some simple, but interesting, string theory to give you some idea of what this theory is all about. Then I want to discuss string field theory and suggest that even though we do not have a complete mathematical formulation, we can get an idea of some of its ontological implications. This is one of the philosophically exciting aspects of string theory. However, I also want to ask whether there is, in fact (good) reason to think that string theory may (or will) emerge to replace quantum field theory. Unfortunately, this may dampen our excitement somewhat.

It is time, then, to turn to string theory. Strings – one-dimensional extended objects – come in two kinds, open and closed (Fig. 6.1). Classically, strings can have translational as well as vibrational motion. The vibrational motion can be decomposed into left- and right-moving normal modes. For the open string, boundary conditions require that these left and right movers be identified, while for the closed string they are independent. When we quantize, these different normal modes become states of different mass and spin. The difference between the open and closed string is that the Hilbert space of states for the closed string is a product space $H = H_R \times H_L$, of the left and right movers – roughly speaking, it is a product of two open string Hilbert spaces.

Of special interest are the massless modes. String theory is fundamentally a theory of the very early universe, when the mass scale was the planck mass, $m_p$. This is the natural mass scale of the theory, so that the massive modes of the string will have

Fig. 6.1. Open and closed strings.

masses that are, roughly, multiples of $m_p$. Thus, only the massless modes of the string will correspond to the particles we see at the relatively low energy scale of the laboratory – the masses of observed particles will instead come from symmetry breaking.

The massless states of the open string form a spacetime vector, so the massless states of the closed string form a second-rank spacetime tensor. The significance of this emerges when we try to make string theory into a unified theory of the fundamental interactions of physics. Since the strong and electroweak interactions are mediated by spin-1 vector particles, while gravity is carried by a spin-2 graviton – the quanta of the second-rank metric field $g_{\mu\nu}$ – a stringy unification requires at least that the gauge vector particles and the graviton are massless states of a single string. Our above remarks, however, suggest that the graviton must come from the massless modes of the closed string, while the gauge bosons must be massless states of the open string. As we will see, string theory has the resources to overcome this problem.

So far we have been talking about the bosonic string; the degrees of freedom are the spacetime co-ordinates, $x^\mu$, of the string, and its excitations have integral spin. To add fermions to the theory we put a spinor at each point of the string world sheet (the surface the string sweeps out in spacetime). Let us draw a picture of this situation (Fig. 6.2).

Amazingly, $\psi$ can be either a world sheet spinor (that is, a spinor with respect to transformations of the world sheet co-ordinates $\sigma, t$) or an explicitly spacetime spinor (the Green–Schwartz string). When done properly, in either case we get a theory with the same spectrum of spacetime bosons and fermions. In particular, we can note two things. First, the purely bosonic string contains a tachyon in its spectrum (a state with spacelike four momentum). By requiring the fermionic string to have spacetime supersymmetry the tachyon is eliminated. Second, it turns out that the (interacting) bosonic string can be consistently formulated only in twenty-six-dimensional spacetime. When we add fermions, the number of required spacetime dimensions is reduced to ten (at least for the standard formulations).

Since in everyday life only four dimensions of spacetime are apparent, in either case all but four of the dimensions must be compactified. That is, each of them is rolled up into a 'cylinder' of very small radius, explaining why we normally do not 'see' them. Compactification may seem like an *ad hoc* move designed merely to remove the embarrassment of more than four spacetime dimensions. But in fact (at least in the standard version of the theory) it does much more than hide

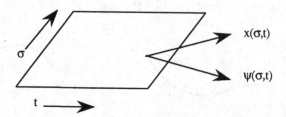

Fig. 6.2. Open string world sheet: $x(\sigma, t)$ are the spacetime co-ordinates of the sheet point, and $\psi(\sigma, t)$ is a spinor field on the sheet.

the extra dimensions. It is meant to explain how the spectrum of particles and interactions we see at our energy scale come from the 'pure' high-energy string theory. We can get an idea of how this is supposed to work by looking at the heterotic string.

But first, a preliminary point. Consider a vector $A^\mu$ in a $D$-dimensional space-time, in which $i = 1 \ldots n$ dimensions are not compactified, the $N = n + 1 \ldots D$ dimensions are compactified. The $A^i$ are thus the components of $A^\mu$ in the $n$ uncompactified dimensions and $A^N$ the components in the compactified dimensions. Then from the point of view of the uncompactified dimensions, since $N$ is not a spacetime index, only the $A^i$ form a vector, while the $A^N$ are $D - n$ scalar fields, which transform into each other according to what the $i$ dimensions regard as an internal symmetry group. In fact, this is a geometrical symmetry of the compactified dimension.

We turn next to the heterotic string. This is a theory of closed strings, in which the left-moving modes are those of a twenty-six-dimensional bosonic string, while the right movers are those of a ten-dimensional fermionic string. This is possible because, as we mentioned, the left and right movers of the closed string are independent. Spacetime therefore has twenty-six dimensions, but the right movers have non-zero components in only ten of them.

We can think of the compactification as occurring in two stages. In stage one we compactify the sixteen purely bosonic dimensions, and then in stage two, six more spacelike dimensions are compactified to give an effectively four-dimensional theory. Let's consider stage one. A vector excitation of the bosonic left movers will be of the form $\alpha^\mu_{-n}|0\rangle$, where $\alpha^\mu_{-n}$ is a creation operator with $n$ labelling the mode created. This will break up into $\alpha^\mu_{-n}|0\rangle = (\alpha^i_{-n}|0\rangle, \alpha^N_{-n}|0\rangle)$. The $i$th components form a vector in the ten (so far) uncompactified dimensions while, as we saw above, the $N$th components will be sixteen scalar fields from the ten-dimensional point of view, with the internal symmetry index $N$.

Similarly, a right-moving vector will be of the form $\beta^i_{-m}|0\rangle$, where the $\beta^i_{-m}$ are suitable creation operators. And note that here we only have the $i$ index since the right movers have only ten components. The closed string states, as we saw, will be products of the left and right movers, so here of the form, $\alpha^\mu_{-n}|0\rangle \times \beta^i_{-m}|0\rangle = (\alpha^j_{-n}|0\rangle \times \beta^i_{-m}|0\rangle, \alpha^N_{-n}|0\rangle \times \beta^i_{-m}|0\rangle)$.

The case of interest to us is $n = m = 1$, the massless vectors. Then there are two important points, since the $\alpha^N_{-1}|0\rangle$ are ten-dimensional scalars, the products $\alpha^N_{-1}|0\rangle \times \beta^i_{-1}|0\rangle$ are still (massless) ten-dimensional vectors. Second, these vectors form a multiplet indexed by $N$, which transforms according to a symmetry determined by the compactification. Thus, we have solved two problems.

We have obtained massless vectors from the closed string, and we have introduced an internal symmetry. (The open string has two distinguished points, the two end points, on which we can put charges. But each point of the closed string is 'the same', so there is no place to put a charge without breaking the symmetry of the closed string.) It turns out that given the proper compactification, we get an $N = 1$ Yang–Mills supermultiplet, whose internal (gauge) symmetry is either $E_8 \times E_8$ or $SO(32)$. In addition, the $\alpha^j_{-n}|0\rangle \times \beta^j_{-n}|0\rangle$ form a second-rank tensor in ten dimensions and, as you would expect, this contains the graviton.

We have obtained the graviton and the Yang–Mills vector bosons, but it has been accomplished in ten, rather than four dimensions. To achieve a realistic theory, we need to complete stage two and compactify from ten down to four dimensions. Here the $E_8 \times E_8$ gauge group offers a possibility. Namely, that the compactification breaks one of the $E_8$s to an $E_6$ symmetry in four dimensions. Unlike $E_8$, $E_6$ has complex representations (which means left- and right-handed particles transform differently under the group), and ones large enough to contain a complete family of the known fermions. The other $E_8$ would remain unbroken and could be the gauge group of the so-called 'shadow' matter – matter that would interact only gravitationally with normal matter. It would be a candidate, then, for the missing mass of the universe.

The heterotic string provides a way of getting massless vector bosons from the closed string. It turns out that conversely, the graviton is contained in the open string. Not as an excitation of the free string, however, but as an intermediate state in the interaction of open strings. To see this, consider the one-loop contribution to an amplitude involving four external strings. There are three kinds of loops that will contribute to this: a planar loop, a non-planar loop (a planar loop with an even number of twists), and a non-orientable loop (a planar loop with an odd number of twists). We can picture these as shown in Fig. 6.3. Because the theory is conformally invariant, the non-planar loop is equivalent to the world sheet pictured in Fig. 6.4, which shows two open strings interacting through a closed string intermediate state! And indeed, exact calculation shows that in $D = 26$, the non-planar loop amplitude contains a closed string pole. Interestingly, here is a place where the necessity of twenty-six dimensions for the bosonic string can be seen. In $D \neq 26$ this simple pole becomes a branch point which destroys the unitarity of the one-loop amplitude.

We see here an important difference between string theory and quantum field theory. We can do quantum field theory without having to have a (quantum) theory

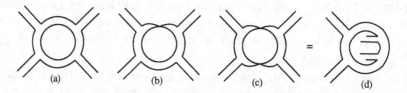

Fig. 6.3. One-loop contributions to four string scattering: (a) a planar loop, (b) a non-orientable loop (the 'pinches' in the figure represent twists) and (c) a non-planar loop; (d) shows the non-planar loop 'untwisted'.

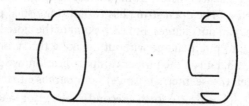

Fig. 6.4. Closed string intermediate state of two interacting open strings.

of gravity. For example, we can do quantum electrodynamics in flat spacetime without the involvement of gravity because the photon field $A^\mu$, and the Dirac fields $\psi$ ($\bar{\psi}$) create only photons and positrons (electrons) respectively. This is fortunate because Einstein's gravity is non-renormalizable as a quantum field theory.

However, since the free closed string already contains the graviton in its spectrum, only the classical open string is consistent without gravity. The minute we quantize, that is, go beyond tree diagrams to the one-loop level, then we have to include gravity to get a consistent theory. Given the non-renormalizability of Einstein gravity, this might have spelled disaster for string theory. But there is a lot of confidence that the superstring is not just renormalizable, but finite. That is, that the different divergences that arise cancel each other, and do not have to be removed by counterterms. Unfortunately, I cannot tell whether this confidence is justified.

Even if we do not have to worry about divergences, however, gravity still poses a problem for string theory. The reason is that at present, string theory provides only a perturbative theory of gravity. Given a background spacetime in which the string propagates, one can calculate graviton scattering to any order in perturbation theory, i.e. up to any number of loops. But the metric of the background spacetime is itself, presumably, a condensate (or coherent state) of some of the massless modes of the strings propagating in it, since, macroscopically, the metric is the gravitational field. At present the background spacetime has to be assumed, but clearly this is at best incomplete. In a deeper and more complete theory, we would expect an explanation of how the metric of spacetime emerges from the string dynamics.

Another aspect of this problem emerges when we realize that so far we have been speaking of first quantized strings. The string is given as the basic object, and when we quantize we get a Hilbert space of states of the string. But this theory can be reformulated as a string field theory, where the fields are functionals of the string co-ordinate functions, $x^\mu(\sigma, t)$, as opposed to being functions of spacetime position as in point particle field theory. But this field theory has two drawbacks. It too is defined only in perturbation theory, and it has resisted a covariant formulation. It must be done in a particular co-ordinate system, such as light cone co-ordinates. Clearly, this too is unsatisfactory.

On the one hand, such non-perturbative objects as monopoles and instantons emphasize that the content of a field theory is not exhausted by its perturbation theory. On the other hand, a covariant formulation of a complete string field theory would (presumably) reveal symmetries of the theory that are hidden by the non-covariant formulation. Therefore, we can see at least two possible benefits from having such a covariant string field theory.

First, we should get an explanation of the fundamental interactions of the theory. For example, the gauge invariance of quantum electrodynamics (QED) explains the origin of the basic trilinear coupling (Fig. 6.5) between electrons and photons. And the non-Abelian gauge symmetry of quantum chromodynamics (QCD) explains why there are four gluon and three gluon self-interaction terms. Unlike the Abelian field strength,

$$F_{\mu\nu} = \partial_\mu A_\nu - \partial_\nu A_\mu,$$

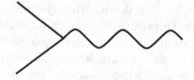

Fig. 6.5. Electron–photon coupling in QED: the coupling is $\bar{\psi}\gamma^{\mu}A_{\mu}\psi$.

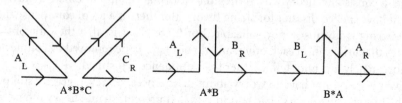

Fig. 6.6. String 'potentials' are associative but not commutative: $A * B \neq B * A$.

whose Lagrangian, $F^{\mu\nu}F_{\mu\nu}$, contains only kinetic terms like $\partial_{\mu}A_{\nu}\,\partial^{\mu}A^{\nu}$ and no self-interaction terms, the non-Abelian field strength is

$$F^a_{\mu\nu} = \partial_{\mu}A^a_{\nu} - \partial_{\nu}A^a_{\mu} + gf^{abc}A^b_{\mu}A^c_{\nu}.$$

This has the extra piece proportional to $[A_{\mu}, A_{\nu}]$, so that it transforms correctly under the gauge group. And it is this extra piece which gives rise to the new trilinear and quartic self-interaction terms in the Lagrangian, $F^a_{\mu\nu}F^{a\mu\nu}$.

Second, we might get an explanation of the metric in terms of some kind of symmetry breaking of the field theory. We can illustrate both of these points, in a stringy context, by looking at a proposal of Witten's (1986) for a covariant field theory of open strings.

Witten's proposal is a generalization of Yang–Mills theory, involving a non-commutative algebra of string forms. If we write Yang–Mills theory in the language of forms, then the gauge-invariant action, $S_{YM}$ is

$$S_{YM} = \int \mathrm{tr}\, F \wedge \hat{F},$$

where $F$ is the gauge field strength 2-form, $F = dA + A \wedge A$, $A$ is the gauge field 1-form, and $\hat{F}$ is the dual field strength. As a stringy generalization of $F$, Witten writes,

$$G = QB + B * B,$$

where $G$ and $B$ are the stringy analogues of $F$ and $A$, and $*$ generalizes the wedge product $\wedge$. It is defined by $(B * C) = (B_L, C_R)\delta(B_R - C_L)$, that is, the left half of $B$ joins to the right hand of $C$. By singling out the midpoint we get an associative product, $(A * B) * C = A * (B * C)$, for parameterized strings, but one that is not commutative, even up to a sign like the wedge product (Fig. 6.6). Finally, $Q$ plays the role of the exterior derivative in string form space. Witten identifies it with the BRST charge.[1]

There is no analogue of the dual of a form (we cannot raise and lower indices) in this formalism, so the closest analogue of $S_{YM}$ for strings would be $\int G * G$ (with $\int$ combining integration and taking the trace). But an arbitrary variation of this vanishes identically, and thus cannot yield any equations of motion. To get an alternative gauge invariant action, Witten turns to the analogue of the Chern Simmons three form $w$, which in Yang–Mills theory obeys, $dw = \text{tr } F \wedge F$. He postulates the action, $S = \int w$,

$$S = \int B * QB + \frac{2}{3} B * B * B,$$

which is invariant under the infinitesimal gauge transformation $\delta B = Q\Lambda + B * \Lambda - \Lambda * B$ ($\Lambda$ is a zero form gauge function), and under an arbitrary variation yields the equation of motion $G = 0$. Unlike Yang–Mills theory, therefore, we have only a cubic interaction term $\frac{2}{3} B * B * B$.

But perhaps of more interest to us is the fact (Horowitz et al. 1986) that we can derive Witten's action from an action containing no kinetic term, by expanding around a background. We postulate action $S_c$, containing only the cubic interaction of $S$, where $S_c$ is invariant under the infinitesimal gauge transformation $\delta_c B = B * \Lambda - \Lambda * B$, and $\delta S_c = 0$ implies the equation of motion $B * B = 0$.

To obtain Witten's form of the action we need the fact that each 1-form $B$ determines a $D_B$ given by,

$$D_B C = B * C - (-1)^{n_c} C * B,$$

where $n_c$ is an integer. If we then split $B$ into a fluctuation $\tilde{B}$, and background $B_o$,

$$B = B_o + \tilde{B},$$

and require that $B_o$ satisfies the equation of motion, $B_o * B_o = 0$, we get,

$$\int B * B * B = \frac{3}{2} \int \left( \tilde{B} * D_{B_o} \tilde{B} + \frac{2}{3} \tilde{B} * \tilde{B} * \tilde{B} \right).$$

If $\delta_c B_o = 0$, then $B$ transforms according to $\delta$ under a gauge transformation and for suitable $B_o$, $D_{B_o} = Q$. Thus, the recovery of Witten's action is complete.

The important point for us is that $Q$ depends on the metric. But if we can define $\int$ and $*$ so that they do not depend on a metric – as in the case of differential forms – then with the simple cubic action we do not have to assume a background gravity field or metric. Instead, the metric of spacetime comes from how the fundamental field $B$ is split (by God?) into a background $B_o$, and fluctuation $\tilde{B}$.

Earlier, when I said that a precise mathematical formulation of string theory did not exist yet, this is exactly what I had in mind. Witten's theory is suggestive, but a complete satisfactory covariant string field theory does not exist yet. Nonetheless, as I have tried to show, I think we can get some idea of the ontological implications of such a theory, namely, that the metric of spacetime is not a fundamental field.

I want to return now to the topics of supersymmetry and extra dimensions, to amplify the remarks I made earlier. After all, we are discussing the superstring.

Supersymmetry is a symmetry of a theory (of the action) under a transformation which changes bosons to fermions, and fermions to bosons. The supersymmetry transformations are closely connected to spacetime transformations. In particular, the result of two supersymmetry transformations is a spacetime translation. In four-dimensional spacetime this is expressed by

$$\{Q_\alpha, Q_\beta\} = (\sigma_\mu)_{\alpha\beta}P^\mu,$$

where $P^\mu$ is the four-momentum operator (the generator of translations), $Q_\alpha$ the supersymmetry generators, and $\sigma_\mu$ the Pauli matrices. This is an important point so let me try to make it plausible, and illustrate supersymmetry generally, with a simple example from quantum mechanics.

Consider a system consisting of an harmonic oscillator and a spin-$\frac{1}{2}$ fermion. Then we know that we can represent the harmonic oscillator by annihilation and creation operators, $a$ and $a^+$, such that $[a, a^+] = 1$, and $[a, a] = [a^+, a^+] = 0$. The Hamiltonian is then, $H_a = N_a + \frac{1}{2}$, where $N_a = a^+ a$ is the number operator. If $|0\rangle$ is the ground state, then

$$a|0\rangle = 0, \qquad (a^+)^n|0\rangle = (n!)^{1/2} \cdot |n\rangle, \ N_a|n\rangle = n|n\rangle.$$

For the fermion, we introduce the operators $b$ and $b^+$, which obey the anti-commutators $\{b, b^+\} = 1$, $\{b, b\} = \{b^+, b^+\} = 0$. Then if $b|0\rangle = 0$, it follows that,

|       | $|\alpha\rangle$ | $|\beta\rangle$ |
|-------|------------------|-----------------|
| $b$   | 0                | $|\alpha\rangle$ |
| $b^+$ | $|\beta\rangle$  | 0               |

and $b$ and $b^+$ are annihilation and creation operators for the states $|\alpha\rangle$ and $|\beta\rangle$. For our analogy we will regard $|\beta\rangle$ as a spin-$\frac{1}{2}$ state, while $|\alpha\rangle$ is the ground state. The states $|n\rangle|\alpha\rangle$ are then regarded as bosonic, and the $|n\rangle|\beta\rangle$ as fermionic.

It follows that for this system, $Q = a^+b$ changes fermions to bosons, while $Q^+ = b^+a$ changes bosons to fermions,

$$Q|n\rangle|\beta\rangle = (n+1)^{1/2}|n+1\rangle|\alpha\rangle,$$

$$Q^+|n\rangle|\alpha\rangle = (n)^{1/2}|n-1\rangle|\beta\rangle,$$

and,

$$\{Q, Q^+\} = (a^+bb^+a + b^+aa^+b) = a^+a + b^+b,$$

by $[a^+, a] = \{b, b^+\} = 1$, and $[a, a] = [a^+, a^+] = \{b, b\} = \{b^+, b^+\} = 0$. So,

$$\{Q, Q^+\} = N_a + N_b = H.$$

The Hamiltonian $H$ is the generator of time translations, and we have the analogue of the relation $\{Q_a, \bar{Q}_\beta\} = (\sigma^\mu)_{ab}P_\mu$ (note that we are taking $H_b = N_b - \frac{1}{2}$).

That the supersymmetry generators form a non-trivial algebra with spacetime generators suggests that we can interpret a supersymmetry transformation as a transformation in some sort of an extension or generalization of spacetime. And this is indeed the case. This extension, called superspace, is obtained by adding anticommuting spinoral co-ordinates $\vartheta^\alpha$ to the usual spacetime co-ordinates $x^\mu$ and then the supersymmetry generator $Q^\alpha$ can be expressed in terms of both the spacetime and spinorial derivatives, $\partial/\partial x^\mu$ and $\partial/\partial \vartheta^\alpha$.

Earlier, we obtained the fermionic string by putting a spinor at each point of the string world sheet. But this was unmotivated. We had no understanding of why there should be a fermionic string. But if superspace is the fundamental manifold, then the co-ordinates of the string will be the superspace co-ordinates $z^m = (x^\mu(\sigma, \tau), \vartheta^\alpha(\sigma, \tau))$. Both $x^\mu$ and $\vartheta^\alpha$ will be degrees of freedom of the string, and quantizing the string will mean quantizing both of them.

Since supersymmetry transformations change bosons and fermions into each other, bosons and fermions occur in the same multiplets (representations) of the supersymmetry transformations. In our simple example, the multiplets have the form, $(|n\rangle|\alpha\rangle, |n-1\rangle|\beta\rangle)$, each state in the multiplet having energy $n$. According to superstring theory, before compactification the fermionic string in ten dimensions has supersymmetry. This implies that every bosonic state (particle) has a fermionic partner of the same mass. But the compactification to four dimensions must break the supersymmetry because observed particles do not form supersymmetry multiplets. Since the supersymmetry generators change the spin by $\frac{1}{2}$, there should be a scalar electron, a massless spin-$\frac{1}{2}$ partner of the photon (this can't be the neutrino which interacts weakly, unlike the photon), and a spin-$\frac{3}{2}$ partner of the graviton. None of these super partners has been observed.

Nonetheless, at a fundamental level, there is supersymmetry, and the question arises as to the ontological significance of this fact. Does the fact that (before symmetry breaking) a boson can be transformed into a fermion by the supersymmetry transformations mean that the two are, at a deeper level, a single particle that can appear as a boson or a fermion? Or does it just mean that physically the two can change into each other?

An often expressed point is that matter is composed of fermions (quarks and leptons), while the forces which hold matter together are bosonic, that is, the interactions are mediated by gauge bosons. The division is not absolutely sharp because bosonic gluons can self-interact and form glue balls. But that is a minor exception. If nature is supersymmetric, then force and matter appear in the same multiplets. But is this really ontological unification?

The answer appears to hinge on how realistically we can (or should) interpret superspace.[2] The $2n + 1$ $z$-components of a particle with spin $n$ (in four dimensions) form a representation of the three-dimensional rotation group. Therefore, for example, the difference between a spin $z = \frac{1}{2}$ electron and $z = -\frac{1}{2}$ electron is that of the orientation of a single particle. And this is so even if the symmetry is broken by a magnetic field and the two have different energies. But such an interpretation does not work when the symmetry transformations are merely 'rotations' in an internal space like isospin space. Because such an internal space is, apparently, just a mathematical space, the symmetry transformations just represent the fact that the particles can be transformed into each other. In the case of supersymmetry, the appropriate space, as we have seen, is superspace. Whether superspace can (or should) be interpreted realistically as an extension of spacetime, ontologically on a par with extending spacetime by adding more (regular) spacetime dimensions, is a question I find it hard to deal with clearly. In as much as I can, the anticommutator connecting $Q_\alpha$ and $P_\mu$ argues that it can. Perhaps another relevant point is the fact, emphasized by Green and Schwartz (1984), that the superstring field is not a finite polynomial of the $\vartheta^\alpha$ as in the case of ordinary (point particle) super fields.

This question about superspace is similar to the question I raised at the beginning about the geometrical significance of the gauge connection. These are difficult interpretive questions, but I think they are important.

Finally, I want to turn to the question I raised earlier, about whether there is any reason to think string theory may (or will) emerge to replace quantum field theory as the framework for fundamental physics. It is often pointed out, even by the most enthusiastic string theorists, that there is absolutely no experimental evidence for string theory. Partly, no one knows how to calculate precisely anything observable at conceivable laboratory energies. But also, there is not even qualitative evidence, such as jets provide for QCD.

But I do not want to dwell on this here. It seems to me that one could be optimistic about a developing theory, even without such evidence, if there were at least some clues in either theory or experiment to suggest that the new theory was on the right track. Unfortunately, even this is lacking in string theory. Let me give some examples of what I mean here by a 'clue' that the theory is on the right track.

I want to mention briefly two examples: general relativity and the $SU(2) \times U(1)$ electroweak theory. In both cases we have a problem which gives a clue to its solution. For the case of general relativity, a clue emerges when we consider the problem of trying to fit gravity into flat spacetime (into special relativity). The natural move would be to consider the gravitational field to be a flat spacetime tensor field on analogy with the electromagnetic field. But then, gravity would not affect the null cone structure of spacetime, and either light would not be affected by gravity (since light worldlines would remain null geodesics in the presence of gravity), or light would be affected by gravity and light worldlines would be timelike in the presence of gravity.

However, if we think the connection between special relativity and local inertial frames is correct, then it would follow both that light is affected by gravity (the equivalence principle), and that light travels null geodesics. These can be reconciled only if we give up flat spacetime and postulate that spacetime in the presence of a gravitational field is curved but locally flat. Even if we did not have the ability to determine whether or not light is affected by gravity, we would have a clue, or suggestion to an important part of general relativity.

In the case of the weak interactions, the problem was that the $V - A$ theory of Feynman and Gell–Mann is non-renormalizable. According to this theory, the weak interaction has the current–current form $J_\mu^- J^{+\mu}$, an example of which is pictured in Fig. 6.7. The current $J_\mu^-$ is a (electric) charge-lowering operator, here changing the neutral electron neutrino ($\nu$) into a negatively charged electron (e), while $J_\mu^+$

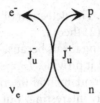

Fig. 6.7. $V - A$ weak interaction.

is a charge-raising operator, here changing the neutral neutron (n) to the positive proton (p). And in each case the currents change the electric charge by one unit (the charge on the electron).

Now the fact that the weak currents are electric charge-raising and lowering operators certainly suggests a connection between the weak and electromagnetic interactions. But the electromagnetic interaction is not a current–current point interaction. Instead, QED is a renormalizable theory in which a gauge vector field mediates the interaction. So given that there is a connection between the two interactions, a natural thought is to try to gauge the weak interaction as well, hopefully getting a renormalizable theory. But how do we do this? What is an appropriate gauge group, and how do we fit the relationship between the two interactions into this theory? Here is where the weak currents provide another clue.

We know that any continuous group $G$, with $n$ generators, which is a symmetry of the weak interactions (of the action) will generate $n$ conserved currents, $J_\mu^a$, $a = 1, \ldots, n$. And the spatial integral of the time component of $J_\mu^a$ will be a generator, $Q^a$, of the group $G$,

$$Q^a = \int J_o^a \, d^3 x.$$

If $G$ is a Lie group, the $Q^a$ will satisfy the the Lie algebra of the group. Of relevance here, we know that $SU(2)$ has the generators $L^a$, where $L^{+(-)} = L^1 \pm iL^2$ are raising (+) and lowering (−) operators with respect to $L^3$. That is, if the charge of $\psi$ is $m$, so that $L^3 \psi = m\psi$, the charge of $L^{+(-)}\psi$ is $m \pm 1$.

This suggests that $SU(2)$ is a symmetry of the weak interaction, with

$$J_\mu^1 = \tfrac{1}{2}(J_\mu^+ + J_\mu^-), \qquad J_\mu^2 = \tfrac{1}{2}i(J_\mu^+ - J_\mu^-),$$

forming two of the conserved currents of that symmetry. And there should be a third conserved neutral current whose time component $J_o^3$ gives us the charge operator, $L^3 = \int J_o^3 \, d^3 x$. We would like this to be the electromagnetic current. However, when we use the known particles to form the currents, the generators do not obey the $SU(2)$ commutation relations.

There are at least two moves we can make, while staying with the basic idea of $SU(2)$ symmetry. We can postulate the appropriate additional particles so that the generators close under commutation to form $SU(2)$, as in the theory of Georgi and Glashow (1972). Or we can try a minimal enlargement of the symmetry group by adding a $U(1)$ factor, which gives a weak neutral current as well as the

149

electromagnetic neutral current. The detection of the weak neutral current in 1973 therefore ruled out the pure $SU(2)$ theory.

This latter example is interesting for us because two things that are often cited in support of string theory are that it provides (hopefully) a finite theory of quantum gravity, and that it unifies the fundamental interactions. And if correct, these do make string theory tremendously interesting. But one thing we have learned from the many attempts at unified field theories in this century, I would argue, is that unification, in itself, is not a guide to a successful theory.

Consider the attempts by Einstein, Weyl, Schrödinger, Misner and Wheeler, and others to fashion a unified geometrical theory of gravity and electromagnetism. All of these were essentially mathematical investigations, because unlike the two cases we have just discussed, there were no empirical or theoretical clues as to how such a unification should take place. Indeed, there was no indication, classically, that such a unification was needed. After all, electromagnetism, conceived as an antisymmetric second-rank tensor field in curved spacetime is perfectly compatible with Einstein's gravity.

True, Einstein's gravity is non-renormalizable, while string theory promises to be a finite theory of quantum gravity. But again, there is not even a hint in what we know about physics, that this is the right way to get a workable quantum theory of gravity. Unless we have some reasons for thinking this, I do not think either of our reasons – providing a unified theory of the fundamental forces, and a finite theory of gravity – can, by themselves, give us (much) reason for thinking string theory may (or will) emerge as the fundamental framework for physics.

Undoubtedly, this is because, as I mentioned earlier, string theory at the fundamental level is a theory of very high energies, many orders of magnitude greater than are obtainable in the laboratory. And this raises an interesting question. If there are no clues in our low-energy world to the high-energy world of strings, how was string theory discovered? Why did anyone think of it in the first place? A two-paragraph history follows.

In the late 1960s it was proposed that the strong interactions obeyed duality. Duality involves the ideas of $s$-channel and $t$-channel processes, which are pictured in Fig. 6.8. In the $s$-channel the incoming particles combine to form an intermediary particle, while in the $t$-channel, the incoming particles exchange an intermediate particle. If $M(s, t)$ is the amplitude for $a + b \rightarrow c + d$, then duality says that we can regard $M$ as a sum of $s$-channel amplitudes or a sum of $t$-channel amplitudes.

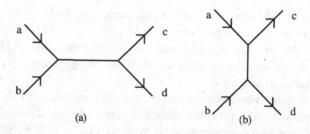

Fig. 6.8. (a)$s$-channel scattering ($s = p_a + p_b$) and (b)$t$-channel scattering ($t = p_a - p_b$).

The two are just different ways of describing the same process. (Compare this with quantum field theory where we have to include both types of diagrams to get the correct amplitude.)

In 1968, G. Veneziano wrote down a formula for $M(s, t)$ that actually exhibited duality. From the analysis of this formula and its generalizations developed the theory of dual models. Then in the early 1970s, it was shown that all the properties of the model could be derived from an underlying Lagrangian theory – that of the relativistic string! However, remember that this was supposed to be a theory of the strong interactions. By then, QCD was emerging and it became clear that dual models do not give a correct account of the strong interaction. But in 1974, Scherk and Schwartz showed that the scattering amplitudes of massless spin two states of the closed string equals the graviton–graviton scattering amplitudes of quantum gravity (in the lowest order of approximation). They therefore proposed reinterpreting string theory as a theory of all interactions, but at the mass scale of the Planck length, rather than that of the strong interaction.

Thus was string theory born. To my mind, this is the most amazing thing about string theory – that it exists at all!

### Notes

Editors' note: Tragically, our mentor Robert Weingard died in 1996.

Reprinted with permission from Weingard, R. 'A Philosopher Looks at String Theory', in *Proceedings of the Philosophy of Science Association 1988: Volume 2*, Fine and Leplin (eds), 1989, pp. 95–106. Copyright 1989, Philosophy of Science Association: permission granted by University of Chicago Press.

1. Using the *BRST* charge operator $Q$ we can define a ghost-free spectrum (i.e. no negative norm states). Namely, $Q|\psi\rangle = 0$ implies $\langle\psi|\psi\rangle \geq 0$. However, since $Q^2 = 0$, there are zero norm states. For example, $Q|\phi\rangle$ has zero norm if $|\phi\rangle$ is not a physical state (so $|\phi\rangle$ is not annihilated by $Q$).
2. This has been discussed by Redhead (1983) and Weingard (1984).

# 7  Black holes, dumb holes, and entropy

## William G. Unruh

It has now been 25 years since Hawking (Hawking 1974, 1975, Bardeen, Carter, and Hawking 1973) first surprised the world of physics with his analysis of quantum fields near black holes. Black holes, as their name implies, were believed to be objects into which things could fall, but out of which nothing could come. They were the epitome of black and dark objects. However, Hawking's analysis predicted that black holes should radiate, the radiation should be continuous and thermal, and the temperature should be inversely proportional to the mass of the black hole. Since black holes can also be said to have an energy proportional to their mass, this result led to opening of a whole new field of black hole thermodynamics.

That black holes could behave like thermodynamic objects had been intimated by results over the the previous five years. Christodolou (1970), Hawking and Ellis (1973, especially Lemma 9.2.2), Misner, Thorne, and Wheeler (1973) and Bekenstein (1973, 1974) had shown that there were certain formal similarities between black holes and thermodynamic objects. In particular, if one assumed positive energy for matter (an uncontested assumption), then – as Hawking most clearly showed – the area of a black hole horizon does not decrease. However, this formal similarity with entropy, which also does not decrease for an isolated system, did not seem to have any real relation with thermodynamics. The entropy of a body does not decrease only if the body is isolated, and not in interaction with any other system. In interaction with other systems, it can, and often does decrease. Furthermore, for ordinary systems, entropy and thermodynamics are primarily of interest in situations in which the system is able to traverse a closed cycle. Black holes on the other hand, have a surface area which always increases when in interaction with other systems. They could never engage in a closed thermodynamic cycle.

Hawking's discovery overturned that picture of black holes. The mass, and thus also the surface area (which is a function only of the mass) of a black hole could decrease, by the emission of thermal radiation into the outside world. That thermal

152

radiation carries away energy, and thus mass, from the black hole. At least formally, the emission of energy was accompanied by the absorption of negative energy by the black hole. This negative energy violated the assumptions of the Christodolou–Hawking theorems, and could therefore result in the shrinkage of the area of the black hole. Suddenly the formal analogy of area to entropy seemed far from simply formal.

This interpretation of the area as an entropy also fits together with the thermal nature of the emission. Since the black hole has a mass, and thus an energy, and since the radiation emitted was thermal with a universal temperature which depended only on the properties of the black hole, and not of the matter fields which were radiated, one could use the standard thermodynamic relations to derive an entropy:

$$T \, dS = dE \tag{7.1}$$

or

$$\frac{dS}{dM} = \frac{1}{T}. \tag{7.2}$$

Since (in the simplest cases) $T$ was only a function of $M$, one could (up to a constant of integration) obtain an expression for the entropy. Surprisingly (or not) this entropy is exactly proportional to the surface area of the black hole. Working in units where $G = \hbar = c = k = 1$, this relationship turned out to be just

$$S = \frac{1}{4\pi} A, \tag{7.3}$$

where $A$ is the area of the horizon of the black hole.

One thus had the astonishing situation that an entropy (in statistical mechanics, a measure of microscopic plenitude for a given macroscopic state) arising from purely quantum considerations was determined by a purely classical macroscopic non-quantum feature of the black hole, namely its area. The past twenty-five years have seen a persistent attempt to understand this puzzle, and a persistent failure to do so. This chapter will outline some of the directions which the attempt to do so has taken.

This chapter makes no pretense to be an exhaustive catalogue of the many ways in which people have tried to understand black hole thermodynamics. Instead, it treads a path which I have followed in trying to attack the problem. As a personal odyssey, it is thus as idiosyncratic as could be expected. It also will neglect to mention other approaches, not because I think they should be ignored – they certainly should not – but because of space limitations, and because I am often not sufficiently expert to present them here. I thus apologize in advance to both the reader and to my colleagues for my sins of omission.

In trying to understand black hole thermodynamics, there are at least two separate things one must try to understand: what causes the radiation which the black hole gives off and why is that radiation thermal; and how does the entropy fit in with the standard view we have of the statistical origin of entropy for all systems other than black holes. My own efforts have concentrated on the former question, and thus this chapter will also concentrate on that same question, coming back to the latter problem in the later sections.

One note about units. I will work throughout in a natural system of units called Planck units. In these units, the fundamental constants – the velocity of light, Planck's constant, Newton's gravitational constant, and Boltzmann's constant – all have the value of unity. In this system of units all dimensioned quantities are expressed in terms of fundamental units. Thus distances are expressed in units of $1.6 \times 10^{-33}$ cm, time in units of approximately $5 \times 10^{-44}$ s, mass in terms of $2.1 \times 10^{-5}$ g, and temperature in terms of $1.4 \times 10^{32}$ K.

## 7.1 Thermodynamics of black holes

With Hawking's calculation, we were faced with the necessity of taking seriously the idea that black holes were true thermodynamic objects. One of the key uses of thermodynamics, and the reason for its invention in the nineteenth century, was its explanation of the operation of heat engines. In particular, the second law of thermodynamics proved to be very powerful in its predictions concerning what kinds of engines could be built, and what the ultimate efficiency of any heat engine was. The limitations imposed on the operations of engines by the second law were so powerful that they led throughout the past two centuries to persistent attempts to violate the second law, to create heat engines which were far more efficient in their conversion of heat to work than was allowed by the strictures of this law. It has of course been one of the minor embarrassments of physics that no complete fundamental proof of the second law exists. However, thousands of examples of the failure of attempts to violate it have given us great confidence in its validity, as do very strong statistical plausibility arguments.

If black holes are thermodynamic objects, we now have another system which one can introduce into the operation of a heat engine, and we have to ask once again whether or not machines which now make use of black holes in a fundamental way also obey the second law, or whether closed thermodynamic cycles can be found in the operation of heat machines with black holes which violate that second law.

This investigation began with Bekenstein and Geroch[1] well before Hawking's discovery. When Bekenstein suggested that the Christodolou–Hawking results could be interpreted as showing that black holes had an entropy, Geroch suggested an example of a heat engine which could then be used to violate the second law (including black holes). Fill a box full of high-entropy radiation. Slowly lower the box toward the black hole by means of a rope. As the box is lowered, the tension in the rope far away from the hole can be used to perform work. In an ideal case, the energy extracted from the box goes as

$$W = E \cdot \left(1 - \sqrt{1 - \frac{2M}{R}}\right), \tag{7.4}$$

where $R$ is the radius away from the black hole to which the box has been lowered, $M$ is the mass of the black hole, and $E$ is the energy in the box. In principle, it would be possible to lower the box arbitrarily close to the black hole. It would thus be possible to extract arbitrarily much of the energy of the radiation in the box in the form of useful work by lowering the box very very near to the horizon. (The requirements on the rope that it not break and on the box that it not to crack open are severe, but since this is an 'in-principle' argument, they will be ignored.) Now, once the box is very near the black hole horizon, open it and allow the radiation to fall into the black hole. The entropy of the radiation within the box is thus lost to the outside world. However, the black hole's entropy increases by an amount proportional to the energy which falls into the black hole, i.e. the energy of the radiation which has not been extracted as useful work. Since that remaining energy can be made as small as desired, the increase in entropy of the black hole is as small as desired, and in particular it can be made smaller than the entropy lost down the black hole when the radiation fell in. Thus, the second law could be violated.

Geroch's argument was basically irrelevant at the time, since black holes could not engage in a cyclic processes; the black hole always grew bigger. Also, since classical black holes had no temperature, they could be regarded as zero temperature heat baths, and it is well known that with a zero temperature heat bath, one can convert all of the heat energy in any system to useful work.

However, with the advent of Hawking's discovery of black hole temperatures, this sanguine dismissal of the problem failed. Black holes could participate in closed cycles within a heat engine. You just needed to wait until the black hole had radiated the tiny bit of energy you had deposited into it. Black holes had a finite temperature, and the laws of thermodynamics dictated a maximum efficiency for the conversion of energy to work. Geroch's argument thus became critical. Could this process be used (in principle) to violate the second law of thermodynamics?

Bekenstein (1981) suggested that the reason that the Geroch argument failed was because the assumption that one could extract an arbitrary amount of work by lowering the box arbitrarily near the black hole was wrong. Boxes have finite sizes. One cannot lower a box closer to the black hole than the vertical dimension of the box. He postulated that there existed another law of nature, namely that the ratio of energy to entropy of the box was limited by the dimension of the box, i.e. that for any system,

$$S/E < \frac{H}{2\pi}, \tag{7.5}$$

where $H$ is the minimum dimension of the box. Since the box cannot be lowered to closer than $H$ from the black hole, we have a limit on the maximum amount of work which can be extracted during the lowering of the box. Since $S$ is also limited by that same $H$ we have a limit on the maximum amount of entropy that can be deposited into the black hole. The factor of $2\pi$ is chosen to ensure that this maximum deposited entropy is less than the increase in area due to the residual unextracted energy left in the box. It is astonishing that the mass of the black hole does not enter into this equation. Bekenstein has examined a number of systems, and has found that most (if not all realistic) in fact obey a bound very similar to this one.

However, his argument ignores the fact that the radiation would tend to gather at the bottom of the box in the arbitrarily large gravitational field felt by the box when near the horizon, i.e. the radiation would be nearer the horizon than the

William G. Unruh

dimension of the box when it was released. Furthermore, I was always disturbed by this line of argument. It implies that one must posit a new law of nature (the entropy to energy bound) in order to rescue the second law of thermodynamics for black holes. Furthermore, it would seem that there are cases in which one could imagine this law to be violated. For example, the entropy of a system with given energy contained within a box, but with $N$ different species of massless particles, is that of a single species times $\ln(N)$. By making the number of species of massless particles sufficiently large, one could thus violate the bound. The fact that the real world does not have this sort of a proliferation of massless particles is surely irrelevant to the validity of the second law. It seems unlikely that black hole thermodynamics could be used to place a bound on the number of species of particles.

The framework for an answer came one day at a Texas Symposium in a lunch lineup, when Kip Thorne and Freeman Dyson posed the Geroch argument to me. After I had told them of my objections to Bekenstein's argument, they asked what a correct answer would be. The glimmering of an answer which came to me then was developed into a complete argument by Bob Wald and myself (Unruh and Wald 1982, 1983) the following summer. At the time of Hawking's work, I had independently discovered (Unruh 1976) a closely related feature of quantum fields in flat spacetime (i.e. in the absence of any gravitational field). I had been worried about how one could define particles in a quantum field theory. Fields are not particles, and the particle aspects of quantum fields arise as secondary features of those fields in their interactions with other matter. Due to the quantization of the energy levels of localized matter, the interaction of that matter with the fields takes place by the absorption and emission of discrete quanta of energy. It is that process which makes the field behave as though its excitations were composed of particles.

The question I posed to myself was how one could interpret the excitations of quantum fields in terms of particles in the presence of strong gravitational fields. The principle I came up with was that particles should be defined to be that which particle detectors detect. Although this definition of particles seems to be a tautology, it focusses attention on a productive aspect of the problem. It is easy to design a system (at least in principle) which will detect particles when the state of the field is such that we usually consider it to have particles. One can now take that same detector and ask how it will respond in novel situations. One can then use that response to define what one means by particles in that novel situation. If it reacts in the same way that it would in detecting particles in well-understood situations, we can say that the field acts as though it were composed of particles in that novel situation.

The novel situation I examined was to accelerate a model of a particle detector in the vacuum state of the field. Now, an unaccelerated detector in that same state will see nothing. The detector will not click: it will respond as though there are no particles present. However, I found that if I examined an accelerated detector with the field left in exactly that same state, it would respond exactly as though that state of the field were populated by an isotropic thermal bath of particles. If the detector is left in contact with that state of the field (and continues in its state of constant acceleration), it will come into thermal equilibrium with a temperature just proportional to the acceleration of the detector. That is, an accelerated object immersed into the vacuum state of the field behaves as though it was in a thermal bath of particles of that field, with the temperature equal to

$$T = \frac{a}{2\pi} \frac{\hbar}{ck}, \tag{7.6}$$

where I have in this case reintroduced Planck's constant, the velocity of light, and Boltzmann's constant. This temperature is such that it requires an acceleration of $10^{24}$ cm/s$^2$ to achieve a temperature of 1 K.

This phenomenon of acceleration radiation can be used to explain why the Geroch argument does not work. As one lowers the box full of radiation toward the black hole, one requires a tension in the rope to keep the box from following a geodesic (straight line) and falling into the black hole: the box must be accelerated to hold it outside the black hole. This accelerated box thus sees the field outside the box as though it is in a thermal state with a temperature proportional to the acceleration. In addition, because it is accelerated, the box also sees itself as being in a gravitational field. Finally, in order that the box hold the radiation inside the box, it must also exclude the radiation outside. Thus, we have exactly the situation analyzed by Archimedes in a preprint over 2000 years ago.[2] A body immersed in a fluid will suffer a buoyant force equal to the weight of the displaced fluid. Thus, the box, immersed in the fluid of that thermal radiation and in the apparent gravitational field due to its own acceleration, will feel a buoyant force proportional to the weight (in that apparent gravitational field) of the displaced thermal fluid. Therefore, the tension in the rope holding the box from falling into the black hole is less than it would be if that thermal fluid were absent. Since the tension in the rope is less, the energy extracted by the work done by that tension far away from the black hole as the box is lowered is also less. In fact, the maximum amount of work extracted occurs not when the box has been lowered to the horizon (where the acceleration is infinite and the resultant buoyant force is thus infinite), but when it has been lowered to the 'floating point' where the weight of the displaced thermal fluid is just equal to the weight of the radiation inside the box.

Our calculation showed that the energy remaining to fall down the black hole (i.e. the total original energy in the box minus the work extracted by lowering the box) was precisely what was needed to increase the mass of the black hole, and thus increase its entropy to a value larger than or equal to the entropy contained in the box. That is, the entropy falling into the black hole in the form of radiation was always less than the entropy increase of the black hole due to its increase in surface area. The second law of thermodynamics was valid, even if one included black holes in the operation of the heat engine. This argument – unlike the Bekenstein argument – is universally valid, no matter how many species of field one imagines, since the buoyant force also increases if the box encloses, and thus excludes, more species of field. Black holes are truly thermodynamic objects: they obey the second law of thermodynamics.

## 7.2 A problem in the derivation of black hole evaporation

Despite this beautiful constancy of the black hole thermodynamics, it was clear from the beginning that the derivation that Hawking gave for the thermal emission from

black holes was flawed at a very fundamental level. The derivation took the background gravitational field of a black hole as a given. On this background spacetime,
one looked at the evolution of quantized fields. The first calculations were for scalar
fields, and that is what I will use in this chapter, but the calculations for other fields
quickly followed and gave exactly the same results.

Let us look at how Hawking actually carried out his calculation. While this may
seem to be somewhat technical, it is important to understand why I made the
statement above that his derivation was clearly wrong (even though, as I will argue,
the results are almost certainly correct). A field (a concept first introduced, with
much controversy, into physics by Faraday) is something which exists at all places
in space at all times. It is characterized, not like particles by positions, but rather by
the whole set of values that the field takes at each point in space and at each instant
in time. Those values at different places and times are related to each other via the
equations of motion that the field obeys. The field thus does not move or travel,
but one can talk about excitations – places where the field has non-zero values, or
certain patterns in the values of that the field takes at different points – travelling.
The field, whether classical or quantum, is a deterministic system, in which the field
values now are determined by the values at an earlier time. Since the disturbances
in the field travel, the state of those disturbances is determined by the state of the
disturbances at an earlier time in different places. We are interested in understanding
where the radiation given off by the black hole came from. To do so, we can propagate
those disturbances back in time to see what aspect of the early state of the field must
have caused the disturbance we are interested in. In particular, disturbances which
are now, at late times, coming out of the black hole must have been caused by
disturbances which came in toward the black hole at a much earlier time.

Although the above is a classical description, in which the magnitude of the fields
at each point has some definite value, the same is true of a quantum field where the
amplitude of the field is a quantum operator with the set of all such operators being in
a certain quantum state. The state of the quantum field can be divided into in-going
and out-going parts, the state of the out-going parts having been determined by the
state of the in-going components far in the past. The gravitational field for a black
hole is such that the spacetime becomes flat far from the black hole, and one can thus
use the usual interpretation of the behaviour of quantized fields in a flat spacetime,
as long as one concentrated on the behaviour of the fields far from the black hole.
The in-going part of the fields were assumed to be in the vacuum state, the state of
lowest energy, or the state in which, under the conventional picture, there exist no
particles corresponding to that field. One then uses the field equations to propagate
those fields into the region of the black hole and then back out to become out-going
components of the field in a region far from the black hole again. Far from the black
hole those out-going fields could again be analysed in the conventional way and the
particle content of the resultant out-going state could be determined.

The result was that if one began with the in-going components of the field in its
vacuum state and with some (non-quantized) matter collapsing together to form a
black hole, then very quickly (on a time scale of the light travel time across a distance
corresponding to the circumference of the black hole) the out-going portions of
the quantum field would settle down to a state such that the out-going field looked

like a thermal state with a temperature of $1/8\pi M$ (for a spherically symmetric black hole).

However, one could, for any particle emitted out of the region of the black hole (or for any mode of the field), trace back, by use of the equations of motion of the field, the disturbances which corresponded to that particle to the in-going aspects of the field which must have created that particle. It was in this determination that one obtains nonsensical answers. Consider a particle which was emitted from the black hole at a time $t - t_0$ after the formation of the black hole at time $t_0$. That particle would have a typical energy of order of $1/M$ and a wavelength of order $M$, the radius of the black hole. However, the initial fluctuations in the field which must have caused this particle would have had an energy of order $(1/M)e^{(t-t_0)/M}$ and a wavelength of order $Me^{-(t-t_0)/M}$. Let us consider a solar mass black hole, for which the typical wavelength of the emitted particles would be a few kilometres, and typical frequency of the order of kilohertz, and consider a particle emitted one second after the black hole had formed. This emitted excitation of the field must have had its origin in the incoming fluctuations in the field with a frequency of $e^{10^4}$ and wavelength $e^{-10^4}$ (units are irrelevant since any known system of units would make only a negligible change in that exponent of $10^4$). It is clear that at these scales the very assumptions which went into the derivation are nonsense. At these frequencies, a single quantum of the field has an energy not only larger than that the energy of the black hole, but also inconceivably larger than the energy of the whole universe. At these scales, the assumption that the quantum field is simply a small perturbation, which does not affect the gravitational field of the black hole, is clearly wrong: Hawking's calculation contains its own destruction. If it is correct, all of the assumptions which were necessary to make his derivation are clearly wrong. But it is equally clear that his results are far too appealing to be wrong. But how can the derivation be nonsense, and the results still be correct? This is a puzzle which has bothered me ever since Hawking's original paper. The problem of course is that this derivation (or essentially equivalent ones, like certain analyticity arguments) were the only evidence for the thermal nature of black holes. Black holes have never been seen to evaporate, nor are we likely to get any such experimental evidence for the validity of this result.

Thus, the picture of the origin of black hole radiation painted by the standard derivation is that we have a mode of the field with such an absurdly high energy. This mode is in its lowest energy state, the vacuum state and is travelling toward the region in which the black hole will form. It enters the collapsing matter just before it forms a black hole. It crosses through the matter, and emerges from the matter just (exponentially 'just' – $e^{-10^4}$ s or less) before the horizon forms. It is trapped against the horizon of the future black hole, exponentially red-shifting, and being torn apart by the horizon, until finally its frequency and wavelength are of the characteristic scale of the black hole. At this point part of this mode escapes from the black hole, leaving part behind to fall into the black hole. The part that escapes is no longer in its ground, or lowest energy state. Instead it is excited and behaves just like an excitation which we would associate with a particle coming out of the black hole. Because of the correlations that existed between the various parts of this mode in the initial state, there are correlations between the part of the mode which escapes from the black hole, and the part left behind inside the black hole. These correlations ensure that the excitation of the field coming out of the black hole is precisely thermal in nature.

The natural question to ask is whether the predication of thermal radiation really does depend on this exponential red-shifting, or whether this apparent dependence is only a result of the approximations made in the derivation. In particular, could an alteration in the theory at high frequencies and energies destroy the prediction of thermal evaporation? That is, is the prediction of thermal radiation from black holes robust against changes to the theory at those high energies which we know must occur? There seemed to be nothing in that original derivation which would indicate that the prediction was robust, except for the beauty of the result. (Of course physics is not supposed to use the appeal of a result as an argument for the truth of that result.)

### 7.3 Dumb holes

Is there some way of testing the idea that the truth of the thermal emission from black holes is independent of the obvious absurdity of assuming that the quantum field theory was valid at arbitrarily high frequencies and short distances? Although any ultimate answer must depend on our being able to solve for the emission of radiation in some complete quantum theory of gravity, one can get clues from some analogous system. One such analogue is what I have called dumb holes, the sonic equivalent to black holes (Unruh 1980).

Let me introduce the analogy by means of a story. Consider a world in which fish swimming in an ocean have become physicists. These fish experience their world, not through sight (they are blind) but through their ears: through sound alone. In this world there exists a particularly virulent waterfall, in which the water, as it runs over the falls, achieves a velocity greater than the velocity of sound. This waterfall will act as an attractor to fish and to sound. Sound waves travelling by this waterfall, will be pulled toward the waterfall, just as light passing by a star is pulled toward the star and bent. Fish swimming near this waterfall will be attracted to it. However, as long as a fish swims outside the boundary surface at which the velocity of water equals the velocity of sound, its voice and the shouts it makes will travel out to any other fish also swimming outside that surface. However, the closer the fish comes to the surface, the longer it will take the sound to get out, since its velocity out is partially cancelled by the flow of the water over the waterfall; the sound waves are swept over the falls along with anything else. Now consider a fish which falls over those falls and calls for help as it does so. The closer to the surface where the velocity of the water equals the velocity of sound, the slower the effective speed of sound getting out is, and the longer the sound takes to get out. A sound wave emitted just at the brink of that surface will take a very long time to escape. Because the sound takes longer and longer to get out, the frequency of the sound emitted by that fish will be bass-shifted to lower and lower frequencies. Any sound emitted after the fish passed through the surface will never get out, and will be swept onto the rocks below along with the fish itself (for a picture and a description of this process, see Susskind 1997).

This description is very similar to that of light around a black hole. The surface where the water speed exceeds the velocity of sound is an analogue of the horizon of a black hole. The bass-shifting of the sound emitted by a fish falling through that surface is analogous to the red-shifting of the light emitted by an object falling through the horizon of a black hole. These dumb holes (holes unable to speak or to emit sound) are thus an analogue to black holes and, since the physics of fluid flow is believed to be well understood, an analogue which one could hope to understand in a way that black holes are not yet understood.

These physicist fish will develop a theory of their surroundings, and in particular will develop a theory of sound waves very similar to our theory of light within General Relativity, Einstein's theory of gravity. This theory will furthermore have in it structures, these waterfalls, which share many of their features with black holes in our world.

It is important to point out that there are features of these dumb holes which are not analogous. If a fish stays stationary outside the waterfall, the frequencies of its sound are not bass-shifted, while the light from such a stationary source is red-shifted. The flow of time is not altered by the presence of the waterfall as is the flow of time in the gravitational field of the black hole. The size of the waterfall, of the sonic horizon, is not a function of how much matter or energy has fallen over the falls, while the black hole size is a direct measure of the amount of energy which has fallen into the black hole. Despite these differences however, the dumb holes still turn out to be ideal subjects for understanding the quantum behaviour of black holes.

Let us consider sound waves in slightly more detail. They are small perturbations in the density and velocity of the background flow of the fluid. As is well known, if one examines only the lowest-order perturbations around a background fluid flow, perturbations which represent sound waves in that fluid, then those perturbations obey a second-order linear partial differential equation. If we furthermore assume inviscid flow, so that the perturbations are conservative (do not lose energy due to dissipation) then the equations are homogeneous second-order equations in space and time derivatives. This can be cast into the form

$$\partial_\mu a^{\mu\nu} \, \partial_\nu \Phi = 0, \tag{7.7}$$

where the tensor $a^{\mu\nu}$ depends on the background fluid flow, and $\Phi$ is some scalar such as the density perturbation or the velocity potential. But by defining

$$g = \det(a^{\mu\nu}), \tag{7.8}$$

$$g^{\mu\nu} = \frac{1}{\sqrt{g}} a^{\mu\nu}, \tag{7.9}$$

this is cast into the form of the perturbation of a scalar field $\Phi$ on a background metric $g_{\mu\nu}$: the equations of motion for the sound waves on a background fluid flow obey exactly the same equations as a massless field on a background spacetime metric. Furthermore, if the fluid flow has a surface on which it becomes supersonic, the metric associated with that fluid flow is the metric of a black hole with that supersonic surface as the horizon of the black hole.

Those small perturbations of sound waves in the fluid flow can be quantized just as can the scalar field on a background metric (quantized sound waves are often called phonons). Furthermore, one can use Hawking's argument to conclude that such a dumb hole should emit radiation in the form of phonons in exactly the same way as a black hole will emit thermal radiation. The fish physicists would come up with a theory that their dumb, sonic, holes would emit thermal radiation, just as black holes do.

Of course in the case of dumb holes, there is no analogue of the mass or energy of the dumb hole. The structure of the sonic horizon is just a consequence of the peculiarities of the flow set-up, and not of the amount of energy that has gone down the dumb hole. There is no analogy to the energy of a black hole, and the relationship between that energy and the temperature that occurs with a black hole does not occur for a dumb hole. Dumb holes would not be thermodynamic objects as black holes are. They are however hot objects. The temperature of dumb holes is an exact analogue of the temperature of black holes.

Because of the pivotal role that the temperature plays in black hole thermodynamics, the existence of a dumb hole temperature makes them a sufficient analogue of black holes to, one hopes, give us clues as to the origin of the radiation from black holes. In particular, in the case of black holes we have no idea as to the form that a correct theory at high energies would have, or at least no idea as to how we could calculate the effects that the alterations of any such theory at high frequencies would have on the thermal emission process. However, for dumb holes we have a good idea as to the correct theory.

Real fluids are made of atoms. We know that once the wavelengths of the sound waves are of order the interatomic spacing then the fluid picture becomes inapplicable. Of course, calculating the behaviour of $10^{23}$ atoms in a fluid is as far out of our reach as is calculating the effects of any putative theory of quantum gravity on the emission process around black holes. This fact kept me from being able to use dumb holes to understand black hole thermal emission for over ten years. However, Jacobson (1991) realized that there are approximations that can be made. One of the key effects of the atomic nature of matter on the behaviour of sound waves is to alter the dispersion relation (the relation between wavelength and frequency) of the sound waves. While sound waves at lower frequencies have a direct proportionality between wavelength and the inverse frequency, just given by a fixed velocity of sound, at higher frequencies this relation can become much more complex. The effective velocity of sound at high frequencies can increase or decrease from its value at low frequencies. Thus, the location of the sonic horizon, which is determined by where the background flow equals the velocity of sound, can be different at different frequencies of the sound waves.

This allows one to ask, and answer, the more limited question: 'If we alter the dispersion relation of the sound waves at high frequencies, does this alter the prediction of thermal radiation at low frequencies?' The calculation proceeds in precisely the same way as it does for a black hole. Consider a wave packet travelling away from the dumb hole in the future, and at a time when it is far from the dumb hole. In order to represent a real particle, we choose that wave packet so that it is made purely of waves with positive frequencies (i.e. their Fourier transform in time at any position is such that only positive and no negative frequencies non-zero). We now propagate that wave backward in time, to see what configuration of the field would have created that wave packet in the future. Going backward in time, that wave packet approaches closer and closer to the (sonic) horizon of the dumb hole. In order to have come from infinity in the past, and if the velocity of sound does not change with frequency, that wave must remain outside the horizon (nothing from inside can escape). But in this propagation backward in time, the wave is squeezed up closer and closer to the horizon, in a process which grows exponentially with time. It is only when we get to the time at which the sonic hole first formed that those waves can escape and propagate (backwards in time) back out to infinity. By that time that exponential squeezing against the horizon has produced a wave with with an incredibly short wavelength, and thus incredibly high frequency. One can now analyse that resultant wave packet to see what its frequency components are. In general they are very high. However, now one can find that that packet which began with purely positive-frequency components, also has negative-frequency components. Those negative-frequency components are a direct measure of the number of particles which were produced in that mode by the process of passing by the dumb hole. In particular, Hawking showed for black holes (and the same follows for dumb holes) that the ratio of amplitudes for those negative- and positive-frequency components is given by $e^{\omega/T}$, where $T$ is the Hawking Temperature of the black hole, proportional to the inverse mass of the black hole, or, in our case, is the temperature of the dumb hole, proportional to the rate of change of velocity of the fluid as it passes through the dumb hole horizon.

What happens if we change the dispersion relation at high frequencies for the fluid? In this case, squeezing against the horizon as one propagates the wave packet back in time no longer occurs to the same extent. As the packet is squeezed, its frequency goes up and the velocity of sound for that packet changes. What was the horizon – the place where the fluid flow equals the velocity of sound – is no longer the horizon for these high-frequency waves. They no longer get squeezed against the horizon, but rather can once again travel freely and leave the vicinity of that low-frequency horizon long 'before' (remember we are actually going backward in time in this description so 'before' means after in the conventional sense) we get back to the time at which the dumb hole formed. Again, we can wait until this wave propagates back to a regime in which the velocity of the water is constant, and we can take the Fourier transform of these waves. Again, we can measure the ratio of the positive and negative frequencies in the incoming wave, and determine the particle creation rate. The astonishing answer I found (Unruh 1995) (and which has been amply confirmed by for example Jacobson and Corley (1996) in a wide variety of situations) is that this ratio is again, to a very good approximation, a thermal factor, with precisely the same temperature as one obtained in the naive case. The history of that packet is entirely different (for example, it never came near the time of formation of the dumb hole, as it did in the naive analysis), and yet the number of particles produced by the interaction with the dumb hole is the same: i.e. such dumb holes produce thermal radiation despite the drastic alteration of the behaviour of the fields at high frequencies. This has been calculated for a wide variety of situations, both where the high-frequency velocity of sound decreases over its low-frequency value, and where it increases sufficiently that there is no longer any horizon at high frequencies. The prediction of thermal radiation from a dumb hole appears to be remarkably robust and independent of the high-frequency nature of the theory.

How can this be understood? The calculation, done for example by Hawking, used exponential red-shifting of the radiation in an essential way. The ultra high-frequency aspects of the theory appeared to be crucial to the argument. Yet we find here that it is not crucial at all. All that appears to be crucial is the the low-frequency behaviour of the theory and the nature of the horizon at low frequencies.

Although the best way of understanding the particle production process by a black hole or a dumb hole has not yet been found, there are at least hints of a possible answer. If we examine the history of such a wave packet from the past to the future, we find that the packet, while near the horizon, suffers an exponential red-shifting. However, the time scale of the red-shifting is of the order of the inverse temperature associated with the hole, $\hbar/kT$. When the packet has a very high frequency, that time scale is very long compared with the inverse frequency of the packet. The packet sees the effects of this red-shifting as occurring very slowly – adiabatically. Now, we know that a quantum system undergoing adiabatic change remains in the state it started in. If the packet started in its ground state, it remains in its ground state during this adiabatic phase. If the packet began in the vacuum, it remains in the vacuum. Thus, during most of the red-shifting that occurs near the horizon, or during the approach of the packet to the horizon, the packet does not notice the fact that something is happening around it. It gets stretched (red-shifted), but so slowly that it remains in its vacuum state (zero particle state). However, eventually it is stretched sufficiently that its frequency is now low enough that the time scale of red-shifting is of the same order as its inverse frequency. The surroundings now change rapidly over the timescale of oscillation of the packet, and it begins to change its state. It is excited from its ground state to a many particle state. Furthermore, because of the correlations across the horizon at these low frequencies, the particles which are emitted out toward infinity are in an incoherent, thermal, state.

Appealing as this scenario is, it still needs to be fleshed out via compelling calculations. However, even at this point it strongly supports one conclusion, namely that the thermal emission process of a black or dumb hole is a low-frequency, low-energy, long-wavelength phenomenon. Despite the apparent importance of high frequencies in the naive calculation, it is only aspects of the theory on scales of the same order as those set by the temperature of the hole (e.g. the inverse mass for black holes) which are important to the emission of thermal radiation. The thermal emission is a low-energy, low-frequency, long-wavelength process, and is insensitive to any short-scale, high-energy or high-frequency physics. This conclusion will play an important role in the following.

## 7.4 Entropy and the 'information paradox'

Having discussed the temperature of black holes, and come to the conclusion that the thermal radiation is created at low frequencies and long wavelengths outside the horizon, let us now return to the issue of the entropy of the black hole. All of our arguments about black holes in the above have been thermodynamic arguments. Black holes have certain macroscopic attributes of mass (energy) and temperature.

This is not the place to enter into a detailed discussion of what entropy really is, but it is important to state the usual relationship of entropy to the internal states of the system under consideration. For all material objects, the work of Boltzmann, Maxwell, Gibbs, etc., showed that this thermodynamic quantity of a system could be related to a counting of the number of microscopic states of a system under the macroscopic constraints imposed by the mode of operation of the heat engine which wished to use the system as part of its 'working fluid'.

This classical analysis of the nature of entropy is closely coupled to the notion of the deterministic evolution of the system. In the evolution of a system, unique initial states evolve to unique final states for the 'universe' as a whole. In quantum physics, this is the unitary evolution of the state of the system: orthogonal initial states must evolve to orthogonal final states. The number of distinct initial states must be the same as the number of distinct intermediate states, and must equal the number of final states. These states may (and will in general) differ in the extent to which a heat engine can differentiate between them. But on a fundamental level the extent to which each state is different from each other state is conserved throughout the evolutions. The law of the increase of entropy then arises when one considers a highly differentiated initial state (as far as the heat engine is concerned). In general this will evolve to a much less differentiated final state. But in principle, if the heat engine were sufficiently accurate in its ability to differentiate between the various states of the system, all systems would evolve with a constant, zero entropy.

However, black holes, as classical objects, have a small number of parameters describing the exterior spacetime (the spacetime in which a heat engine can operate), namely the mass, angular momentum, and the variety of long-range charges which a black hole can support. All aspects of the matter that falls into the black hole fall into the singularity at its centre, taking the complexity of the structure and state of that matter with them. The horizon presents a one-way membrane, impervious to any prying hands, no matter how sophisticated. Arbitrarily complicated initial states all evolve, if the matter falls into the black hole, into simple final states of a black hole.

Does the situation change in a quantum theory of black holes? One difficulty is that no such theory exists, making the question difficult to answer definitively, but one can at least try to understand why the question has proven as difficult to answer as it has.

Ever since black hole thermodynamics was discovered in the 1970s, attempts have been made to try to understand black hole entropy in a statistical mechanical sense. At the semiclassical level, black holes evaporate giving off maximal entropy thermal radiation, no matter how the black hole was formed. They can however be formed in a wide variety of ways. Is the emission from black holes truly thermal, and independent of the formation of the black hole, or are there subtleties in the outgoing radiation which carry off the information as to how the black hole was formed? If one forms a black hole in two distinct (orthogonal) ways, are the final states, after the black hole in each case has evaporated, also distinct? The fact that the radiation in the intermediate times looks thermal is no barrier to this being true. A hot lump of coal looks superficially as though it is emitting incoherent thermal radiation. However, we strongly believe that the final states of the radiation field would be distinct if the initial states which heated up the lump of coal had been distinct.

How does the situation proceed for that lump of coal? If the coal were heated by a pure, zero entropy, state initially, how does it come about that finally the radiation field again has zero (fine-grained) entropy despite the fact that at intermediate times the coal was emitting apparently thermal incoherent radiation? The answer is that the coal has an internal memory, in the various states which the constituents of the coal can assume. When the coal is originally heated by the incoming radiation, the precise state in which those internal degrees of freedom of the coal are left is completely and uniquely determined by the state of the heating radiation. The coal will then emit radiation, in an incoherent fashion: i.e. the radiation emitted will be probabilistically scattered over a huge variety of ways in which it could be emitted. However, each way of emitting the radiation will leave the internal state of the coal in a different, completely correlated state. The coal remembers both how it was heated, and exactly how it has cooled since that time. Thus as the coal cools, its state will help determine how the radiation at late times is given off, and that radiation will then be correlated with what was given off at early times. The state of the emitted radiation, after the coal has cooled, will look – if looked at in a limited region or a limited time – as though it were completely incoherent and thermal. However, there will exist subtle correlations between the radiation in the various regions of space and at various times. It is those correlations which will make the final state a unique function of the initial state of the heating radiation. And the possibility of that final state being uniquely determined is predicated on that coal's internal state being uniquely determined by the initial state and the state of the radiation emitted up to the point of time in question. It is entirely predicated on the coal's having a memory (those possible internal states) and on it remembering how it was formed and how it radiated.

If black holes also behave this way then they also must have a memory, and retain a memory of how they were formed. Furthermore, that memory must affect the nature and state of the radiation which is subsequently given off by the black hole. The radiation given off by the black hole at any time must depend on both the state of the matter which originally formed the black hole, and the state of the radiation which has already been given off by the black hole.

Of course the alternative is simply to accept that black holes are different from other bodies, in that the out-going radiation is unaffected either by how the hole was formed or by how it has decayed since that time. Black holes would then differ fundamentally from all other forms of matter. The final incoherent state of the radiation emitted by a black hole would not be determined by the initial state of the matter that formed the black hole. All initial states would produce the same final, incoherent, mixed state. Of course this is of no practical importance, since those subtle correlations in the emitted radiation responsible for the maintenance of unitarity from even a small lump of coal are entirely unmeasurable.

In the 1980s Banks, Susskind, and Peskin (1984) presented a *reductio ad absurdum* argument that black holes must have a memory, and must preserve unitary evolution. Their argument was that if the black hole really was a memoryless system, then the out-going radiation would be what is technically known as Markovian. They then showed that such a Markovian evaporation process would lead to energy non-conservation, and furthermore that the energy non-conservation would be extreme for small black holes. For black holes of the order of the Plank mass $(10^{-5}\,\mathrm{g})$ one would expect radiation to be emitted in quanta whose energy was also of the order of that same Plank mass. Furthermore, one might expect virtual quantum processes to create such Plank mass black holes which would then evaporate creating a severe energy non-conservation in which one might expect energies of that order to be liberated. Since that is of the order of a ton of TNT, one could argue that such non-conservation has been experimentally ruled out, as physicists all might have noticed the presence of an extra ton of TNT in energy liberated in their labs (or indeed even non-physicists might have noticed this). Therefore, they concluded that black holes could not be memoryless: that they must remember how they were formed. However this argument assumed that black holes remembered nothing, not even how much energy went into their formation and evaporation. But black holes clearly have a locus for the memory of the energy, angular momentum, and charge of their formation, namely in the gravitational and other long-range fields which can surround a black hole (Unruh and Wald 1995). There is thus no reason in principle or in experiment why black holes could not be objects very different from other matter, lacking all memory of their formation (apart from those few bits of memory stored outside the horizon).

However, many people have found this conclusion too radical to countenance. While I do not share their horror of black holes being afflicted by this ultimate in Alzheimer's disease, it is certainly important to examine the alternatives.[3]

Various suggestions have been made through the years for a statistical mechanical origin for the black hole entropy. One suggestion has been that the entropy of a black hole is equal to the logarithm of the number of distinct ways that the black hole could have been formed (Thorne, Zurek, and Price 1986). However, this clearly does not identify any locus of memory for the black hole. It is, in fact, simply a statement that the black hole obeys the second law of thermodynamics. If the black hole had many more ways of being formed than the exponential of its entropy, then one could prepare a state which was an incoherent sum of all of the possible ways of forming the black hole. The entropy of this state would just be the logarithm of that number of states. Once the black hole had formed and evaporated, then the entropy of the resulting matter radiation would just be the entropy of the black hole. Thus the entropy of matter, under this process, would have decreased, leading to a violation of the second law of thermodynamics. Similarly, if the entropy of the black hole were larger than the logarithm of the number of all the possible states which went into its formation, the entropy of the world would increase under the evaporation process. However, one of the ways of increasing the size of a black hole would simply be to reverse the radiation which was emitted back to the black hole: the time reverse of the emitted radiation is one of the ways of creating the black hole. Thus the entropy of the emitted radiation also forms a lower bound on the logarithm of the number of ways of creating the black hole. Thus the statement that the number of ways of forming a black hole gives a measure of its entropy is simply a thermodynamic consistency condition, and gives no clues as to the form of the memory (internal states) of a black hole.

Within the past four years, however, string theorists (originally Strominger and Vaffa 1996) have suggested that they may have found a statistical origin for the black hole entropy, a description of those internal states which would form the black hole's memory. These suggestions are as yet only that – suggestions – but they carry a powerful conviction and hope that they may offer a way of describing black hole entropy. This hope has been adequately broadcast, so what I wish to do in the remainder of this chapter is to raise some potential difficulties which may beset this explanation.

Let me give a cartoon review of the string theory argument. At long distances, and low energy, one can regard string theory as leading to an effective field theory, with the various excitations of the strings corresponding to various types of fields. That low-energy field theory includes gravity, and has, in certain cases, black holes as solutions. In the case where one demands that these solutions to the low-energy field theory retain some of the super-symmetries of the string theory, the fields must be chosen carefully so as to obtain a black hole solution, rather than some singular solution of the field equations. These super-symmetric black holes have, in addition to their gravitational field, a combination of other massless fields. These other fields, like the electromagnetic field, imply that the black hole must also carry certain charges. The mass of these black holes is furthermore a given function of those charges. It is not an independent parameter, as it would be for a generic black hole.

If one wishes to create a model for these black holes within string theory, one requires that the model have sources for those fields, that the configuration of the strings carries those charges. For some of the fields, such sources were no problem. They arise in the perturbative form of string theory. However for other fields, the so-called 'RR fields', there was nothing within standard perturbative theory which carried these RR charges. Polchinski introduced additional structures, called 'D-Branes', into string theory precisely to carry those RR charges: to be sources for the associated fields. D-Branes are higher dimensional surfaces on which open strings could have their endpoints.[4] It is assumed that they will arise in string theory from non-perturbative effects. However, if super-symmetric black holes were to have a string theory analogue, that analogue would have to contain not only the standard strings, but also D-Branes to act as the origin for the charges carried by the black hole.

D-Branes are extended surfaces without edges. In order that the black hole be a localized object, it is assumed that our ordinary four dimensions (three space and one time) are all orthogonal to these D-Brane surfaces. In order that the extra dimensions demanded of string theory not embarrass us by not having been observed, these extra dimensions (in which the D-Branes stretch) are assumed to be 'curled up' so that travelling a (very small) distance along one of these extra dimensions always brings one back to the origin. Thus to us, these D-Branes would look as though they were located at a point (or at least a very small region) of our observable three dimensions of space.

In a certain limit, one could carry out a calculation of how such an analogue structure would behave. If one adjusted one of the parameters in the string theory – the so-called 'string coupling constant' – so that the couplings between strings were turned off, then those D-Branes and their attached strings would constitute a free gas of particles. The D-Branes would be held together by a coherent collection of strings which connected them to each other, preserving the localized nature of this black hole analogue.

Now one could ask in how many ways the D-Branes be could distributed in order to produce the same macroscopic total charges. And in how many ways could the open strings connect the various D-Branes together and to each other? In the weak coupling limit, since the D-Branes and their attached open strings have a perturbative description, the number of states of the system could be calculated by solving the low-energy super-gravity conformal field theory to count the degeneracy of these states, and thus their entropy. The answer was that the entropy, as a function of the charges, was precisely the same as the entropy one would have for a super-symmetric, maximal black hole with those same charges.

Now the string theory analogue is *not* a black hole. The only way in which the calculations can be carried out is by assuming that the gravitational constant (proportional to the string coupling constant) is essentially zero. In that case the string analogue would be spread out over a region much larger than the radius of the horizon of the black hole which those charges would form. The string theory calculation is a calculation in which everything is embedded in a flat spacetime.

As a thermodynamic object, a super-symmetric black hole is anomalous. Although it can be said to have both entropy and energy, its temperature is identically zero. It does not evaporate. However, one (Maldecena and Strominger 1997) can also carry out calculations for the D-Brane model if one goes very slightly away from the exactly super-symmetric model. In this case both the black hole and the D-Brane models radiate. The spectrum of the radiation, assuming that the D-Brane is in thermal equilibrium, is again identical in both cases, and in neither case is it thermal. In the case of the black hole, this non-thermality is caused by the non-zero albedo of the black hole. If the wavelength of the radiation is much longer than the diameter of the black hole, most of the radiation scatters from the curvature around the black hole, and is not absorbed. By detailed balance, the emitted radiation is suppressed for exactly those components which would have been scattered if they were incoming radiation. In the D-Brane calculation, the emitted radiation is caused by the collision of open strings of differing temperatures on the D-Branes. The astonishing result is that both calculations give the same, very non-trivial result.

Both of these results, the entropy and the low-temperature emission, make one feel that surely string theory is giving us an insight into the behaviour of black holes. In the case of D-Branes, the entropy has its origin in the standard way, as a measure of the number of ways that the internal configuration (fluctuations of the open strings and configurations of the D-Branes) of the system can vary, constrained by the macroscopic parameters (the charges). The entropy of the D-Branes is an entropy of the same type as the entropy of a lump of coal. The D-Brane analogue has a locus for its memory, that locus being the various configurations of D-Branes and strings which make up the object.

However, it is again important to emphasize that these D-Brane bound states are just analogues to black holes, they are not themselves black holes. As one increases the gravitational constant to a value commensurate with the structure forming a black hole, perturbative calculations in string theory become impossible.[5] Thus the crucial question becomes 'does this D-Brane analogue give us an insight into the entropy of black holes?' Or is the equality of the entropy calculations – one a thermodynamic one from the temperature of the radiation given off by the black hole, and the other a statistical one from the degeneracy of D-Brane bound states and strings – either an accident or an indication that in any theory the black hole entropy is the maximum entropy that any system could have? One could then argue that since the D-Branes also form such a maximal entropic system, they must have the same entropy as does the black hole.

The key problem is that if string theory is going to give a statistical origin for the entropy of a black hole, it must identify the locus of the memory of the black hole. How can the black hole remember how it was formed and what the state of the radiation was which was previously emitted by the black hole during its evaporation? If one believes that some sort of D-Brane structures are responsible for this memory, where in the black hole are they located? Let us examine some of the possibilities.

(i) *Inside the black hole*: The interior of the black hole, as a classical object, contains a singularity. While the existence of singularities in string theory is unknown, the suspicion is that the theory will not be singular. However, in the field theory description of a charged black hole, it is the singularity which carries the charges associated with the black hole. It is at the singularity that one would therefore expect to find the D-Branes, the carriers of the RR charges, in the string theory formulation. While the presence of the D-Branes might well smooth out the singularity, or make irrelevant the notion of a spacetime in which the singularity lives, one would expect them to live in the region of the spacetime where the curvatures and field strengths approach the string scale. In regions where the curvatures and strengths are much less, the field theory approximation to string theory should be good, and it contains no sources for the charges of the fields making up the super-symmetric black hole.

Could D-Branes living at or near the singularity be the location for the black hole memory (Horowitz and Marolf 1997)? The singularity certainly forms a place which one would expect to be affected by any matter falling into the black hole or by the radiation emitted by the black hole. (That emitted radiation, in the standard calculations, is correlated with radiation which is created inside the horizon and falls into the singularity.) Thus, the formation and evaporation of the black hole could certainly affect the memory of the black hole if this is where it was located. However, to be effective as a source of entropy the memory cannot simply remember, like the unread books of an academic historian, but must also affect the future behaviour of the radiation emitted by the black hole. The radiation emitted late in the life of the black hole must be exquisitely and exactly correlated with the radiation emitted early in its life, in order that the state of the total radiation field be precisely determined by the state of the matter which formed the black hole. The memory must affect, in detail, the radiation which the black hole emits.

But here the low-frequency nature of the emitted radiation gets in the way. As I argued from the behaviour of the dumb hole, the radiation created by black holes is insensitive to the high-frequency behaviour and interactions of the radiation fields. It is a product of the behaviour of the spacetime at the low frequencies and long-distance scales typical of the black hole (milliseconds and kilometres for a solar mass black hole). At these scales the classical spacetime picture of the black hole is surely an accurate one, and at these scales the interior of the black hole is causally entirely separate from the exterior. Anything happening inside the black hole, at its centre, can have no effect on the detailed nature of the radiation emitted. The D-Brane memory, if it exists at the centre of a black hole, must be sterile.

(ii) *Outside the horizon*: I have argued above that black holes are not completely forgetful. They do have a memory for energy, charges, and angular momentum. Since this memory is encoded in the fields outside the black hole, it would not be expected to be sterile, but can (and we believe does) have a determining effect on the emitted radiation. Could not the rest of the memory be encoded in the same way? Could not the memory of a black hole be in the curvature of the gravitational field or the structures of the other fields outside the black hole? As stated it would be hard to use D-Branes, as a black hole carries no RR charges outside the horizon. However, other string-like structures perhaps could have a wide variety of states for any macroscopic distribution of the gravitational field.

Let me first give some completely disparate evidence in favour of this possibility, before expressing my reservations. In the late 1970s, Paul Davies, Steve Fulling, and I (Davies, Fulling, and Unruh 1976) were looking at the stress–energy tensor of a massless scalar field in a two-dimensional model spacetime. Now, as a massless conformal field, the stress–energy tensor obeys the relation that its trace is zero:

$$g_{\mu\nu} T^{\mu\nu} = 0. \tag{7.10}$$

However, when we tried to calculate the expectation value of the tensor it was, as expected, infinite. By regularizing the value (in that case via 'point splitting' – taking the fields which form $T^{\mu\nu}$ at slightly different locations and looking at the behaviour of $T^{\mu\nu}$ as the points were brought together) we found that we were faced with a choice. We could either take the trace to be zero, and then either lose the conservation equation

$$T^{\mu\nu}{}_{;\nu} = 0, \tag{7.11}$$

or demand that the conservation equation be satisfied and lose the trace-free nature of $T^{\mu\nu}$. The most natural choice which follows from the regularization is to keep the trace-free nature and get

$$T^{\mu\nu}{}_{;\nu} = R_{,\mu}. \tag{7.12}$$

That is, the stress–energy tensor is not conserved. However the standard attitude to this problem is to maintain conservation and allow the trace to be non-zero. This is the so-called 'conformal' or 'trace anomaly' (rather than the 'divergence anomaly').

However, if we take the other route, namely abandoning conservation, then we can interpret this equation as saying that in the quantum regime, stress–energy is created by the curvature of spacetime. This interpretation is especially interesting because this conservation equation is really all we need in two dimensions to derive the Hawking radiation. This interpretation would say that the radiation in black hole evaporation is caused by the curvature of spacetime.

The chief argument against this view is that the expectation of the stress–energy tensor is supposed to form the right side of Einstein's equations, and the left side, $G^{\mu\nu}$ satisfies the conservation law as an identity. However, if the metric $g^{\mu\nu}$ is taken as an operator, then that identity depends on the commutation of various derivatives of the metric and the metric itself: $G^{\mu\nu}$ itself does not necessarily obey the conservation law (the Bianchi identity).

This is not the place to follow this line of reasoning any further (especially since I am not sure that I have anything else to say). The suggestive point is that this interpretation would indicate that the source of the black hole radiation is curvature outside the black hole. Short-scale fluctuations in the curvature, due to the strings and perhaps D-Branes, could then affect the radiation given off by the black hole.

The problem with this locus for the memory is that it is very hard to see how this memory could be affected by the matter which formed the black hole, or which is emitted by the black hole. In particular, the matter which falls into the black hole would have to leave behind all of the quantum correlations and information which distinguishes one state from the other. In the case of the long-range fields (coupled to the mass, charges, and angular momentum), the natural equations of motion of the fields can retain this information in the external fields. However, in the case of all of the other aspects of the fields falling in, there is no known mechanism to strip each of the physical systems of all of their information before they fall into the hole. (The suggestion that, perhaps, because someone looking at matter falling into the black hole never sees it ever crossing the horizon, the information never enters the black hole either, I find very difficult to understand. It would be similar to claiming that because the fish outside the waterfall never hear the fish actually cross the sonic horizon of a dumb hole, the fish and all of the information which constitute the fish, never cross the sonic horizon either.)

We are thus left with the uncomfortable situation that if the memory is located where it can be affected, it cannot affect the out-going radiation, while if it is located where it can affect, it is not affected. In either case it is not efficacious as memory for allowing the black hole entropy to be explainable as having a statistical origin.

(iii) *Inside* and *Outside*: There is of course one other possibility, namely that the memory is located at both the centre and outside at the same time, i.e. that the memory is non-local, in two causally separated (at least at low frequencies and wavelengths) places at once (Susskind and Uglum 1994). For a solar mass black hole this would require a non-locality over distances of kilometres. The structure of the stringy world, acting on macroscopic low-energy phenomena, would have to be such that a single item could exist in two places, separated by a kilometre, and have macroscopic effects (although very subtle and long time scale) over those distances. This is very hard to accept, and would require a radical alteration of our views of how the world operates.

### 7.5 Conclusion

Black hole evaporation presents us still, twenty-five years after its discovery, with some fascinating and frustrating problems. What causes the radiation? Why do black holes behave like thermodynamic objects? What is the link, if any, between the thermodynamics of black holes and the statistical origin of all other thermodynamic systems?

Dumb holes, the sonic analogue of black holes, suggest that any ultimate understanding of the thermal radiation must concentrate on its creation as a low-energy, large length–scale, phenomenon. The radiation is created on scales of the order of the size of the black hole, and at time scales of the order of the light crossing time for the black hole, by the non-adiabatic nature of the black hole spacetime seen by the fields at such scales.

On the other hand, the only candidate we have for a statistical origin for the black hole thermodynamics, D-Branes and strings, are phenomena at the string scale (the Planck scale). If they are, as would naturally be expected, located inside the horizon, there seems no way that they could affect the low-energy radiation given off outside the horizon. If they correspond to structures lying outside the black hole, it is difficult to see how they could strip off all of the information from matter falling into the black hole.

As with all situations in which alternative explanations for a phenomenon lead to contradictory conclusions and where we do not really understand the physics, this paradox holds within itself the chance that it will uncover fundamental alterations in our view of physics. Black holes are a far from dead subject, and their understanding will lead to some crucial changes in our understanding of nature and our philosophical stance toward what a physical explanation can look like. Truly, black holes form one of the key frontiers where philosophy and physics meet at the Planck scale.

### Notes

I would like to thank a number of people with whom I have discussed these issues. In addition to my collaborators (I would especially single out Bob Wald) through the years, I would thank Gary Horowitz for explaining to me much of what little I know about string theory, both in discussions and in his papers. I also thank Lenny Susskind for being a goad. We often disagree, but I at least have learned a lot in trying to understand why I disagree. Finally, I would like to thank N. Hambli for reading earlier versions and preventing me from making at least some mistakes in string theory I would otherwise have done. (All remaining mistakes are my fault, not his.) I would thank the Canadian Institute for Advanced Research for their salary support throughout the past 15 years, and the Natural Science and Engineering Research Council for their support of the costs of my research.

1. Geroch presented a model for extracting all of the energy of a system by lowering it into a black hole in a colloquium at Princeton University in 1972.
2. 'Οχουμένωνά'. For an English translation of a Latin translation (by William of Moerbeck in 1269) see 'On Floating Bodies' in Heath (1897).
3. For an extensive analysis of the possibilities for the 'loss of information' by black holes, see the review paper by Page (1994). He makes many of the points I do (though pre D-Branes) and gives a calculation of how the correlations in the out-going radiation could contain details of the initial state, despite the apparent thermal and random character of emission process.
4. For a pedagogical (though technical) discussion of D-Branes, see Polchinski (1998).
5. Horowitz and Polchinski (1997) argue that one would expect the entropy of an excited string to (of order of magnitude) smoothly join to the entropy of a black hole as one increased the string coupling, and thus the gravity, from zero: even in the strongly coupled regime, the string entropy and the black hole entropy should be comparable.

# Part III

# Topological Quantum Field Theory

# 8     Higher-dimensional algebra and Planck scale physics

John C. Baez

## 8.1 Introduction

At present, our physical worldview is deeply schizophrenic. We have, not one, but two fundamental theories of the physical universe: general relativity, and the Standard Model of particle physics based on quantum field theory. The former takes gravity into account but ignores quantum mechanics, while the latter takes quantum mechanics into account but ignores gravity. In other words, the former recognizes that spacetime is curved but neglects the uncertainty principle, while the latter takes the uncertainty principle into account but pretends that spacetime is flat. Both theories have been spectacularly successful in their own domain, but neither can be anything more than an approximation to the truth. Clearly some synthesis is needed: at the very least, a theory of *quantum gravity*, which might or might not be part of an overarching 'theory of everything'. Unfortunately, attempts to achieve this synthesis have not yet succeeded.

Modern theoretical physics is difficult to understand for anyone outside the subject. Can philosophers really contribute to the project of reconciling general relativity and quantum field theory? Or is this a technical business best left to the experts? I would argue for the former. General relativity and quantum field theory are based on some profound insights about the nature of reality. These insights are crystallized in the form of mathematics, but there is a limit to how much progress we can make by just playing around with this mathematics. We need to go back to the insights behind general relativity and quantum field theory, learn to hold them together in our minds, and dare to imagine a world more strange, more beautiful, but ultimately more *reasonable* than our current theories of it. For this daunting task, philosophical reflection is bound to be of help.

However, a word of warning is in order. The paucity of experimental evidence concerning quantum gravity has allowed research to proceed in a rather unconstrained manner, leading to divergent schools of opinion. If one asks a string theorist about quantum gravity, one will get utterly different answers than if one asks someone working on loop quantum gravity or some other approach. To make matters worse, experts often fail to emphasize the difference between experimental results, theories supported by experiment, speculative theories that have gained a certain plausibility after years of study, and the latest fads. Philosophers must take what physicists say about quantum gravity with a grain of salt.

To lay my own cards on the table, I should say that as a mathematical physicist with an interest in philosophy, I am drawn to a strand of work that emphasizes 'higher-dimensional algebra'. This branch of mathematics goes back and reconsiders some of the presuppositions that mathematicians usually take for granted, such as the notion of equality and the emphasis on doing mathematics using one-dimensional strings of symbols. Starting in the late 1980s, it became apparent that higher-dimensional algebra is the correct language to formulate so-called 'topological quantum field theories'. More recently, various people have begun to formulate theories of quantum gravity using ideas from higher-dimensional algebra. While they have tantalizing connections to string theory, these theories are best seen as an outgrowth of loop quantum gravity.

The plan of this chapter is as follows. In Section 8.2, I begin by recalling why some physicists expect general relativity and quantum field theory to collide at the Planck length. This is a unit of distance concocted from three fundamental constants: the speed of light $c$, Newton's gravitational constant $G$, and Planck's constant $\hbar$. General relativity idealizes reality by treating Planck's constant as negligible, while quantum field theory idealizes it by treating Newton's gravitational constant as negligible. By analysing the physics of $c$, $G$, and $\hbar$, we get a glimpse of the sort of theory that would be needed to deal with situations where these idealizations break down. In particular, I shall argue that we need a *background-free quantum theory with local degrees of freedom propagating causally*.

In Section 8.3, I discuss 'topological quantum field theories'. These are the first examples of background-free quantum theories, but they lack local degrees of freedom. In other words, they describe imaginary worlds in which everywhere looks like everywhere else! This might at first seem to condemn them to the status of mathematical curiosities. However, they suggest an important analogy between the mathematics of spacetime and the mathematics of quantum theory. I argue that this is the beginning of a new bridge between general relativity and quantum field theory.

In Section 8.4, I describe one of the most important examples of a topological quantum field theory: the Turaev–Viro model of quantum gravity in three-dimensional spacetime. This theory is just a warm-up for the four-dimensional case that is of real interest in physics. Nonetheless, it has some startling features which perhaps hint at the radical changes in our worldview that a successful synthesis of general relativity and quantum field theory would require.

In Section 8.5, I discuss the role of higher-dimensional algebra in topological quantum field theory. I begin with a brief introduction to categories. Category theory can be thought of as an attempt to treat processes (or 'morphisms') on an equal footing with things (or 'objects'), and it is ultimately for this reason that it serves as a good framework for topological quantum field theory. In particular, category theory allows one to make the analogy between the mathematics of spacetime and the mathematics of quantum theory quite precise. But to fully explore this analogy one must introduce '$n$-categories', a generalization of categories that allows one to speak of processes between processes between processes ... and so on to the $n$th degree. Since $n$-categories are purely algebraic structures but have a natural relationship to the study of $n$-dimensional spacetime, their study is sometimes called 'higher-dimensional algebra'.

Finally, in Section 8.6 I briefly touch upon recent attempts to construct theories of four-dimensional quantum gravity using higher-dimensional algebra. This subject is still in its infancy. Throughout the chapter, but especially in this last section, the reader must turn to the references for details. To make the bibliography as useful as possible, I have chosen references of an expository nature whenever they exist, rather than always citing the first paper in which something was done.

## 8.2 The Planck length

Two constants appear throughout general relativity: the speed of light $c$ and Newton's gravitational constant $G$. This should be no surprise, since Einstein created general relativity to reconcile the success of Newton's theory of gravity, based on instantaneous action at a distance, with his new theory of special relativity, in which no influence travels faster than light. The constant $c$ also appears in quantum field theory, but paired with a different partner: Planck's constant $\hbar$. The reason is that quantum field theory takes into account special relativity and quantum theory, in which $\hbar$ sets the scale at which the uncertainty principle becomes important.

It is reasonable to suspect that any theory reconciling general relativity and quantum theory will involve all three constants $c$, $G$, and $\hbar$. Planck noted that apart from numerical factors there is a unique way to use these constants to define units of length, time, and mass. For example, we can define the unit of length now called the 'Planck length' as follows:

$$\ell_P = \sqrt{\frac{\hbar G}{c^3}}.$$

This is extremely small: about $1.6 \times 10^{-35}$ m. Physicists have long suspected that quantum gravity will become important for understanding physics at about this scale. The reason is very simple: any calculation that predicts a length using only the constants $c$, $G$, and $\hbar$ must give the Planck length, possibly multiplied by an unimportant numerical factor like $2\pi$.

For example, quantum field theory says that associated to any mass $m$ there is a length called its Compton wavelength, $\ell_C$, such that determining the position of a particle of mass $m$ to within one Compton wavelength requires enough energy to create another particle of that mass. Particle creation is a quintessentially quantum-field-theoretic phenomenon. Thus, we may say that the Compton wavelength sets the distance scale at which quantum field theory becomes crucial for understanding the behaviour of a particle of a given mass. On the other hand, general relativity says that associated to any mass $m$ there is a length called the Schwarzschild radius, $\ell_S$, such that compressing an object of mass $m$ to a size smaller than this results in the formation of a black hole. The Schwarzschild radius is roughly the distance scale at which general relativity becomes crucial for understanding the behaviour of an object of a given mass. Now, ignoring some numerical factors, we have

$$\ell_C = \frac{\hbar}{mc}$$

and

$$\ell_S = \frac{Gm}{c^2}.$$

These two lengths become equal when $m$ is the Planck mass. And when this happens, they both equal the Planck length!

At least naively, we thus expect that *both* general relativity and quantum field theory would be needed to understand the behaviour of an object whose mass is about the Planck mass and whose radius is about the Planck length. This not only explains some of the importance of the Planck scale, but also some of the difficulties in obtaining experimental evidence about physics at this scale. Most of our information about general relativity comes from observing heavy objects like planets and stars, for which $\ell_S \gg \ell_C$. Most of our information about quantum field theory comes from observing light objects like electrons and protons, for which $\ell_C \gg \ell_S$. The Planck mass is intermediate between these: about the mass of a largish cell. But the Planck length is about $10^{-20}$ times the radius of a proton! To study a situation where both general relativity and quantum field theory are important, we could try to compress a cell to a size $10^{-20}$ times that of a proton. We know no reason why this is impossible in principle, but we have no idea how to actually accomplish such a feat.

There are some well-known loopholes in the above argument. The 'unimportant numerical factor' I mentioned above might actually be very large, or very small. A theory of quantum gravity might make testable predictions of dimensionless quantities like the ratio of the muon and electron masses. For that matter, a theory of quantum gravity might involve physical constants other than $c$, $G$, and $\hbar$. The latter two alternatives are especially plausible if we study quantum gravity as part of a larger theory describing other forces and particles. However, even though we cannot prove that the Planck length is significant for quantum gravity, I think we can glean some wisdom from pondering the constants $c$, $G$, and $\hbar$ – and more importantly, the physical insights that lead us to regard these constants as important.

What is the importance of the constant $c$? In special relativity, what matters is the appearance of this constant in the Minkowski metric

$$ds^2 = c^2\, dt^2 - dx^2 - dy^2 - dz^2$$

which defines the geometry of spacetime, and in particular the lightcone through each point. Stepping back from the specific formalism here, we can see several ideas at work. First, space and time form a unified whole which can be thought of geometrically. Second, the quantities whose values we seek to predict are localized.

That is, we can measure them in small regions of spacetime (sometimes idealized as points). Physicists call such quantities 'local degrees of freedom'. And third, to predict the value of a quantity that can be measured in some region $R$, we only need to use values of quantities measured in regions that stand in a certain geometrical relation to $R$. This relation is called the 'causal structure' of spacetime. For example, in a relativistic field theory, to predict the value of the fields in some region $R$, it suffices to use their values in any other region that intersects every timelike path passing through $R$. The common way of summarizing this idea is to say that nothing travels faster than light. I prefer to say that a good theory of physics should have *local degrees of freedom propagating causally*.

In Newtonian gravity, $G$ is simply the strength of the gravitational field. It takes on a deeper significance in general relativity, where the gravitational field is described in terms of the curvature of the spacetime metric. Unlike in special relativity, where the Minkowski metric is a 'background structure' given *a priori*, in general relativity the metric is treated as a field which not only affects, but also is affected by, the other fields present. In other words, the geometry of spacetime becomes a local degree of freedom of the theory. Quantitatively, the interaction of the metric and other fields is described by Einstein's equation

$$G_{\mu\nu} = 8\pi G T_{\mu\nu},$$

where the Einstein tensor $G_{\mu\nu}$ depends on the curvature of the metric, while the stress–energy tensor $T_{\mu\nu}$ describes the flow of energy and momentum due to all the other fields. The role of the constant $G$ is thus simply to quantify how much the geometry of spacetime is affected by other fields. Over the years, people have realized that the great lesson of general relativity is that a good theory of physics should contain no geometrical structures that affect local degrees of freedom while remaining unaffected by them. Instead, all geometrical structures – and in particular the causal structure – should themselves be local degrees of freedom. For short, one says that the theory should be *background-free*.

The struggle to free ourselves from background structures began long before Einstein developed general relativity, and is still not complete. The conflict between Ptolemaic and Copernican cosmologies, the dispute between Newton and Leibniz concerning absolute and relative motion, and the modern arguments concerning the 'problem of time' in quantum gravity – all are but chapters in the story of this struggle. I do not have room to sketch this story here, nor even to make more precise the all-important notion of 'geometrical structure'. I can only point the reader towards the literature, starting perhaps with the books by Barbour (1989) and Earman (1989), various papers by Rovelli (1991, 1997), and the many references therein.

Finally, what of $\hbar$? In quantum theory, this appears most prominently in the commutation relation between the momentum $p$ and position $q$ of a particle:

$$pq - qp = -i\hbar,$$

together with similar commutation relations involving other pairs of measurable quantities. Because our ability to measure two quantities simultaneously with complete precision is limited by their failure to commute, $\hbar$ quantifies our inability to simultaneously know everything one might choose to know about the world. But there is far more to quantum theory than the uncertainty principle. In practice, $\hbar$ comes along with the whole formalism of complex Hilbert spaces and linear operators.

There is a widespread sense that the principles behind quantum theory are poorly understood compared to those of general relativity. This has led to many discussions about interpretational issues. However, I do not think that quantum theory will lose its mystery through such discussions. I believe the real challenge is to better understand why the mathematical formalism of quantum theory is precisely what it is. Research in quantum logic has done a wonderful job of understanding the field of candidates from which the particular formalism we use has been chosen. But what is so special about this particular choice? Why, for example, do we use complex Hilbert spaces rather than real or quaternionic ones? Is this decision made solely to fit the experimental data, or is there a deeper reason? Since questions like this do not yet have clear answers, I shall summarize the physical insight behind $\hbar$ by saying simply that a good theory of the physical universe should be a *quantum theory* – leaving open the possibility of eventually saying something more illuminating.

Having attempted to extract the ideas lying behind the constants $c$, $G$, and $\hbar$, we are in a better position to understand the task of constructing a theory of quantum gravity. General relativity acknowledges the importance of $c$ and $G$, but idealizes reality by treating $\hbar$ as negligibly small. From our discussion above, we see that this is because general relativity is a background-free classical theory with local degrees of freedom propagating causally. On the other hand, quantum field theory as normally practised acknowledges $c$ and $\hbar$ but treats $G$ as negligible, because it is a background-dependent quantum theory with local degrees of freedom propagating causally.

The most conservative approach to quantum gravity is to seek a theory that combines the best features of general relativity and quantum field theory. To do this, we must try to find a *background-free quantum theory with local degrees of freedom propagating causally*. While this approach may not succeed, it is definitely worth pursuing. Given the lack of experimental evidence that would point us towards fundamentally new principles, we should do our best to understand the full implications of the principles we already have!

From my description of the goal one can perhaps see some of the difficulties. Since quantum gravity should be background-free, the geometrical structures defining the causal structure of spacetime should themselves be local degrees of freedom propagating causally. This much is already true in general relativity. But because quantum gravity should be a quantum theory, these degrees of freedom should be treated quantum-mechanically. So at the very least, we should develop a quantum theory of some sort of geometrical structure that can define a causal structure on spacetime.

String theory has not gone far in this direction. This theory is usually formulated with the help of a metric on spacetime, which is treated as a background structure rather than a local degree of freedom like the rest. Most string theorists recognize that this is an unsatisfactory situation, and by now many are struggling towards a background-free formulation of the theory. However, in the words of two experts, 'it seems that a still more radical departure from conventional ideas about space and time may be required in order to arrive at a truly background independent formulation' (Helling and Nicolai 1998).

Loop quantum gravity has gone a long way towards developing a background-free quantum theory of the geometry of space (Ashtekar 1999, Rovelli 1998b), but not so far when it comes to spacetime. This has made it difficult to understand dynamics, and in particular the causal propagation of degrees of freedom. Work in earnest on these issues has begun only recently. One reason for optimism is the recent success in understanding quantum gravity in three spacetime dimensions. But to explain this, I must first say something about topological quantum field theory.

### 8.3 Topological quantum field theory

Besides general relativity and quantum field theory as usually practised, a third sort of idealization of the physical world has attracted a great deal of attention during the past decade. These are called topological quantum field theories, or 'TQFTs'. In the terminology of the previous section, a TQFT is a *background-free quantum theory with no local degrees of freedom.*[1]

A good example is quantum gravity in three-dimensional spacetime. First let us recall some features of *classical* gravity in three-dimensional spacetime. Classically, Einstein's equations predict qualitatively very different phenomena depending on the dimension of spacetime. If spacetime has four or more dimensions, Einstein's equations imply that the metric has local degrees of freedom. In other words, the curvature of spacetime at a given point is not completely determined by the flow of energy and momentum through that point: it is an independent variable in its own right. For example, even in the vacuum, where the energy–momentum tensor vanishes, localized ripples of curvature can propagate in the form of gravitational radiation. In three-dimensional spacetime, however, Einstein's equations suffice to completely determine the curvature at a given point of spacetime in terms of the flow of energy and momentum through that point. We thus say that the metric has no local degrees of freedom. In particular, in the vacuum the metric is flat, so every small patch of empty spacetime looks exactly like every other.

The absence of local degrees of freedom makes general relativity far simpler in three-dimensional spacetime than in higher dimensions. Perhaps surprisingly, it is still somewhat interesting. The reason is the presence of 'global' degrees of freedom. For example, if we chop a cube out of flat three-dimensional Minkowski space and form a three-dimensional torus by identifying the opposite faces of this cube, we get a spacetime with a flat metric on it, and thus a solution of the vacuum Einstein equations. If we do the same starting with a larger cube, or a parallelipiped, we get a different spacetime that also satisfies the vacuum Einstein equations. The two spacetimes are *locally* indistinguishable, since locally both look just like flat Minkowski spacetime. However, they can be distinguished *globally* – for example, by measuring the volume of the whole spacetime, or studying the behaviour of geodesics that wrap all the way around the torus.

Since the metric has no local degrees of freedom in three-dimensional general relativity, this theory is much easier to quantize than the physically relevant four-dimensional case. In the simplest situation, where we consider 'pure' gravity without matter, we obtain a background-free quantum field theory with no local degrees of freedom whatsoever: a TQFT.

I shall say more about three-dimensional quantum gravity in Section 8.4. To set the stage, let me sketch the axiomatic approach to topological quantum field theory proposed by Atiyah (1990). My earlier definition of a TQFT as a 'background-free quantum field theory with no local degrees of freedom' corresponds fairly well to how physicists think about TQFTs. But mathematicians who wish to prove theorems about TQFTs need to start with something more precise, so they often use Atiyah's axioms.

An important feature of TQFTs is that they do not presume a fixed topology for space or spacetime. In other words, when dealing with an $n$-dimensional TQFT, we are free to choose any $(n-1)$-dimensional manifold to represent space at a given time[2]. Moreover given two such manifolds, say $S$ and $S'$, we are free to choose any $n$-dimensional manifold $M$ to represent the portion of spacetime between $S$ and $S'$, as long as the boundary of $M$ is the union of $S$ and $S'$. Mathematicians call $M$ a 'cobordism' from $S$ to $S'$. We write $M : S \rightarrow S'$, because we may think of $M$ as the process of time passing from the moment $S$ to the moment $S'$.

For example, in Fig. 8.1 we depict a two-dimensional manifold $M$ going from a one-dimensional manifold $S$ (a pair of circles) to a one-dimensional manifold $S'$ (a single circle). Crudely speaking, $M$ represents a process in which two separate spaces collide to form a single one! This may seem outré, but these days physicists are quite willing to speculate about processes in which the topology of space changes with the passage of time. Other forms of topology change include the formation of a wormhole, the appearance of the universe in a 'big bang', or its disappearance in a 'big crunch'.

There are various important operations that one can perform on cobordisms, but I will only describe two. First, we may 'compose' two cobordisms $M : S \rightarrow S'$ and $M' : S' \rightarrow S''$, obtaining a cobordism $M'M : S \rightarrow S''$, as illustrated in Fig. 8.2. The idea here is that the passage of time corresponding to $M$ followed by the passage of time corresponding to $M'$ equals the passage of time corresponding to $M'M$. This is analogous to the familiar idea that waiting $t$ seconds followed by waiting $t'$ seconds is the same as waiting $t + t'$ seconds. The big difference is that in topological quantum

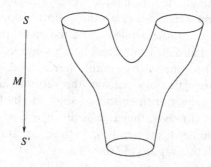

Fig. 8.1. A cobordism.

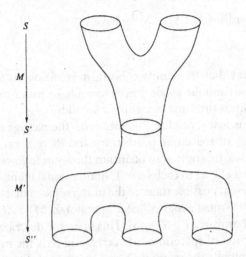

Fig. 8.2. Composition of cobordisms.

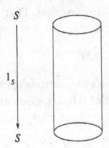

Fig. 8.3. An identity cobordism.

field theory we cannot measure time in seconds, because there is no background metric available to let us count the passage of time! We can only keep track of topology change. Just as ordinary addition is associative, composition of cobordisms satisfies the associative law:

$$(M''M')M = M''(M'M).$$

However, composition of cobordisms is not commutative. As we shall see, this is related to the famous non-commutativity of observables in quantum theory.

Second, for any $(n-1)$-dimensional manifold $S$ representing space, there is a cobordism $1_S : S \rightarrow S$ called the 'identity' cobordism, which represents a passage of time without any topology change. For example, when $S$ is a circle, the identity cobordism $1_S$ is a cylinder, as shown in Fig. 8.3. In general, the identity cobordism $1_S$ has the property that for any cobordism $M : S' \rightarrow S$ we have

$$1_S M = M,$$

185

while for any cobordism $M : S \to S'$ we have

$$M1_S = M.$$

These properties say that an identity cobordism is analogous to waiting 0 seconds: if you wait 0 seconds and then wait $t$ more seconds, or wait $t$ seconds and then wait 0 more seconds, this is the same as waiting $t$ seconds.

These operations just formalize the notion of 'the passage of time' in a context where the topology of spacetime is arbitrary and there is no background metric. Atiyah's axioms relate this notion to quantum theory as follows. First, a TQFT must assign a Hilbert space $Z(S)$ to each $(n-1)$-dimensional manifold $S$. Vectors in this Hilbert space represent possible states of the universe given that space is the manifold $S$. Second, the TQFT must assign a linear operator $Z(M) : Z(S) \to Z(S')$ to each $n$-dimensional cobordism $M : S \to S'$. This operator describes how states change given that the portion of spacetime between $S$ and $S'$ is the manifold $M$. In other words, if space is initially the manifold $S$ and the state of the universe is $\psi$, after the passage of time corresponding to $M$ the state of the universe will be $Z(M)\psi$.

In addition, the TQFT must satisfy a list of properties. Let me just mention two. First, the TQFT must preserve composition. That is, given cobordisms $M : S \to S'$ and $M' : S' \to S''$, we must have

$$Z(M'M) = Z(M')Z(M),$$

where the right-hand side denotes the composite of the operators $Z(M)$ and $Z(M')$. Second, it must preserve identities. That is, given any manifold $S$ representing space, we must have

$$Z(1_S) = 1_{Z(S)},$$

where the right-hand side denotes the identity operator on the Hilbert space $Z(S)$.

Both these axioms are eminently reasonable if one ponders them a bit. The first says that the passage of time corresponding to the cobordism $M$ followed by the passage of time corresponding to $M'$ has the same effect on a state as the combined passage of time corresponding to $M'M$. The second says that a passage of time in which no topology change occurs has no effect at all on the state of the universe. This seems paradoxical at first, since it seems we regularly observe things happening even in the absence of topology change. However, this paradox is easily resolved: a TQFT describes a world quite unlike ours, one without local degrees of freedom. In such a world, nothing local happens, so the state of the universe can only change when the topology of space itself changes.[3]

The most interesting thing about the TQFT axioms is their common formal character. Loosely speaking, they all say that a TQFT maps structures in differential topology – by which I mean the study of manifolds – to corresponding structures in quantum theory. In devising these axioms, Atiyah took advantage of a powerful analogy between differential topology and quantum theory, summarized in Table 8.1.

I shall explain this analogy between differential topology and quantum theory further in Section 8.5. For now, let me just emphasize that this analogy is exactly the

Table 8.1. Analogy between differential topology and quantum theory

| Differential topology | Quantum theory |
| --- | --- |
| $(n-1)$-dimensional manifold (space) | Hilbert space (states) |
| cobordism between $(n-1)$-dimensional manifolds (spacetime) | operator (process) |
| composition of cobordisms | composition of operators |
| identity cobordism | identity operator |

sort of clue we should pursue for a deeper understanding of quantum gravity. At first glance, general relativity and quantum theory look very different mathematically: one deals with space and spacetime, the other with Hilbert spaces and operators. Combining them has always seemed a bit like mixing oil and water. But topological quantum field theory suggests that perhaps they are not so different after all! Even better, it suggests a concrete programme of synthesizing the two, which many mathematical physicists are currently pursuing. Sometimes this goes by the name of 'quantum topology' (Baadhio and Kauffman 1993, Turaev 1994).

Quantum topology is very technical, as anything involving mathematical physicists inevitably becomes. But if we stand back a moment, it should be perfectly obvious that differential topology and quantum theory must merge if we are to understand background-free quantum field theories. In physics that ignores general relativity, we treat space as a background on which states of the world are displayed. Similarly, we treat spacetime as a background on which the process of change occurs. But these are idealizations which we must overcome in a background-free theory. In fact, the concepts of 'space' and 'state' are two aspects of a unified whole, and likewise for the concepts of 'spacetime' and 'process'. It is a challenge, not just for mathematical physicists, but also for philosophers, to understand this more deeply.

## 8.4 Three-dimensional quantum gravity

Before the late 1980s, quantum gravity was widely thought to be just as intractable in three spacetime dimensions as in the physically important four-dimensional case. The situation changed drastically when physicists and mathematicians developed the tools for handling background-free quantum theories without local degrees of freedom. By now, it is easier to give a complete description of three-dimensional quantum gravity than most quantum field theories of the traditional sort!

Let me sketch how one sets up a theory of three-dimensional quantum gravity satisfying Atiyah's axioms for a TQFT. Before doing so, I should warn the reader that there are a number of inequivalent theories of three-dimensional quantum gravity (Carlip 1998). The one I shall describe is called the Turaev–Viro model (Turaev 1994). While in some ways this is not the most physically realistic one, since it is a quantum theory of Riemannian rather than Lorentzian metrics, it illustrates the points I want to make here.

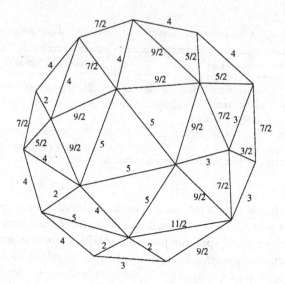

Fig. 8.4. A state in the preliminary Hilbert space for three-dimensional quantum gravity.

To get a TQFT satisfying Atiyah's axioms we need to describe a Hilbert space of states for each two-dimensional manifold and an operator for each cobordism between two-dimensional manifolds. We begin by constructing a preliminary Hilbert space $\tilde{Z}(S)$ for any two-dimensional manifold $S$. This construction requires choosing a background structure: a way of chopping $S$ into triangles. Later we will eliminate this background-dependence and construct the Hilbert space of real physical interest.

To define the Hilbert space $\tilde{Z}(S)$, it is enough to specify an orthonormal basis for it. We decree that states in this basis are ways of labelling the edges of the triangles in $S$ by numbers of the form $0, \frac{1}{2}, 1, \frac{3}{2}, \ldots, k/2$. An example is shown in Fig. 8.4, where we take $S$ to be a sphere. Physicists call the numbers labelling the edges 'spins', alluding to the fact that we are using mathematics developed in the study of angular momentum. But here these numbers represent the *lengths* of the edges as measured in units of the Planck length. In this theory, length is a discrete rather than continuous quantity!

We then construct an operator $\tilde{Z}(M) : \tilde{Z}(S) \rightarrow \tilde{Z}(S')$ for each cobordism $M : S \rightarrow S'$. Again, we do this with the help of a background structure on $M$: we choose a way to chop it into tetrahedra, whose triangular faces must include among them the triangles of $S$ and $S'$. To define $\tilde{Z}(M)$ it is enough to specify the transition amplitudes $\langle \psi', \tilde{Z}(M)\psi \rangle$ when $\psi$ and $\psi'$ are states in the bases given above. We do this as follows. The states $\psi$ and $\psi'$ tell us how to label the edges of triangles in $S$ and $S'$ by spins. Consider any way to label the edges of $M$ by spins that is compatible with these labellings of edges in $S$ and $S'$. We can think of this as a 'quantum geometry' for spacetime, since it tells us the shape of every tetrahedron in $M$. Using a certain recipe we can compute a complex number for this geometry, which we think of as its 'amplitude' in the quantum-mechanical sense. We then sum these amplitudes over all such geometries to get the total transition amplitude from $\psi$ to $\psi'$. The reader familiar with quantum field theory may note that this construction is a discrete version of a 'path integral'.

Now let me describe how we erase the background-dependence from this construction. Given an identity cobordism $1_S : S \to S$, the operator $\tilde{Z}(1_S)$ is usually not the identity, thus violating one of Atiyah's axioms for a topological quantum field theory. However, the next best thing happens: this operator maps $\tilde{Z}(S)$ onto a subspace, and it acts as the identity on this subspace. This subspace, which we call $Z(S)$, is the Hilbert space of real physical interest in three-dimensional quantum gravity. Amazingly, this subspace does not depend on how we chopped $S$ into triangles. Even better, for any cobordism $M : S \to S'$, the operator $\tilde{Z}(M)$ maps $Z(S)$ to $Z(S')$. Thus, it restricts to an operator $Z(M) : Z(S) \to Z(S')$. Moreover, this operator $Z(M)$ turns out not to depend on how we chopped $M$ into tetrahedra. To top it all off, it turns out that the Hilbert spaces $Z(S)$ and operators $Z(M)$ satisfy Atiyah's axioms.

In short, we started by chopping space into triangles and spacetime into tetrahedra, but at the end of the day nothing depends on this choice of background structure. It also turns out that the final theory has no local degrees of freedom: all the measurable quantities are global in character. For example, there is no operator on $Z(S)$ corresponding to the 'length of a triangle's edge', but there is an operator corresponding to the length of the shortest geodesic wrapping around the space $S$ in a particular way. Even better, if we take this theory and consider the limit as $\hbar \to 0$, we recover the classical version of Einstein's equations for Riemannian metrics in three dimensions. These miracles are among the main reasons for interest in quantum topology. They only happen because of the carefully chosen recipe for computing amplitudes for spacetime geometries. This recipe is the real core of the whole construction. Sadly, it is a bit too technical to describe here, so the reader will have to turn elsewhere for details (Kauffman 1993, Turaev 1994). I can say this, though: the reason this recipe works so well is that it neatly combines ideas from general relativity, quantum field theory, and a third subject that might at first seem unrelated – higher-dimensional algebra.

### 8.5 Higher-dimensional algebra

One of the most remarkable accomplishments of the early twentieth century was to formalize all of mathematics in terms of a language with a deliberately impoverished vocabulary: the language of set theory. In Zermelo–Fraenkel set theory, everything is a set, the only fundamental relationships between sets are membership and equality, and two sets are equal if and only if they have the same elements. If in Zermelo–Fraenkel set theory you ask what sort of thing is the number $\pi$, the relationship 'less than', or the exponential function, the answer is always the same: a set! Of course one must bend over backwards to think of such varied entities as sets, so this formalization may seem almost deliberately perverse. However, it represents the culmination of a worldview in which things are regarded as more fundamental than processes or relationships.

More recently, mathematicians have developed a somewhat more flexible language, the language of category theory. Category theory is an attempt to put processes and relationships on an equal status with things. A category consists of a collection of 'objects', and for each pair of objects $x$ and $y$, a collection of 'morphisms' from $x$ to $y$. We write a morphism from $x$ to $y$ as $f : x \rightarrow y$. We demand that for any morphisms $f : x \rightarrow y$ and $g : y \rightarrow z$, we can 'compose' them to obtain a morphism $gf : x \rightarrow z$. We also demand that composition be associative. Finally, we demand that for any object $x$ there be a morphism $1_x$, called the 'identity' of $x$, such that $f1_x = f$ for any morphism $f : x \rightarrow y$ and $1_x g = g$ for any morphism $g : y \rightarrow x$.

Perhaps the most familiar example of a category is Set. Here, the objects are sets and the morphisms are functions between sets. However, there are many other examples. Fundamental to quantum theory is the category Hilb. Here, the objects are complex Hilbert spaces and the morphisms are linear operators between Hilbert spaces. In Section 8.3 we also met a category which is important in differential topology, the category $n$Cob. Here, the objects are $(n - 1)$-dimensional manifolds and the morphisms are cobordisms between such manifolds. Note that in this example, the morphisms are not functions! Nonetheless, we can still think of them as 'processes' going from one object to another.

An important part of learning category theory is breaking certain habits that one may have acquired from set theory. For example, in category theory one must resist the temptation to 'peek into the objects'. Traditionally, the first thing one asks about a set is: what are its elements? A set is like a container, and the contents of this container are the most interesting thing about it. But in category theory, an object need not have 'elements' or any sort of internal structure. Even if it does, this is not what really matters! What really matters about an object is its morphisms to and from other objects. Thus, category theory encourages a relational worldview in which things are described, not in terms of their constituents, but by their relationships to other things.

Category theory also downplays the importance of equality between objects. Given two elements of a set, the first thing one asks about them is: are they equal? But for objects in a category, we should ask instead whether they are isomorphic. Technically, the objects $x$ and $y$ are said to be 'isomorphic' if there is a morphism $f : x \rightarrow y$ that has an 'inverse': a morphism $f^{-1} : y \rightarrow x$ for which $f^{-1}f = 1_x$ and $ff^{-1} = 1_y$. A morphism with an inverse is called an 'isomorphism'. An isomorphism between two objects lets turn any morphism to or from one of them into a morphism to or from the other in a reversible sort of way. Since what matters about objects are their morphisms to and from other objects, specifying an isomorphism between two objects lets us treat them as 'the same' for all practical purposes.

Categories can be regarded as higher-dimensional analogues of sets. As shown in Fig. 8.5, we may visualize a set as a bunch of points, namely its elements. Similarly, we may visualize a category as a bunch of points corresponding to its objects, together with a bunch of one-dimensional arrows corresponding to its morphisms. (For simplicity, I have not drawn the identity morphisms in Fig. 8.5.)

We may use the analogy between sets and categories to 'categorify' almost any set-theoretic concept, obtaining a category-theoretic counterpart (Baez and Dolan 1998). For example, just as there are functions between sets, there are 'functors' between categories. A function from one set to another sends each element of the

Table 8.2. Analogy between set theory and category theory

| Set theory | Category theory |
| --- | --- |
| elements | objects |
| equations between elements | isomorphisms between objects |
| sets | categories |
| functions between sets | functors between categories |
| equations between functions | natural isomorphisms between functors |

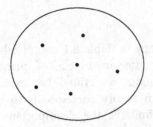 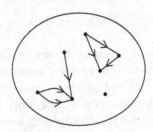

Fig. 8.5. A set and a category.

first to an element of the second. Similarly, a functor $F$ from one category to another sends each object $x$ of the first to an object $F(x)$ of the second, and also sends each morphism $f : x \rightarrow y$ of the first to a morphism $F(f) : F(x) \rightarrow F(y)$ of the second. In addition, functors are required to preserve composition and identities:

$$F(f'f) = F(f')F(f)$$

and

$$F(1_x) = 1_{F(x)}.$$

Functors are important because they allow us to apply the relational worldview discussed above, not just to objects in a given category, but to categories themselves. Ultimately what matters about a category is not its 'contents' – its objects and morphisms – but its functors to and from other categories!

We summarize the analogy between set theory and category theory in Table 8.2. In addition to the terms already discussed, there is a concept of 'natural isomorphism' between functors. This is the correct analogue of an equation between functions, but we will not need it here – I include it just for the sake of completeness.

The full impact of category-theoretic thinking has taken a while to be felt. Categories were invented in the 1940s by Eilenberg and MacLane for the purpose of clarifying relationships between algebra and topology. As time passed, they became increasingly recognized as a powerful tool for exploiting analogies throughout mathematics (MacLane 1988). In the early 1960s they led to revolutionary – and still controversial – developments in mathematical logic (Goldblatt 1979). It gradually became clear that category theory was a part of a deeper subject, 'higher-dimensional algebra' (Brown 1992), in which the concept of a category is generalized to that of an

'$n$-category'. But only by the 1990s did the real importance of categories for physics become evident, with the discovery that higher-dimensional algebra is the perfect language for topological quantum field theory (Lawrence 1993).

Why are categories important in topological quantum field theory? The most obvious answer is that a TQFT is a functor. Recall from Section 8.3 that a TQFT maps each manifold $S$ representing space to a Hilbert space $Z(S)$ and each cobordism $M : S \rightarrow S'$ representing spacetime to an operator $Z(M) : Z(S) \rightarrow Z(S')$, in such a way that composition and identities are preserved. We may summarize all this by saying that a TQFT is a functor

$$Z : n\text{Cob} \rightarrow \text{Hilb}.$$

In short, category theory makes the analogy in Table 8.1 completely precise. In terms of this analogy, many somewhat mysterious aspects of quantum theory correspond to easily understood facts about spacetime! For example, the non-commutativity of operators in quantum theory corresponds to the non-commutativity of composing cobordisms. Similarly, the all-important 'adjoint' operation in quantum theory, which turns an operator $A : H \rightarrow H'$ into an operator $A^* : H' \rightarrow H$, corresponds to the operation of reversing the roles of past and future in a cobordism $M : S \rightarrow S'$, obtaining a cobordism $M^* : S' \rightarrow S$.

But the role of category theory goes far beyond this. The real surprise comes when one examines the details of specific TQFTs. In Section 8.4 I sketched the construction of three-dimensional quantum gravity, but I left out the recipe for computing amplitudes for spacetime geometries. Thus, the most interesting features of the whole business were left as unexplained 'miracles': the background-independence of the Hilbert spaces $Z(S)$ and operators $Z(M)$, and the fact that they satisfy Atiyah's axioms for a TQFT. In fact, the recipe for amplitudes and the verification of these facts make heavy use of category theory. The same is true for all other theories for which Atiyah's axioms have been verified. For some strange reason, it seems that category theory is precisely suited to explaining what makes a TQFT tick.

For the past ten years or so, various researchers have been trying to understand this more deeply. Much remains mysterious, but it now seems that TQFTs are intimately related to category theory because of special properties of the category $n$Cob. While $n$Cob is defined using concepts from differential topology, a great deal of evidence suggests that it admits a simple description in terms of '$n$-categories'.

I have already alluded to the concept of 'categorification' – the process of replacing sets by categories, functions by functors and so on, as indicated in Table 8.2. The concept of '$n$-category' is obtained from the concept of 'set' by categorifying it $n$ times! An $n$-category has objects, morphisms between objects, 2-morphisms between morphisms, and so on up to $n$-morphisms, together with various composition operations satisfying various reasonable laws (Baez 1997). Increasing the value of $n$ allows an ever more nuanced treatment of the notion of 'sameness'. A 0-category is just a set, and in a set the elements are simply equal or unequal. A 1-category is a category, and in this context we may speak not only of equal but also of isomorphic objects. Unfortunately, this careful distinction between equality and isomorphism breaks down when we study the morphisms. Morphisms in a category are either the same or different; there is no concept of isomorphic morphisms. In a 2-category this is remedied by introducing 2-morphisms between morphisms. Unfortunately, in a 2-category we cannot speak of isomorphic 2-morphisms. To remedy this we must introduce the notion of 3-category, and so on.

We may visualize the objects of an $n$-category as points, the morphisms as arrows going between these points, the 2-morphisms as two-dimensional surfaces going between these arrows, and so on. There is thus a natural link between $n$-categories and $n$-dimensional topology. Indeed, one reason why $n$-categories are somewhat formidable is that calculations with them are most naturally done using $n$-dimensional diagrams. But this link between $n$-categories and $n$-dimensional topology is precisely why there may be a nice description of $n$Cob in the language of $n$-categories.

Dolan and I have proposed such a description, which we call the 'cobordism hypothesis' (Baez and Dolan 1995). Much work remains to be done to make this hypothesis precise and prove or disprove it. Proving it would lay the groundwork for understanding topological quantum field theories in a systematic way. But beyond this, it would help us towards a purely algebraic understanding of 'space' and 'space-time' – which is precisely what we need to marry them to the quantum-mechanical notions of 'state' and 'process'.

## 8.6 Four-dimensional quantum gravity

How important are the lessons of topological quantum field theory for four-dimensional quantum gravity? This is still an open question. Since TQFTs lack local degrees of freedom, they are at best a warm-up for the problem we really want to tackle: constructing a background-free quantum theory with local degrees of freedom propagating causally. Thus, even though work on TQFTs has suggested new ideas linking quantum theory and general relativity, these ideas may be too simplistic to be useful in real-world physics.

However, physics is not done by sitting on one's hands and pessimistically pondering the immense magnitude of the problems. For decades, our only insights into quantum gravity came from general relativity and quantum field theory on space-time with a fixed background metric. Now we can view it from a third angle, that of topological quantum field theory. Surely it makes sense to invest some effort in trying to combine the best aspects of all three theories!

And indeed, during the past few years various people have begun to do just this, largely motivated by tantalizing connections between topological quantum field theory and loop quantum gravity. In loop quantum gravity, the preliminary Hilbert space has a basis given by 'spin networks' – roughly speaking, graphs with edges labelled by spins (Baez 1996, Smolin 1997). We now understand quite well how a spin network describes a quantum state of the geometry of space. But spin networks are also used to describe states in TQFTs, where they arise naturally from considerations of higher-dimensional algebra. For example, in three-dimensional quantum gravity the state shown in Fig. 8.4 can also be described using the spin network shown in Fig. 8.6.

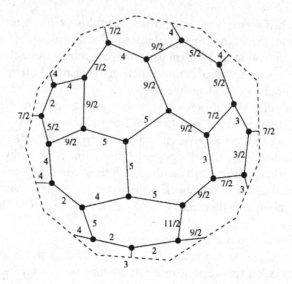

Fig. 8.6. A spin network.

Using the relationships between four-dimensional quantum gravity and topological quantum field theory, researchers have begun to formulate theories in which the quantum geometry of spacetime is described using 'spin foams' – roughly speaking, two-dimensional structures made of polygons joined at their edges, with all the polygons being labelled by spins (Baez 1998, Barrett and Crane 1998, Freidel and Krasnov 1998, Reisenberger 1996, Reisenberger and Rovelli 1997). The most important part of a spin foam model is a recipe assigning an amplitude to each spin foam. Much as Feynman diagrams in ordinary quantum field theory describe processes by which one collection of particles evolves into another, spin foams describe processes by which one spin network evolves into another. Indeed, there is a category whose objects are spin networks and whose morphisms are spin foams! And like nCob, this category appears to arise very naturally from purely n-categorical considerations.

In the most radical approaches, the concepts of 'space' and 'state' are completely merged in the notion of 'spin network', and similarly the concepts of 'spacetime' and 'process' are merged in the notion of 'spin foam', eliminating the scaffolding of a spacetime manifold entirely. To me, at least, this is a very appealing vision. However, there are a great many obstacles to overcome before we have a full-fledged theory of quantum gravity along these lines. Let me mention just a few of the most pressing. First, there is the problem of developing quantum theories of Lorentzian rather than Riemannian metrics. Second, and closely related, we need to better understand the concept of 'causal structure' in the context of spin foam models. Only the work of Markopoulou and Smolin (1998) has addressed this point so far. Third, there is the problem of formulating physical questions in these theories in such a way that divergent sums are eliminated. And fourth, there is the problem of developing computational techniques to the point where we can check whether these theories approximate general relativity in the limit of large distance scales – i.e. distances much greater than the Planck length. Starting from familiar territory we have sailed into strange new waters, but only if we circle back to the physics we know will the journey be complete.

### Notes

Conversations and correspondence with many people have helped form my views on these issues. I cannot list them all, but I especially want to thank Abhay Ashtekar, John Barrett, Louis Crane, James Dolan, Louis Kauffman, Kirill Krasnov, Carlo Rovelli, and Lee Smolin.

1. It would be nicely symmetrical if TQFTs involved the constants $G$ and $\hbar$ but not $c$. Unfortunately I cannot quite see how to make this idea precise.
2. Here and in what follows, by 'manifold' I really mean 'compact oriented smooth manifold', and cobordisms between these will also be compact, oriented, and smooth.
3. Actually, while perfectly correct as far as it goes, this resolution dodges an important issue. Some physicists have suggested that the second axiom may hold even in quantum field theories *with* local degrees of freedom, so long as they are background-free (Barrett 1995). Unfortunately, a discussion of this would take us too far afield here.

# Part IV

# Quantum Gravity and
# the Interpretation of General Relativity

# 9    On general covariance and best matching

Julian B. Barbour

## 9.1 Introduction

This chapter addresses issues raised by Christian (Chapter 14), Penrose (Chapter 13), and others in this volume: What is the physical significance, if any, of general covariance? Norton (1993) has given a valuable historical survey of the tangled history of this question. His subtitle 'Eight decades of dispute' is very apt! Many people attribute the difficulties inherent in the quantization of gravity to the general covariance of Einstein's general theory of relativity, so clarification of its true nature is important.

Christian gives a clear account of what may be called the current orthodoxy with regard to the status of general covariance, and I disagree with little of what he says. However, in my view that merely draws attention to a problem – it does not say how the problem is overcome. I believe general relativity does resolve the problem, but that this has escaped notice. If this view is correct – and most of this chapter will be devoted to arguing that it is – then I believe it has important implications for the problem that Penrose is trying to solve in Chapter 13. I think he is trying to solve a problem that has already been solved. Nevertheless, I do feel all attempts to construct viable models of physical collapse are valuable, since they constitute one of the few alternatives to many-worlds interpretations of quantum mechanics. Though I personally incline to such interpretations, it is always important to seek alternatives. I also welcome Christian's attempt to give concrete form to the notion of transience, since I think an explicit theory of it would be most interesting.

## 9.2 Einstein's understanding of general covariance

### 9.2.1 The historical background

Up to and even after he created general relativity, Einstein was convinced that the requirement of general covariance had deep physical significance. At the same time, he got into quite a muddle with his hole argument. The argument and its resolution have been discussed by many people since Stachel brought it to widespread notice (Stachel 1989, 1993). Christian gives an excellent historical account of the hole argument, but ends with Einstein's recognition of his error and realization that general covariance is not incompatible with unique statements of the physical content of a generally covariant theory. However, to form an accurate picture of the complete question we need to consider some of the earlier history and some of the later history.

The root of the issue goes back to Newton's famous scholium on absolute and relative motion and his failure to make good a promise at its end ('For it was to this end that I composed it [the *Principia*]'). The promise was to show how his absolute motions could be deduced from observed relative motions. Ironically, this problem was completely and most beautifully solved in 1771 for the three-body problem in a celebrated prize-winning essay by Lagrange. Lagrange showed how Newton's equations could be reformulated in such a way that they contained only the time and the relative co-ordinates of the system (the three sides of the triangle formed by the three gravitationally interacting bodies). The most significant thing about his equations is that they contain two constants of the motion, the total energy $E$ and the square of the intrinsic angular momentum $M^2$. In addition, one of his equations contains the third derivatives of the sides of the triangle with respect to the time, while the two remaining equations contain only second derivatives. Lagrange also showed how, from knowledge of only the relative quantities (the sides of the triangle), the co-ordinates in absolute space (or rather, in the center-of-mass inertial frame of reference) could be deduced. Dziobek (1888) gives the best account of this work that I know.

Ironically, Lagrange was not much interested in this aspect of his work, which has considerable implications for the absolute–relative debate that raged in Newton's time. Eighty-five years on from the publication of the *Principia*, the existence of absolute space was no longer a contentious issue. Even more ironic was the subsequent history. In 1872 and again in 1883, Mach raised very forcibly the issue of whether absolute space existed and through his writings prompted Ludwig Lange in Germany to attack the problem of how inertial frames of reference could be determined from known relative data (Lange 1884). Independently, James Thomson (the brother of Lord Kelvin) and Peter Tait addressed the same problem in Edinburgh (Thomson 1883, Tait 1883). None of these authors seems to have been aware of Lagrange's work, or at least aware of its significance for their proposals. The point is that they only attempted to establish the inertial frames of reference from the motions of mass points known to be moving inertially, whereas Lagrange had already completely solved the vastly more difficult problem with interacting particles! The solutions found by Thomson and Lange are mechanical and ungainly, whereas Tait's is elegant and recommended to the reader. It is unfortunate that his paper passed almost completely unnoticed, because he showed how not only the absolute motions but also the absolute times could be constructed from the relative motions of three inertial particles. In contrast, by sheer persistence Lange ensured that his work became reasonably widely known, and he is rightly credited with introducing the concept of an inertial system (or inertial frame of reference). Mach brought his work to wide attention (Mach 1960).

It is nevertheless a strange fact that most theoretical physicists simply adopted the notion of an inertial system as, so to speak, God given and never bothered themselves with the details of Lange's construction, which contains some subtle issues. Among the relativists of stature, only von Laue (von Laue 1951) drew attention to Lange's achievement. In a late letter to von Laue, Einstein argued (incorrectly) that Lange's contribution was trivial (Einstein 1952a).

The most important fact that emerges from Tait's superior treatment is as follows. If one knows *three* instantaneous relative configurations of $n$ inertially moving point masses at certain unknown times, then (up to the unit of time) one can (if $n > 5$) determine the family of inertial frames in which they satisfy Newton's laws of motion and construct a clock that measures absolute time. For $n = 3$ and 4, more than three configurations are needed. But the really important thing is that two relative configurations are never enough to solve the Lange–Tait problem. This exactly matches the fact that one of Lagrange's equations must necessarily contain a third time derivative. The underlying reason for this state of affairs is that two relative configurations give no information about the rotational state of the system and so contain no information about the angular momentum, which however has a profound effect on the dynamics.

### 9.2.2 Einstein and the determination of inertial frames

For reasons that I shall not go into here (see Barbour 1999, 1999a, in preparation), when Einstein (1905) and Minkowski (1908) created special relativity they both completely side-stepped the issue of how inertial frames of reference can be determined from observable relative quantities. They simply assumed that they *already* had access to one. The central point of their papers is that, if one already has one such frame of reference, then there exist transformation laws leading to other similar frames of reference moving uniformly relatively to the first in which the laws of nature take the identical form. Einstein says virtually nothing (in any of his papers) about how one is to obtain the all-important first inertial frame of reference. Minkowski says a bit more:

> 'From the totality of natural phenomena it is possible, by successively enhanced approximations, to derive more and more exactly a system of reference $x, y, z, t$, space and time, by means of which these phenomena then present themselves in agreement with definite laws.'

He then points out that one such reference system is by no means uniquely determined and that there are transformations that lead from it to a whole family of others, in all of which the laws of nature take the same form. However, he never says what he means by the totality of natural phenomena or what steps must be taken to perform the envisaged successive approximations. But how is it done? This is a perfectly reasonable question to ask. We are told how to get from one such reference system to another, but not how to find the first one. Had either Einstein or Minkowski asked this question explicitly and gone through the steps that must be taken, the importance of extended relative configurations of matter would have become apparent. (When bodies interact, determination of the inertial frame of reference involves in principle all matter in the universe

(Barbour, in preparation). This will be important later in the discussion of Penrose's proposal.)

The curious circumstance that two such configurations are never sufficient to find the inertial frames might have made them more aware of another issue. For the fact is that, rather remarkably, Einstein never said what he meant by the laws of nature. How are they to be expressed? However, it is implicit in the manner in which he proceeded that he took them to have the form of differential equations formulated in any of the allowed inertial frames of reference. But what determines the inertial frames of reference? Does some law of nature do that? If so, what form does it take? Einstein's persistent failure to ask let alone answer these questions directly has been the cause of great confusion.

### 9.2.3 Einstein, Kretschmann, and the relativity principle

Omitting again the reasons (given in Barbour 1999, 1999a, in preparation) for his approach, let me simply point out that Einstein believed that he would arrive at a satisfactory theory of gravitation and inertia if he could generalize his restricted principle of relativity to encompass absolutely all possible co-ordinate transformations. To the arguments in support of this standpoint that he had advanced prior to the autumn of 1915 (when he saw the flaw in his hole argument), he then added another based on the argument he had correctly used to defuse it. I need to give his argument in full (Einstein 1916):

> 'All our space-time verifications invariably amount to a determination of space-time coincidences. If, for example, events consisted merely in the motion of material points, then ultimately nothing would be observable but the meetings of two or more of these points. Moreover, the results of our measurings are nothing but verifications of such meetings of the material points of our measuring instruments with other material points, coincidences between the hands of a clock and points on the clock dial, and observed point-events happening at the same place at the same time. The introduction of a system of reference serves no other purpose than to facilitate the description of the totality of such coincidences.'

It was this argument that Kretschmann picked up. What Einstein says above about the nature of (objective) verifications is unquestionably true. It is a fact of life that the physics is in the coincidences, not in any co-ordinates that we may choose to lay out over them. It follows from this that general covariance is not a physical principle, but a formal necessity. When Kretschmann pointed this out, Einstein immediately conceded the point. This is his complete response in 1918:

> 'Relativity Principle: The laws of nature are merely statements about space-time coincidences; they therefore find their only natural expression in generally covariant equations. . . .
> Kretschmann has commented that, formulated in this way, the relativity principle is not a statement about physical reality, i.e. not a statement about the *content* of the laws of nature but only a requirement on their mathematical *formulation*. For since our entire physical experience rests solely on coincidences, it must always be possible to represent the laws that connect these coincidences in a generally covariant form. He therefore believes it is necessary to associate a different meaning with the principle of relativity. I consider Kretschmann's argument to be correct, but do not feel that the new proposal he makes is to be recommended. Even

though it is correct that one must be able to cast every empirical law into a generally covariant form, the relativity principle (as formulated above) still possesses a significant heuristic power that has already been brilliantly demonstrated in the problem of gravitation and is based on the following. From two theoretical systems that are compatible (reading the printed *vereinbarten* as a misprint for *vereinbaren*) with experimental data, one should choose the one that is simpler and more transparent from the point of view of the absolute differential calculus. If one were to attempt to bring Newton's gravitational calculus into the form of (four-dimensional) absolutely covariant equations, one would surely be persuaded that the resulting theory is admittedly theoretically possible but in practice ruled out!'

Einstein never wavered from this 'Kretschmann-corrected' position until the end of his life. He repeatedly said that general covariance (and the relativity principle!) has no physical significance. This conclusion seems to be in direct conflict with the positions of Christian and Penrose, both of whom (Christian explicitly) seem to regard general covariance as an important property that general relativity possesses and other theories do not. Christian, drawing on a paper of Stachel (1993), seeks to pin down the difference to the presence or absence of *a priori* individuation of the points of the spacetime manifold used to express the theory.

## 9.3 General covariance is not a physical principle

### 9.3.1 Newtonian mechanics in generally covariant form

In this chapter I want to argue that Einstein's revised position was quite right. No theory that has any hope of describing nature can fail to be generally covariant. As Kretschmann argued (and Christian denies), general covariance is physically vacuous. I believe that the physically significant issue is not whether or not points have *a priori* individuation, but the relative complexity of rival theories when expressed in generally covariant form. We shall see that this is closely related to the determination of inertial frames of reference and also an issue central to Penrose's proposal.

Following Stachel, Christian says that 'if there are any non-dynamical structures present, such as the globally specified Minkowski metric tensor field ... of special relativity then the impact of general covariance is severely mitigated.' This is because they can be used to introduce what Stachel calls 'inertial individuating fields', which can be used 'to set apart a point $q$ from a point $p$ of a manifold, bestowing *a priori* spatio-temporal individuality to the points of the manifold'. In further support of this statement, Christian cites Wald (1984).

Such views arise, in my opinion, because theoretical physicists – and above all Einstein – have paid insufficient attention to the work of Lagrange and Tait and the actual determination of inertial frames of reference. It is widely held that Newton's theory contains a non-dynamical structure and thus points with *a priori* individuality. However, Lagrange's formulation of the three-body problem (which can be expanded to the general $n$-body problem) is generally covariant as regards the treatment of the spatial variables (I shall come to time later). It uses the distances between the particles and nothing else. No points of the space in which the particles reside are singled out. There is no *a priori* individuation. Stachel's attempt (followed by Christian) to distinguish between genuinely covariant theories and theories with individuating fields must ultimately fail for this reason. I shall come back to this point. Einstein was much closer to the truth when he said that what counts is the relative simplicity and transparency of a theory when cast in generally covariant form. Interestingly, what Lagrange's work does show is that Newtonian theory, when cast in generally covariant form, is actually several different theories at once. This is because, as Poincaré (1902) noted (without, unfortunately referring explicitly to Lagrange's paper which would have made things clearer and strengthened the impact of his remarks), the constants $E$ and $M^2$ that appear in Newton's equations for $n$ bodies *regarded as representing the universe* can be interpreted, not as the reflection of contingent initial conditions, but as universal constants. We shall see shortly that there is a very close parallel between $E$ and the cosmological constant in Einstein's equation. The point is that the equations and solutions of Newton's theory are characteristically and qualitatively different for different numerical values of $E$ and $M^2$. Especially interesting in this connection are the theories that arise if $E$ or $M^2$ (especially the latter) is zero. The corresponding theories for these special cases are then very different, as I shall shortly show.

### 9.3.2 Poincaré and the initial-value problem

The most distinctive property of Newton's equations when cast into the generally covariant form found by Lagrange is the presence of the two freely disposable constants $E$ and $M^2$ and the third time derivative. As Poincaré (1902) noted in the same penetrating remarks, this makes Newton's theory rather unsatisfactory. If only relative quantities count in the world, one would have hoped that specification of the relative separations and the time rates of change of these separations would suffice to predict the future uniquely. However, even if one knew the values of $E$ and $M^2$, this hope is thwarted by the presence of the third time derivatives in Lagrange's equations. However, Poincaré failed to note that if $M^2 = 0$ the form of Lagrange's equations changes qualitatively and the third derivatives disappear.

### 9.3.3 Mechanics without background structures

In fact, during the 1980s the special status of Newtonian motions with vanishing intrinsic angular momentum was discovered independently by Bertotti and myself in work on Mach's principle (Barbour and Bertotti 1982), by Guichardet (1984) in molecular dynamics, and by Shapere and Wilczek (1987, 1989) in a study of how micro-organisms swim in viscous fluids! Littlejohn and Reinsch (1997) have given a beautiful review of the work that developed from the discoveries of Guichardet and Shapere and Wilczek. Among other things, this makes elegant use of a connection with associated geometric (Berry) phase that is defined on the relative configuration space of any $n$-body system by the condition that its motions are constrained to have vanishing intrinsic angular momentum in absolute space.

Perhaps the most important thing about these results, which has a close bearing on Penrose's proposal, is that there exists a way to derive motions of the universe that have vanishing intrinsic angular momentum and makes no use at all of any external Newtonian structures like absolute space and time. A metric is defined on the relative configuration space of the universe by a 'best-matching' procedure as follows. Take one relative configuration of $n$ point masses and another that differs from it slightly. Hold the first fixed, and try to move the second around relative to it so that it is brought into the closest fit in the following sense. For any relative placing of the second configuration, there will be some distance $dx_i$ between the position of particle $i$ with mass $m_i$ in the two relative configurations. Now consider the quantity

$$\left[\sum_i m_i(dx_i)^2\right]^{1/2}.$$ (9.1)

It is a measure of the difference between the two configurations but is arbitrary as yet because the placing of the second configuration relative to the first is arbitrary. However, this placing can be varied until eqn. 9.1 is extremalized (minimized in this case). That will then give a completely co-ordinate-free and objective measure of the difference between the two relative configurations. A non-trivial geodesic principle on the relative configuration space is then defined by the action principle

$$\delta \int \left[(E-V)\sum_i (m_i/2)(Dx_i)^2\right]^{1/2} = 0,$$ (9.2)

where $Dx_i$ is the displacement in the 'best-matching' position, $E$ is a constant, and $V = \sum_i m_i m_j / r_{ij}$.

It is easy to show (Barbour and Bertotti 1982) that the sequence of relative configurations through which a universe that satisfies eqn. 9.2 passes is identical to a sequence of relative configurations through which a Newtonian universe of the same point masses passes, provided that it has the total energy $E$ and vanishing intrinsic angular momentum. It should be noted that eqn. 9.2, which is a Machian form of Jacobi's principle for the timeless orbit of a dynamical system in its configuration space (Lanczos 1949, Barbour 1994, 1994a), makes no use of either absolute time or absolute space. Nothing but relative configurations appears in it. It is manifestly general covariant.

In the light of the earlier discussion, eqn. 9.2 represents a distinct physical theory for each value of $E$. It is worth mentioning in this connection that Einstein's general relativity can be cast in a form very like eqn. 9.2 with the cosmological constant playing the role of $E$ (see, for example, Brown and York 1989). Everyone accepts that the forms of general relativity with different values of that constant represent different physical theories, so it is obvious that the same should be done here. It is important here that eqn. 9.2 is meant to describe the complete universe, not subsystems of it. Even though the complete universe described by eqn. 9.2 will have vanishing total angular momentum and energy, subsystems within it can have all values of these quantities. In addition, the complete family of theories represented by eqn. 9.2 with different values of $E$ are to be seen as theories distinct from the general Newtonian solutions with non-vanishing intrinsic angular momentum. For the three-body problem these are described in generally covariant form by Lagrange's equations, which are much more complicated than the ones that follow from eqn. 9.2. For the $n$-body problem, the equations have only been found (to the best of my knowledge) in implicit form (Zanstra 1924).

It is important to note that the generally covariant principle eqn. 9.2 comes with an automatic prescription that makes it possible to cast it into a form in which it seems to be formulated with an external framework. Namely, if one has found a sequence of $n$-body relative configurations by means of eqn. 9.2 and its underlying best-matching prescription to determine the $Dx_i^2$, the same best matching can be used to 'stack the configurations horizontally'. One starts with the first configuration, which one supposes to be placed anywhere in absolute space. The second configuration is then brought into the best-matching position relative to the first and thereby acquires a definite position in absolute space. The third is then placed by best-matching relative to the second, and so on. At the end, the configurations have exactly the positions in absolute space that the corresponding solution with vanishing intrinsic angular momentum has. All this has been done using the directly observable matter, not any 'individuating' marks in non-dynamical frames.

This has an important bearing on my contention that Stachel's attempt to introduce inertial individuating fields will not solve the problem of establishing the true content of general covariance. The point is that in all the different generally covariant theories that correspond to Newtonian mechanics in its standard form one can introduce distinguished frames of reference. All the theories seem to contain non-dynamical elements. (Even general relativity has such elements, since spacetime is always Minkowskian in the small – that is what one could call an absolute element.) Nevertheless, the theories differ in their complexity and predictive power: the simpler the theory, the greater its predictive power. Most arguments about the formulation and content of general covariance are mere shadow boxing. Since all viable theories must be generally covariant, the only question of interest is their relative simplicity and the nuts and bolts of their construction.

Returning to the question of the construction of (Newtonian) spacetime from purely relative configurations, the times at which the positions are occupied in the constructed 'absolute space' has not yet been determined. This is, however, readily done without recourse to any knowledge of an external time. The equations of motion that follow from eqn. 9.2 have the form (for an arbitrary parameter $\lambda$ that labels the configurations along the path in the configuration space)

$$(d/d\lambda)\left\{\left[(E-V)/\sum_i (m_i/2)(Dx_i)^2\right]^{1/2} m_i(Dx_i/d\lambda)\right\}$$

$$= -\left[\sum_i (m_i/2)(Dx_i/d\lambda)^2/(E-V)\right]\partial V/\partial x_i. \tag{9.3}$$

There is a unique choice of the parameter that simplifies eqn. 9.3 dramatically. It is the one for which

$$E - V = \sum_i (m_i/2)(Dx_i/d\lambda)^2. \tag{9.4}$$

Then eqn. 9.3 becomes identical to Newton's second law, and in Newtonian terms eqn. 9.4 becomes the statement that the total energy is given by $E$.

The parameter $\lambda$ is then indistinguishable from Newton's absolute time, while in the 'horizontal stacking' achieved by the best-matching construction $Dx_i/d\lambda$ simply becomes the Newtonian velocity in absolute space. From a variational principle formulated solely with relative configurations and no external framework, we have constructed a framework identical to Newton's absolute space and time.

However, this is a framework that also contains whole families of solutions that belong to *different* generally covariant theories when the Lagrangian variables are employed. The Newtonian solutions with different intrinsic (centre-of-mass) values of $E$ and $M^2$ belong to different theories.

### 9.3.4 A modification of Einstein's simplicity argument

This is one way in which one can examine the manner in which Newtonian mechanics is generally covariant. It was used unconsciously by Lagrange and Jacobi, and it tells one how a four-dimensional framework of space and time can be put together from *three-dimensional relative configurations*. There are characteristically different ways in which this can be done. The Machian theory corresponding to eqn. 9.2 is manifestly simpler than the families of theories corresponding to all the different non-vanishing values of $M^2$.

It is very tempting to modify and make somewhat more precise Einstein's standpoint as expressed in his (1918) paper. He seems to be implying that the choice that has to be made is between general relativity in the generally covariant form in which he found it and Newtonian theory when it is cast in (four-dimensional) generally covariant form. But this is a somewhat artificial and unsatisfactory choice. The two theories have very different underlying ontologies (curved space and point particles in Euclidean space respectively), and Newton's theory was already then known to be no longer fully compatible with observations.

It is much more illuminating to consider the many different (three-dimensionally) generally covariant theories hidden within general Newtonian theory and apply to them Einstein's simplicity requirement. It will immediately rule out the theories with $M^2 \neq 0$, which have a vastly more complicated generally covariant structure, and, less emphatically, suggest the choice $E = 0$ among the remaining family of theories (avoiding the specification of an arbitrary constant). We shall see shortly that the family of theories with $M^2 = 0$ and $E \neq 0$ ($E$ fixed) should be seen as the true non-relativistic generally covariant analogues of general relativity.

### 9.3.5 Cartan's reformulation misses the point

Because this rich substructure of generally covariant theories within Newtonian theory escaped wide notice, no one has hitherto thought about the meaning of general covariance in such terms. Instead, Cartan (1923), Friedrichs (1927), and many others took up Einstein's challenge to cast Newtonian theory into a four-dimensional generally covariant form. This is done purely formally by the introduction of general co-ordinates and two distinct (and rather bizarre) vector fields on four-dimensional Newtonian spacetime. It is clear that the physical content of Newton's theory is completely unaffected by this treatment. In fact, even then only imperfect analogy with Einstein's theory is achieved.

Despite the absence of change, the Cartan form of Newtonian theory is held to cast light on the conceptual problems associated with general covariance. I believe this is incorrect. Since no new physical principle has been added, no new physical insight can be extracted. Far from clarifying the situation with regard to covariance, it has muddied the waters. (However, I think the Cartan formulation might well cast useful light, as I think Christian argues, on quantization of connections.)

The problem is that the general covariance of a dynamical theory raises two quite different issues. The first has been clearly recognized and well understood ever since Einstein saw the flaw in his hole argument. This is that the objective content of a four-dimensional theory cannot be changed merely by changing the co-ordinates used to describe it. Equally, using different and more modern terminology, it cannot be changed by an active diffeomorphism. Essentially, when Einstein saw the flaw in his hole argument he actually rediscovered Leibniz's principle of the identity of indiscernibles. No one will disagree on this, and for further interesting discussion see Chapter 13 in this volume.

It is the second issue that has not been grasped. It is this: Can four-dimensional structures like Newtonian spacetime or Einstein's spacetime be conceived as being put together from three-dimensional structures? If so, what does the recognition of the truth first stated for four-dimensional structures imply for the task of putting together three-dimensional structures? For they too are structures whose objective content cannot be changed by painting co-ordinates on them. Equally significantly, mere co-ordinate values can in no way be used to identify points belonging to objectively different three-dimensional structures.

## 9.4 Penrose's argument

This point is central to the arguments that Penrose develops. For example, he says:

> 'It is clear from the principles of general relativity that it is not appropriate, in general, to make a precise identification between points of one space-time and corresponding points of the other ... all that we can expect will be some kind of *approximate* pointwise identification. .... There is no canonical way of identifying the individual *points* of a section of one space-time with corresponding points of the other.'

Penrose argues that exactly the same thing applies to the comparison of three-dimensional spaces and for this reason is critical of things that are 'common in discussions of quantum gravity':

> 'According to the sorts of procedure that are often adopted in quantum gravity, the superposition of different space-times is indeed treated in a very formal way, in terms of complex functions on the space of 3-geometries (or 4-geometries), for example, where there is no pretence at a pointwise identification of the different geometries under consideration.'

In disagreement with the thrust of what Penrose says here, I believe that there is a canonical way in which one could achieve exact pairwise identification of points between slightly differing spacetimes if, inappropriately, one wanted to (in order, perhaps, to construct some five-dimensional structure) and more significantly that there is an equally well-defined canonical procedure for doing the same with 3-geometries. This canonical procedure is nothing more nor less than the Hilbert variational principle that leads to Einstein's field equations in general relativity. I shall amplify this in a moment.

Moreover, the actual superpositions of 3-geometries envisaged in the discussions in quantum gravity are far from formal. Quite the opposite, the coefficients that appear in them are determined by exactly the same variational procedure that leads to Einstein's field equations and solves the problem that Penrose says is insoluble in any exact sense: the pairwise identification of points in different 3-geometries. The superpositions that Penrose distrusts are simply Einstein's field equations in a different guise. This is why John Wheeler correctly calls the Wheeler–DeWitt equation the Einstein–Schrödinger equation. Just as the ordinary Schrödinger equation is really Newton's laws in quantum clothes, the two constraints of quantum gravity are simply Einstein's equations in a quantum guise. They match exactly. (However, it would be wrong to think that the superpositions of 3-geometries allowed by the Wheeler–DeWitt equation are to be thought of as somehow residing in the same space. They are simply there in their separate totalities and in their own right.)

Before I spell out how this happens, it is worth noting that Leibniz's principle of the identity of indiscernibles only revealed to Einstein the reason why his hole argument was wrong. But it could also have told him something much more exciting. It could have explained why there is something very right and appropriate about the way an Einsteinian spacetime is constructed.

### 9.5 The significant issue

Penrose correctly identifies a deep problem. Given two objects that are intrinsically different, is it possible in any objective sense to say that a certain point in one is at the same position as a certain point in the other? One's first reaction is exactly what Penrose says – it will be only approximately possible to achieve such pairing. But an *exact* pairing is precisely what best matching does. It brings two different objects to what might be called the closest possible approach to congruence (established in the case of point particles by minimizing eqn. 9.1) and then declares that the points paired in this way are 'at the same position' or 'equilocal'. This solution to the problem is clearly meaningful and the outcome is in general unique. It is possible because the two compared objects, while being intrinsically different, are nevertheless generically of the same kind. It will be helpful to say something about how the best matching is achieved in the case of 3-geometries; for a fuller account, see Barbour and Bertotti (1982) and Barbour (1994, 1994a).

Imagine one 3-geometry and lay out co-ordinates on it in some arbitrary manner. These will be kept fixed. Imagine a second 3-geometry that differs from it slightly and lay out co-ordinates on it which are such that the resulting metric tensors on the two spaces differ little at equal values of their arguments. Otherwise the co-ordinates on the second 3-geometry are arbitrary. Even though this is the case, one could say by definition that the points on the two 3-geometries that have the same co-ordinate values are 'at the same position' or 'equilocal'. The two sets of co-ordinates establish a 'trial equilocality relationship' between the two spaces. With respect to it, one can compare the values of the 3-metrics on the two three geometries at equilocal points. Their difference gives a measure of how the geometry has changed between the two

Julian B. Barbour

3-geometries at the points declared to be equilocal by this (as yet arbitrary) procedure. This local difference can then be integrated over all of space to find a global measure of the difference between the two 3-geometries. This too will be arbitrary.

This defect can be eliminated if one systematically seeks out *all possible ways* in which the trial equilocality pairing of points is established. One will say that the best-matching position has been achieved if the global measure of difference is extremal (does not change to first order) with respect to all possible infinitesimal changes of the equilocality pairing. This is the exact analogy of the simpler procedure for point particles described above. Actual equations are given in Barbour (1994, 1994a).

The point of interest in connection with the discussion of the physical significance of general covariance (or diffeomorphism invariance) is the manner in which all possible changes of the equilocality pairing is implemented. It is done by a diffeomorphism. Using the somewhat more old-fashioned language of co-ordinates, one first changes the co-ordinates on the second 3-geometry, obtaining a new metric tensor of new arguments $(g'_{ij}(x'))$, after which one looks at the values of the new tensor at the previous values of the arguments $(g'_{ij}(x))$. This generates what is now called a diffeomorphism. The point on the second 3-geometry with the co-ordinate value $x$ is not the point that has the co-ordinate value $x$ in the untransformed co-ordinates. Since the co-ordinate values are being used to identify points on the two different 3-geometries that are said to be equilocal, the diffeomorphism carried out on the second 3-geometry changes the equilocality pairing between the two geometries.

This, I think, is the key point and shows how different is the conceptual issue that is at stake. The only insight Einstein had when he saw the flaw in his hole argument was that changing the label on a bottle of wine in no way changes the wine within it. A given thing, be it a toadstool, a spacetime, or a 3-geometry is completely unchanged by labels attached to it. But here we are not at all engaged in changing the labelling on a single object just to have different representations of it. We are making diffeomorphisms systematically in order to generate physically different trial pairings between two different things and hence obtain a measure of the difference between them. We are forced to this by Einstein's insight that mere co-ordinate values do not have any physical significance. The diffeomorphisms are being used creatively to overcome this undoubted fact but for a purpose that goes far beyond the obtaining of different representations of one given thing. They are being used to compare *two different things*.

This realization is what has been lacking from the interminable discussions of the significance of general covariance that Norton (1993) has chronicled so well. What distinguishes general relativity from Newtonian theory is not the presence or absence of individuating points in the respective four-dimensional spacetimes corresponding to them. There are none in either case when you consider what is truly observable. It is the manner in which the respective four-dimensional objects are put together out of three-dimensional constituents. General relativity does it by best matching; Newtonian theory does not, and is therefore more complicated. However, the solutions of Newtonian theory with vanishing intrinsic angular momentum form a different theory in which the four-dimensional spacetime is put together by best matching. It is therefore to be regarded as the non-relativistic analogue of general relativity.

210

The most important thing that follows from this is that the idea of best matching makes it possible to define a metric on the space of all the intrinsically different but generically homogeneous structures that one wishes to compare, be they configurations of mass points in Euclidean space or 3-geometries, as happens in general relativity. I should mention that, in the case of general relativity, the pairing of points between the compared 3-geometries results not from some arbitrary pairing established at infinity, as Penrose seems to imply, but from comparison across the whole of both 3-geometries. Both the entire 3-geometries are taken into account. Once a metric has been defined on such a 'superspace', geodesics are defined. Now to determine a geodesic, one needs not only a metric but also some point along the geodesic and a direction at it. This brings me to the connection of all this with Einstein's equations and the constraints into which they are transformed in the Hamiltonian formulation found by Arnowitt, Deser, and Misner (ADM). I shall not give any details (which can be found in Barbour 1994, 1994a), but only the main points. As has been shown by many people but especially Kuchař (1993), the entire content of Einstein's field equations are captured and faithfully reproduced by the three equations of the so-called momentum constraints and the single equation of the Hamiltonian constraint. As explained previously (Barbour 1994, 1994a), the three momentum constraints really do express the guts of Einsteinian dynamics and show that it arises through the creation of a metric on superspace by best-matching comparison of slightly different 3-geometries. The Hamiltonian constraint is really an expression of the fact that the initial condition for a geodesic requires specification of a direction in superspace. (I should mention that the situation is especially subtle in general relativity, and the Hamiltonian constraint is actually infinitely many constraints – one at each space point – each of which defines a 'local direction'.)

### 9.6 Conclusions

In the light of these comments, it does seem to me rather unlikely that nature, having used the precise and exact global best-matching condition to put together 3-geometries into Einsteinian spacetimes, should use some more approximate and essentially local (as Penrose envisages) matching between four-dimensional structures to collapse superpositions of spacetimes. I also feel that Cartan's purely formal rewriting of Newtonian mechanics in four-dimensional generally covariant form is likely to cast less light on the problems of quantum gravity than Lagrange's physical reformulation of the same mechanics in a three-dimensional generally covariant form. Quantum mechanics is, so it seems to me, about superpositions of three-dimensional structures that classical dynamics knits together into four-dimensional structures.

Finally, both Penrose and Christian call for a more 'even-handed' approach to quantum gravity, complaining that Einstein's beautiful theory is expected to cede too much to the quantum formalism. It seems to me that this too is not correct if one takes DeWitt's canonical approach, in which the constraints are transformed into operators and constitute the entire theory (in the case of a spatially closed universe; see DeWitt 1967). At a stroke, this eliminates the external time and inertial frames of reference that form such a characteristic part of ordinary quantum mechanics. In quantum cosmology there are as yet no conventional observables and no Hilbert space. It is to be expected that they will emerge as effective concepts in the WKB regime (Barbour 1994, 1994a, 1999). It is clear that in quantum cosmology there will be a different measurement theory, and none of the usual business with non-quantum classical measuring instruments external to the quantum system under study. In fact, in DeWitt's canonical approach, all that remains of quantum mechanics is the wave function. In contrast, the entire dynamical content of Einstein's theory remains intact, though transfigured into quantum form, through which, in anything but an arbitrary fashion, it determines what superpositions of 3-geometries are allowed. As I have argued previously (Barbour 1994, 1994a, 1999), one should not look to interpret these superpositions as representing physical processes that occur simultaneously in some common space. Rather, each 3-geometry (with matter distribution in it) is to be regarded as an *instant of time*. It is a possible instantaneous state of the universe. Ontologically, it is just the same as in classical physics. Contrary to a widely held view, I do not believe that the most significant difference between quantum and classical physics is in the ontology, but in what is done with it.

## Notes

I am grateful to Domenico Giulini for drawing to my attention von Laue's interest in Lange. It also made me check out the later correspondence between Einstein and von Laue, which is very rewarding. In fact, their entire correspondence, spanning nearly forty years, is both absorbing and poignant (on account of the Nazi period).

**Pre-Socratic quantum gravity**

Gordon Belot and John Earman

## 10.1 Introduction

Physicists who work on canonical quantum gravity will sometimes remark that the general covariance of general relativity is responsible for many of the thorniest technical and conceptual problems in their field.[1] In particular, it is sometimes alleged that one can trace to this single source a variety of deep puzzles about the nature of time in quantum gravity, deep disagreements surrounding the notion of 'observable' in classical and quantum gravity, and deep questions about the nature of the existence of spacetime in general relativity.

Philosophers who think about these things are sometimes sceptical about such claims. We have all learned that Kretschmann was quite correct to urge against Einstein that the 'General Theory of Relativity' was no such thing, since *any* theory could be cast in a generally covariant form, and hence that the general covariance of general relativity could not have any physical content, let alone bear the kind of weight that Einstein expected it to.[2] Friedman's assessment is widely accepted: 'As Kretschmann first pointed out in 1917, the principle of general covariance has no physical content whatever: it specifies no particular physical theory; rather, it merely expresses our commitment to a certain style of formulating physical theories' (Friedman 1983, p. 44). Such considerations suggest that general covariance, as a technically crucial but physically contentless feature of general relativity, simply cannot be the source of any significant conceptual or physical problems.[3]

Physicists are, of course, conscious of the weight of Kretschmann's points against Einstein. Yet they are considerably more ambivalent than their philosophical colleagues. Consider Kuchař's conclusion at the end of a discussion of this topic:

> ' . . . the Einstein–Kretschmann discussion is clearly relevant for the canonical quantization of covariant theories, but, as so many times before in the ancient controversy between the relative and the absolute, it is difficult to decide which of

these two alternative standpoints is correct and fruitful. This leaves the canonical quantization of covariant systems uncomfortably suspended between the relative and the absolute'.

(Kuchař 1988, p. 118)

It becomes clear in the course of Kuchař's discussion that he takes the physical content of the general covariance of general relativity to reside not in the fact that that theory, like every other, *can* be given a generally covariant formulation, but in the fact that it *ought* to be so formulated (see, e.g., pp. 95–6). The idea is that one does some sort of violence to the physical content of general relativity if one breaks its general covariance by introducing preferred co-ordinates, slicings, or other geometrical structure, in a way in which one does not when one moves from a generally covariant formulation of Newtonian mechanics or special relativity to the standard formulations in which inertial co-ordinates play a special role.

Central to this way of thinking about general covariance is the idea that misjudging the physical content of a given theory can lead one astray in attempts to construct new theories – since, e.g. empirically equivalent formulations of a given classical theory may well lead to inequivalent quantum theories, it is important to begin with the *correct* formulation. This link between content and method is the source of the sentiment – which is widespread among physicists working on canonical quantum gravity – that there is a tight connection between the interpretative problems of general relativity and the technical and conceptual problems of quantum gravity.

Our goal in this chapter is to explicate this connection for a philosophical audience, and to evaluate some of the interpretative arguments which have been adduced in favour of various attempts to formulate quantum theories of gravity. We organize our discussion around the question of the extent to which the general covariance of general relativity can (or should) be understood by analogy to the gauge invariance of theories like classical electromagnetism, and the related questions of the nature of observables in classical and quantum gravity and the existence of time and change in the quantum theory.

We provide neither a comprehensive introduction to the formalism of quantum gravity, nor a survey of its interpretative problems (readers interested in the latter should turn to the canonical survey articles: Isham 1991, 1993, Kuchař 1992). We do, however, want the chapter to be both accessible to readers who are unfamiliar with the formalism, and helpful for those who would like to use it as a starting point for a serious study of the field. To this end, we have tried to keep the presentation in the body of the text as intuitive as possible, while relegating technicalities and references to background literature to footnotes and an Appendix (p. 249).[4]

We begin in the next section with a sketch of the formalism of gauge theories, and a brief discussion of their interpretative problems. This is followed in Section 10.3 by a discussion of how general relativity itself may be cast as a gauge theory, and how in this context the hole argument can be viewed as a special case of the general interpretative problem of gauge invariance. In Section 10.4 we bring out some of the potential demerits of reading the general covariance of general relativity as a principle of gauge invariance. Most importantly, we discuss the fact that this reading seems to force us to accept that change is not a fundamental reality in classical and quantum gravity. This sets up the discussion of the following two sections, where we survey a number of proposals for understanding the general covariance of general relativity and discuss the proposals for quantizing gravity which they underwrite. Finally, in Section 10.7, we argue that the proposals canvassed in Sections 10.5 and 10.6 are directly related to interpretative views concerning the ontological status of the spacetime of general relativity. We conclude that problems about general covariance are indeed intimately connected with questions about the correct quantization of gravity and the nature of time and change in physical theory.

## 10.2 Hamiltonian and gauge systems

There are a number of ways to formulate classical physical theories. One of the most straightforward is to proceed as follows. Construct a space whose points represent the physically possible states of the system in which you are interested. Then introduce some further structure which singles out a set of curves in this space that correspond to dynamically possible histories of the system. In the first two subsections we will sketch two implementations of this strategy: the Hamiltonian formalism and the gauge-theoretic formalism. We will see that the notion of a gauge system is a modest generalization of the notion of a Hamiltonian system – one simply weakens the geometric structure which is imposed on the space of states. As will become clear in the third subsection, however, this relatively small difference generates some very interesting interpretative problems: in the context of gauge systems, one is forced to make difficult decisions concerning the nature of the representation relation which holds between the mathematical space of states and the set of physically possible states of the system. We close the section with a brief discussion of the quantization of gauge systems.

### 10.2.1 Hamiltonian systems

Many classical physical systems can be modelled by *Hamiltonian systems*. These are triples of mathematical objects, $(M, \omega, H)$. Here $M$ is manifold, and $\omega$ is a tensor, called a *symplectic form*, which gives $M$ a geometric structure. The pair $(M, \omega)$ is called a *symplectic geometry*; the dension of $M$, if finite, must be even. For our purposes, it is sufficient to note two ways in which the symplectic structure $\omega$ interacts with the set $C^\infty(M)$ of smooth real-valued functions on $M$. The first is that the symplectic structure $\omega$ gives us, *via Hamilton's equations*, a map $f \mapsto X_f$ between smooth functions on $M$ and vector fields on $M$. Given $f \in C^\infty(M)$, one can integrate its vector field, $X_f$, to obtain a unique curve through each point of $M$ (that is, one looks for the family of curves whose tangent vector at $x \in M$ is just $X_f(x)$). Thus, we can associate a partition of $M$ into curves with each smooth function on $M$. The second, and related, important function of the symplectic structure is to endow the set $C^\infty(M)$ with an interesting algebraic structure, the *Poisson bracket*. This is a binary operation which associates a smooth function, denoted $\{f, g\}$, with each pair of functions $f, g \in C^\infty(M)$. Intuitively, $\{f, g\}$ measures the rate of change of $g$ along the set of curves generated by $f$, so that $g$ is constant along the curves generated by $f$ iff (i.e. if and only if) $\{f, g\} = 0$. The Poisson bracket plays a crucial role in quantization.

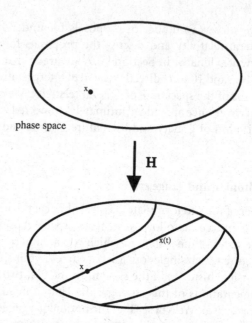

phase space

**H**

Fig. 10.1. Hamiltonian systems.

We construct a Hamiltonian system by supplementing a symplectic geometry $(M, \omega)$ by a choice of a distinguished element $H \in C^{\infty}(M)$, called the *Hamiltonian*. The set of curves on $M$ determined by $\omega$ and $H$ are called the *dynamical trajectories*. Figure 10.1 depicts a Hamiltonian system: at the top, we have a symplectic geometry $(M, \omega)$; specifying a Hamiltonian serves to determine a unique dynamical trajectory through each point.

Taken together, $H$ and $(M, \omega)$ constitute a theory of the behaviour of the system in the following sense. We think of $(M, \omega)$ as being the space of dynamically possible states of some physical system – the *phase space* of the system. Each point of $(M, \omega)$ corresponds to exactly one physically possible state of the system, so a curve in phase space corresponds to a history of physically possible states of the system. To say that there is a unique dynamical trajectory through each point is to say that our theory specifies a unique past and future for every possible present state of the system. It is a complete and deterministic theory.

In the context of classical mechanics, one typically constructs a phase space by beginning with a smaller space, $Q$, the *configuration space*, which is taken to be the space of possible configurations of some set of particles or fields relative to physical space. One then identifies the phase space with the *cotangent bundle*, $T^*Q$, of $Q$. A point of $T^*Q$ is a pair $(q, p)$ where $q \in Q$, and $p$ is a covector at $q$. If $Q$ represents the set of possible positions of some set of particles relative to physical space, then $T^*Q$ can be thought of as the space of possible positions and momenta of these particles. We can tell a similar story about fields. There is a canonical way of endowing $T^*Q$ with a symplectic structure. We can now impose the Hamiltonian, $H$, whose value at a point $(q, p)$ is just the energy of a system with that position and momentum. The dynamical trajectories for this Hamiltonian ought to model the observed behaviour of our system. Examples of this are the free particle and the Klein–Gordon field.

### 10.2.1.1 *The free particle*

If we are dealing with a single particle in Euclidean space, then $Q = \Re^3$ – the space of possible configurations of the particle is just the space of positions of the particle relative to physical space. The phase space is $T^*Q = T^*\Re^3$, and $H$ is just the kinetic energy. More generally, if we have a free particle moving in a physical space which is modelled by a Riemannian geometry $(S, g)$, then the configuration space is $S$ and the phase space is $T^*S$ endowed with the canonical symplectic form, $\omega$. The dynamical trajectories corresponding to the particle moving along the geodesics of $(S, g)$ are again generated by setting the Hamiltonian equal to the kinetic energy $\frac{1}{2}g^{ab}p_a p_b$.

### 10.2.1.2 *The Klein–Gordon Field*

Fix a simultaneity slice, $\Sigma$, in Minkowski spacetime, and let $Q$ be the space of configurations on this slice of the Klein–Gordon field $\phi$ of mass $m$ – thus each point in $Q$ corresponds to a $\phi : \Re^3 \to \Re$. We then look at $T^*Q$, where a point corresponds to a pair $(\phi, \dot\phi)$. Our phase space consists of $T^*Q$ equipped with the canonical symplectic structure, $\omega$. Fixing an arbitrary timeslice, $\Sigma$, we can write:

$$\omega((\phi_1, \dot\phi_1), (\phi_2, \dot\phi_2)) = \int_\Sigma \phi_1 \dot\phi_1 - \phi_2 \dot\phi_2 \, dx^3$$

(here we are dealing with a linear field theory, so that $Q$ and $T^*Q$ are vector spaces, and we may identify vectors on $Q$ with elements of $T^*Q$). For our Hamiltonian, we take:

$$H = \frac{1}{2} \int_\Sigma \left( \dot\phi^2 + \nabla\phi + m\phi^2 \right) dx^3.$$

The equation of motion is just the usual Klein–Gordon equation,

$$\partial^a \partial_a \phi - m\phi = 0.$$

### 10.2.2 Gauge systems

We consider an especially interesting generalization of the Hamiltonian framework: gauge systems.[5] The starting point is to relax one of the conditions imposed upon symplectic forms. This leads to a more general class of geometries, known as *presymplectic geometries*, which serve as the phase spaces of gauge theories. The presymplectic structure, $\sigma$, of a presymplectic geometry $(N, \sigma)$ determines a natural foliation of the manifold $N$ by submanifolds of some fixed dimension – there is one such submanifold through every point of $N$ (see the top half of Fig. 10.2). These submanifolds are called the *gauge orbits* of $(N, \sigma)$.

Since $(N, \sigma)$ is partitioned by gauge orbits, 'being in the same gauge orbit' is an equivalence relation. We denote this relation by $x \sim y$, and denote the gauge orbit of $x$ by $[x]$. We call a diffeomorphism $\Phi : N \to N$ a *gauge transformation* if it preserves gauge orbits – i.e. if $x \sim x'$ implies $\Phi(x) \sim \Phi(x')$. We call a function $f : N \to \Re$ *gauge-invariant* if $f$ is constant on each gauge orbit – i.e. if $x \sim x'$ implies $f(x) = f(x')$.

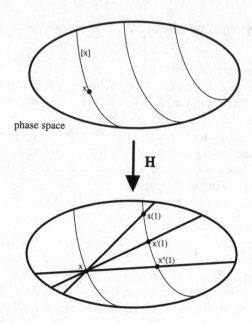

Fig. 10.2. Gauge systems.

If we take a gauge-invariant function, $H$, on $N$ as our Hamiltonian, then we can use the resulting *gauge system*, $(N, \sigma, H)$, to model physical systems. We can again investigate the dynamical trajectories generated by $H$. Whereas in the Hamiltonian case there was a single dynamical trajectory through each point of phase space, we find in the gauge-theoretic case that there are infinitely many trajectories through each point.

The saving grace is that the family of dynamical trajectories through a given point, although they in general disagree radically about which point represents the future state of the system at a given time, do agree about which gauge orbit this point lies in. That is: if $x(t)$ and $x'(t)$ are dynamical trajectories which have their origin at the same point $x(0) = x'(0) = x_0$, then we have that $x(t) \sim x'(t)$ for all $t \in \Re$. Thus, although the presymplectic geometry is not strong enough to determine a unique dynamical trajectory through each point, it *is* strong enough to force all of the dynamical trajectories through a given point to agree about which gauge orbit the system occupies at a given time (Fig. 10.2). In particular, if $f$ is a gauge-invariant function on $N$ then the initial value problem for $f$ is well-posed in the sense that if we fix an initial point $x_0 \in N$, then for any two dynamical trajectories $x(t)$ and $x'(t)$ which have their origin at the same point $x(0) = x'(0) = x_0$, we find that $f(x(t)) = f(x'(t))$ for all $t$. Thus specifying the initial state of the system completely determines the past and future values of any gauge-invariant quantity.

In practice, the most interesting gauge systems arise as *constrained Hamiltonian systems*. This means that our presymplectic phase space $(N, \sigma)$ arises by restricting attention to a regular submanifold, $N$, of a symplectic geometry $(M, \omega)$, where $N$ is equipped with the presymplectic form $\sigma = \omega \mid_N$ (the restriction of $\omega$ to $N$).[6] We introduce the notation $f \cong g$ (read '$f$ is weakly equal to $g$') to indicate $f \mid_N = g \mid_N$, where $f, g \in C^\infty(M)$.

Locally, we can specify $N$ by requiring that some set, $\mathcal{C} = \{C_a\}$ of real-valued functions on $M$ vanish. Such functions are called *constraints*. There are two kinds of constraints: if $C_a \in \mathcal{C}$ is such that for $x \in N$, $X_{C_a}(x)$ is tangent to $[x]$, then $C_a$ is a *first-class constraint*, and is denoted $\gamma_a$; otherwise, $C_a$ is a *second-class constraint*, and is denoted $\chi_a$. Equivalently, the first-class constraints are those which commute with all of the constraints. The first-class constraints, but not the second-class constraints, generate gauge transformations on $N$.[7] That is: following in $N$ the integral curve of a vector field associated with a first-class constraint carries one along gauge orbits of $(N, \sigma)$ (here we are thinking of vector fields as the infinitesimal generators of diffeomorphisms). In fact, at each point $x \in N$, $\{X_{\gamma_a}(x)\}$ is a basis for the tangent space of $[x]$ so that the dimensionality of the gauge orbit of $x$ is just the cardinality of $\{\gamma_a\}$.

A function $f : M \to \mathfrak{R}$ has a gauge-invariant restriction to $N$ iff $\{f, \gamma\} \cong 0$ for all first-class constraints $\gamma$ (here $\{,\}$ are the Poisson brackets on $(M, \omega)$). One describes this result by saying that gauge-invariant functions *commute* with the first-class constraints. In what follows we shall be exclusively concerned with first-class constraints.

### 10.2.2.1 *Trivial gauge-invariant function*

Let $(T^*Q, \omega)$ be a finite dimensional cotangent bundle with its canonical symplectic structure, and let $(q^i, p_i)$ be canonical co-ordinates. Now let $(N, \sigma)$ arise by imposing the first-class constraint $p_1 = 0$.[8] The Hamiltonian vector field of $p_1$ in $(M, \omega)$ generates motions in the $q^1$ direction, so the gauge orbits in $(N, \sigma)$ are of the form $\{(s, q^2, \ldots; p_2, \ldots)$: where $s \in \mathfrak{R}$ and all other $q^i$ and $p_i$ are fixed$\}$. Thus the gauge-invariant functions on $N$ are those which are independent of $q^1$. For any Hamiltonian, a dynamical trajectory is of the form $(q^i(t), p_i(t))$, with $q^i(t)$ and $p_i(t)$ determined uniquely for $i \geq 2$, but with $q^1(t)$ an arbitrary function of time – we call the physically irrelevant $q^1$ a *gauge* degree of freedom. The behaviour of this trivial example is typical: if $(N, \sigma, H)$ is a constrained Hamiltonian system in a finite dimensional symplectic manifold $(M, \omega)$, then we can always find local canonical co-ordinates, $(q^i; p_i)$, on $(M, \omega)$ so that the first-class constraints are of the form $p_i = 0$ for $i \leq k$ and the $q^i(t)$ are arbitrary for $i \leq k$.

### 10.2.2.2 *Vacuum electromagnetism*

Let $(S, g)$ be a three-dimensional Riemannian manifold representing physical space, and let $Q = \{A: S \to \mathfrak{R}^3\}$ be the infinite dimensional space of covector fields on $S$ – that is, each element of $Q$ is a function which maps each point of $S$ to a three-vector. We construct the cotangent bundle, $T^*Q$. A point in $T^*Q$ is a pair $(A, E)$, where $E$, like $A$, is a vector field on $S$ (again, identifying vectors on a linear space with elements of that space). We endow $T^*Q$ with the canonical symplectic structure, $\omega$. In order to construct the phase space of electromagnetism, we restrict attention to those points $(A, E) \in T^*Q$ such that div $E = 0$. This is a first-class constraint. The constraint manifold, $N$, is an infinite dimensional submanifold of $T^*Q$. We equip $N$ with the presymplectic form, $\sigma = \omega \mid_N$. The presymplectic manifold, $(N, \sigma)$ is the phase space of electromagnetism. The gauge orbits of $(N, \sigma)$ are determined by the following equivalence relation: $(A, E) \sim (A', E')$ iff $E' = E$ and $A' = A + \text{grad } \Lambda$ for some $\Lambda : S \rightarrow \Re$. Thus, $[(A, E)] = \{(A', E): A' = A + \text{grad } \Lambda, \Lambda : S \rightarrow \Re\}$. What are the gauge-invariant functions on this phase space? If, for example we fix a point $\xi \in S$, then the function $\xi_E : N \rightarrow \Re^3$ whose value at $(A, E)$ is just $E(\xi)$ is gauge-invariant. On the other hand, the function which returns the value $A(\xi)$ is clearly *not* gauge-invariant (in general, $A(\xi) \neq A'(\xi)$ even if $(A, E) \sim (A', E)$). We can, however, use $A$ to construct gauge-invariant quantities. Of these, the most important is the magnetic field, $B \equiv \text{curl } A$. Since $\text{curl}(A) = \text{curl}(A + \text{grad } \Lambda)$ for any scalar $\Lambda$, we find that $B(\xi) = B'(\xi)$ whenever $(A, E) \sim (A', E)$.

We choose our Hamiltonian to be $H = \int_S (|E|^2 + |\text{curl } A|^2) \, dx$. Hamilton's equations are $\dot{A} = -E$ and $\dot{E} = \text{curl (curl } A)$. These are Maxwell's equations for $E$ and $A$, the electric field and the vector potential. Here we find the behaviour that we expect from a gauge system: specifying an initial point does not serve to determine a unique dynamical trajectory. But we do find that if $(A(t), E(t))$ and $(A'(t), E'(t))$ are solutions of Maxwell's equations for the initial data $(A_0, E_0) \in N$, then for each $t$, $E(t) = E'(t)$ and there is a scalar function on space, $\Lambda(t)$, such that $A'(t) = A(t) + \text{grad } \Lambda(t)$. Equivalently: if $(A(t), E(t))$ and $(A'(t), E'(t))$ are dynamical trajectories with their origins in the same point of $(N, \sigma)$, then we have that $[(A(t), E(t)] = [(A'(t), E'(t)]$ for all $t$. Maxwell's equations do not determine the future value of $A(t)$, but they do determine in which gauge orbit $A(t)$ will lie.

### 10.2.3 Interpreting gauge theories

The interpretation of theories cast in Hamiltonian form is typically quite straightforward. Given a Hamiltonian system, $(M, \omega, H)$, one can always stipulate that it represents a system whose dynamically possible states stand in a one-to-one correspondence with the points of $M$ (call this the *literal* approach to interpreting the theory). Furthermore, in the context of classical mechanics, it often happens that $M = T^*Q$, where $Q$ can be viewed as the space of possible configurations of a set of particles or fields relative to some inertial frame. In this case, one ends up with an extremely attractive interpretation of the theory as a deterministic account of a physically reasonable system.

Unfortunately, interpreting gauge theories is seldom so simple. In the case of electromagnetism, the application of the literal strategy leads to the claim that our system has a distinct dynamically possible state for every pair $(A, E)$. But then one is committed to viewing electromagnetism as an indeterministic theory: specifying the initial dynamical state, $(A_0, E_0)$, fails to determine the future dynamical state, since if $(A, E)$ is a dynamically possible state at time $t$ according to Maxwell's equations, then so is $(A + \text{grad } \Lambda, E)$. The present state of the electromagnetic field fails to determine the future state of the field. Clearly, the same sort of indeterminism will arise whenever a gauge theory is given a literal interpretation.

This flies in the face of common sense: given initial data one can use Maxwell's equations to make highly accurate predictions. So there has to be something wrong with our literal interpretation of the theory. There are two possible diagnoses here. The first is that the interpretation, although essentially correct, needs to be supplemented with an account of measurement which will ensure that the predictions derivable from our gauge theory are perfectly determinate. The second is that the formalism of our gauge theory presented above contains 'surplus structure', which must be eliminated – either at the level of formalism or the level of interpretation – if we are to have a physically sensible understanding of the theory.[9] We consider each of these alternatives in turn.

If we wish to stick with our literal interpretation of our gauge theory, then we have to explain how it is that the theory is used to make determinate predictions despite its indeterminism. The most obvious way of doing so is to claim that some physically real quantities are not measurable. In order to produce determinate predictions, we need to work with physical quantities whose initial value problems are well-posed. A function on phase space has a well posed initial value problem iff it is gauge-invariant. So we will want to stipulate that only gauge-invariant quantities are measurable. This allows us to maintain predictability, even in the face of indeterminism. In the case of electromagnetism, implementing this strategy will mean accepting gauge-invariant quantities like the electric field, $E$, and the magnetic field, $B$, as measurable, while denying the vector potential, $A$, is directly measurable. Nonetheless, the vector potential will be a physically real quantity: since every point of phase space corresponds to a distinct physically possible situation, $(A, E)$ and $(A', E')$ will represent distinct situations in virtue of disagreeing as to the value of the quantity $A$ – even if $[A] = [A']$ so that the two states of affairs are not empirically distinguishable.

This sort of ploy is likely to seem rather desperate, however. It seems far more natural to insist that the *only* physically real quantities are gauge-invariant quantities (call this strategy the adoption of a *gauge-invariant* interpretation). In this case, one need not resort to a tricky account of measurement: one can stick to the orthodox position that every classical physical quantity is (in principle) measurable with arbitrary accuracy. Furthermore, the interpretation renders the theory deterministic, since specifying the initial state determines the future and past values of the physically real quantities. In the case of electromagnetism, for instance, it is natural to maintain that $E$ and $B$ taken together encode all of the structure of the electromagnetic field. When physical space is simply connected, the divergence free magnetic fields are in one-to-one correspondence with the gauge orbits of vector potentials, so that this move is tantamount to taking $[A]$ rather than $A$ as the physically real quantity. The resulting interpretation is fully deterministic, and supports an orthodox account of measurement. Notice that this establishes that determinism cannot be a formal property of theories: to ask whether electromagnetism is deterministic or not is not to ask a technical question about the formalism of the vacuum electromagnetism example; rather it is to ask whether one prefers a literal or gauge-invariant interpretation of this formalism.

There is a formal move which is associated with the interpretative move from literal to gauge-invariant interpretations: *reduction*. As it stands, our formalism is good at predicting which gauge orbit we will end up in, but lousy at predicting which point we will end up at. This suggests that what we really need is a theory of gauge orbits rather than points. Thus we attempt to do the following: we build a new manifold, $\tilde{M}$, whose points are the gauge orbits of $(N, \sigma)$; we then use $\sigma$ to construct a form $\tilde{\omega}$ on $\tilde{M}$; finally, we use $H$ to induce a Hamiltonian $\tilde{H}$ on $(\tilde{M}, \tilde{\omega})$. This is called the *reduced phase space.* It is not always possible to carry out this construction: the set of gauge orbits will not be a manifold, even locally, unless there exist sufficiently many gauge-invariant quantities to fix a gauge orbit. If not, it will of course be impossible to construct a symplectic form on $\tilde{M}$. But when $\tilde{M}$ is well behaved, $\tilde{\omega}$ is a symplectic form so that $(\tilde{M}, \tilde{\omega}, \tilde{H})$ is a genuine Hamiltonian system rather than a gauge system. This Hamiltonian system describes the way in which the trajectories of the original gauge system travel through gauge orbits. Giving a gauge-invariant interpretation of the original gauge theory is the same thing as giving a literal interpretation of the reduced phase space. In a sense, then, it is always easy to find a gauge-invariant interpretation, barring technical difficulties: simply construct the reduced phase space and adopt a literal interpretation. It can happen, however, that the reduced phase space does not admit any physically attractive literal interpretations – it need not, for instance, have the structure of a cotangent bundle over configuration space for reasonable particles or fields. Something like this actually happens in the case of electromagnetism when space is multiply connected.[10]

Gauge systems differ from Hamiltonian systems in that their equations of motion fail to determine the evolution of all of their variables. In the classical context, it is reasonable to regard this fact as reflecting a shortcoming of the formalism (the inclusion of excess variables) rather than a genuine ontological indeterminism. That is: it is preferable to look for interpretations in which only those variables whose evolution is determined by the equations of motion are taken to correspond to physically real quantities. If we can find a large enough set of such quantities to fix the gauge orbit of the system, and which can be taken to correspond to plausible physical quantities, then we have found an acceptable interpretation. In the case of ordinary vacuum electromagnetism on a simply connected spacetime, $E$ and $B$ serve this function admirably. In what follows, we will see that many foundational issues in classical and quantum gravity turn upon the difficulty of finding a complete set of physically reasonable gauge-invariant quantities for general relativity.

### 10.2.4 Quantizing gauge theories

A quantization of a Hamiltonian system $(M, \omega, H)$ consists of a Hilbert space, $\mathcal{H}$, equipped with a Hamiltonian operator, $\hat{H}$, and a representation of an appropriate subalgebra of the Poisson algebra of classical observables as an algebra of self-adjoint operators on $\mathcal{H}$. If $M$ can be written as $T^*Q$ for some natural configuration space $Q$, then one normally chooses $\mathcal{H}$ to be $L^2(Q, \mu)$, the space of complex functions on $Q$ which are square-integrable with respect to some physically relevant measure $\mu$.

How does one quantize a gauge theory? There are two main routes. The first is to construct the reduced phase space and apply canonical technique to the resulting Hamiltonian system. This tends to be impracticable, however – even when the reduced phase space exists, its structure is often difficult to determine. The alternative is to quantize the gauge system directly, employing a technique due to Dirac.[11]

Suppose that one has a gauge system $(N, \sigma, H)$ where $(N, \sigma)$ is the submanifold of a symplectic geometry $(M, \omega)$ determined by the first-class constraint $\gamma$ given by $C \equiv 0$. Then one chooses a set of co-ordinates on $M$, and finds a vector space, $V$, which carries a representation of their Poisson algebra as linear operators: if $(p, q)$

are canonical co-ordinates on $M$, then one looks for operators $\hat{q}$ and $\hat{p}$ satisfying $[\hat{q}, \hat{p}] = -i\hbar$. One then looks for a quantum analogue, $\hat{C}$, of the classical constraint – e.g. if $C = p^2$ then $\hat{C} = \hat{p}^2$.[12] Next, one imposes the quantum constraint to construct the space of physical states: $V_{phys} = \{\psi \in V: \hat{C}\psi = 0\}$. This ensures that the quantum states are gauge-invariant: if a given degree of freedom, $q$, is gauge (i.e. physically irrelevant) at the classical level, then it should be gauge at the quantum level.

Suppose, for example, that the classical constraint is $C = p$. Then we know from the example on p. 219 that the classical degree of freedom, $q$, is gauge. Thus gauge-invariant functions are independent of $q$. Working in the standard Schrödinger representation, we have that

$$\hat{C} = \hat{p} = i\frac{\partial}{\partial q}$$

so that stipulating that $\hat{C}\psi = 0$ amounts to requiring that the quantum wave functions be independent of the gauge degree of freedom, $q$. Similarly, imposing the quantum constraint corresponding to div $E = 0$ forces the states $\psi(A)$ of quantum electrodynamics to be independent of the choice of gauge (i.e. if $A' = A + \text{grad } \Lambda$, then $\psi(A) = \psi(A')$).

Finally, one looks for an appropriate inner product to make $V_{phys}$ into a Hilbert space, and for an appropriate quantum Hamiltonian, $\hat{H}$, which determines the quantum dynamics via Schrödinger's equation.

## 10.3 General relativity as a gauge theory

In its standard version, the hole argument looks something like the following (see Earman and Norton 1987). Let $\mathcal{M} = (M, g)$ be a model of general relativity, and let $d : M \to M$ be a diffeomorphism (called a *hole* diffeomorphism) which differs from the identity only on some small open set, $U$. The general covariance of the theory implies that $\mathcal{M}' = (M, d^*g)$ is also a model. If one views $\mathcal{M}$ and $\mathcal{M}'$ as representing distinct physically possible worlds, then one is committed to believing that general relativity is an indeterministic theory – specifying the state of the gravitational field on a Cauchy surface prior to $U$ fails to determine the state of the field inside $U$. Furthermore, it is claimed, if one is a substantivalist about the spacetime points of general relativity, then, *prima facie*, one *is* committed to viewing $\mathcal{M}$ and $\mathcal{M}'$ as representing distinct states of affairs. The conclusion is that substantivalists are *prima facie* committed to the doctrine that general relativity is an indeterministic theory.

In this section we will show that the hole argument is a special case of the observation made in the previous section: a gauge theory is indeterministic under a literal interpretation. We sketch the formulation of general relativity as a gauge theory and then argue that certain forms of substantivalism are, in fact, literal interpretations of this formalism.

### 10.3.1 Formalism

We begin our search for a gauge-theoretic formulation of general relativity by considering how to represent an instantaneous state of a general relativistic world, since we will want to work with the set of such representations as our phase space. To this end, we fix for of the remainder of this section a compact three manifold, $\Sigma$. Now consider a globally hyperbolic vacuum solution of the Einstein field equations, $(M, g)$, whose Cauchy surfaces are diffeomorphic to $\Sigma$.[13] We embed $\Sigma$ in $(M, g)$ via a diffeomorphism $\phi : \Sigma \to M$ such that $S = \phi(\Sigma)$ is a Cauchy surface of $(M, g)$, and we study the geometry which $g$ induces on $S$. We will take this geometry to represent an instantaneous state of the gravitational field.[14] This geometry is characterized by two symmetric tensors on $S$, $q_{ab}$, and $K_{ab}$. Here $q$ is the Riemannian metric on $S$, called the *first fundamental form*, which results from restricting to $T_x S$ (for $x \in S$) the inner product which $g$ induces on $T_x M$. $K$ is the *second fundamental form*, or *extrinsic curvature*, which encodes information about how $S$ is embedded in $(M, g)$. Very roughly, the extrinsic curvature of $S$ is the time derivative of $q$ (see eqn. 10.2.13 of Wald 1984). We use $\phi$ to pull these tensors back to $\Sigma$, and henceforth regard them as being defined on $\Sigma$ rather than on $S$.

This tells us what sort of geometric structure $\Sigma$ inherits when viewed as a submanifold of $(M, g)$. Now suppose that we imagine $\Sigma$ to come equipped with symmetric tensors $q$ and $K$, with $q$ a Riemannian metric. It is natural to wonder under what circumstances we can view $(\Sigma, q, K)$ as being the geometry of a Cauchy surface of some model $(M, g)$. The answer is that $\Sigma$ may be embedded in some $(M, g)$ in such a way that $q$ and $K$ arise as the first and second fundamental forms of $\Sigma$ iff the following two relations – known as the *Gauss* and *Codazzi* constraints respectively – hold:

$$R + (K_a^a)^2 - K^{ab} K_{ab} = 0$$

$$\nabla^a K_{ab} - \nabla_b K_a^a = 0.$$

Here, the metric $q$ on $\Sigma$ is used to define the scalar curvature, $R$, and the covariant derivative, $\nabla$. Note that these conditions make reference only to $q$ and $K$ – they do not mention $g$.

All of this suggests that we should regard a pair $(q, K)$ as representing the dynamical state of gravitational field at a given time if it satisfies the Gauss and Codazzi constraints. The metric $q$ describes the geometry of a Cauchy surface; the symmetric tensor $K$ describes the embedding of the slice in the ambient spacetime, and corresponds roughly to the time derivative of $q$. Thus, we can regard $q$ as the 'position' of the gravitational field. The natural starting point for writing down general relativity as a constrained Hamiltonian system is Riem($\Sigma$), the space of Riemannian metrics on $\Sigma$. We regard this as the configuration space, $Q$, of our theory of gravity. In order to construct the phase space, we first construct $T^*Q$, and then endow it with the canonical symplectic structure, $\omega$. The momentum canonically conjugate to $q$ is given not by $K$ but by

$$p^{ab} \equiv \sqrt{\det q}(K^{ab} - K_c^c q^{ab}).$$

The phase space of general relativity is the constraint surface $N \subset T^*Q$ given by the following first-class constraints, known as the *scalar* and *vector* constraints (or, alternatively, as the *Hamiltonian* and *momentum* constraints):

$$h \equiv \sqrt{\det q}(p^{ab}p_{ab} - \tfrac{1}{2}(p_a^a)^2 - R) = 0$$

$$h_a \equiv \nabla_b p_a^b = 0.$$

Each of these equations actually determines an infinite dimensional family of constraints, since each of them must hold at every point of $\Sigma$. Notice that the scalar and vector constraints are just the Gauss and Codazzi constraints, rewritten in terms of $p$ rather than $K$. Let $\sigma = \omega|_N$, and let $H \equiv 0$. Then general relativity is the gauge theory $(N, \sigma, H)$.[15]

At each point $x \in N$, the gauge orbit of $(N, \sigma, H)$ is infinite dimensional. These orbits have the following structure. Fix $x = (q, p)$ and $x' = (q', p')$ in $N$. Then $x$ and $x'$ lie in the same gauge orbit iff there is a solution to the Einstein field equations, $(M, g)$, and embeddings, $\phi, \phi' : \Sigma \rightarrow M$, such that: (i) $\phi(\Sigma)$ and $\phi'(\Sigma)$ are Cauchy surfaces of $(M, g)$; (ii) $q$ and $q'$ are the first fundamental forms of $\phi(\Sigma)$ and $\phi'(\Sigma)$; (iii) $p$ and $p'$ are the second fundamental forms of $\phi(\Sigma)$ and $\phi'(\Sigma)$. That is, two points are gauge-related if they describe spatial geometries of the *same* model of general relativity. Thus, each gauge orbit can be viewed as being the space embeddings of $\Sigma$ as a Cauchy surface of some model $(M, g)$ (it could, of course, equally well be viewed as being the space of such embeddings for any other model isometric to $(M, g)$). This means that each dynamical trajectory lies in a single gauge orbit: as the gravitational field evolves, it always stays in the same gauge orbit. This is, in fact, the significance of setting $H \equiv 0$: the vanishing of the Hamiltonian means that the dynamical trajectories are always tangent to the gauge orbits, which is just to say that once a dynamical trajectory is in a given gauge orbit, it never leaves. As we will see below, a zero Hamiltonian is closely related to the lack of a preferred time parameter.

Given a model $\mathcal{M} = (M, g)$, we can find a dynamical trajectory of $(N, \sigma, H)$ corresponding to $\mathcal{M}$ as follows. We first choose a foliation of $M$ by Cauchy surfaces (which are, of course, all diffeomorphic to $\Sigma$). We then choose a time function $\tau : M \rightarrow \mathfrak{R}$, which is compatible with the foliation in the sense that the level surfaces of $\tau$ are the Cauchy surfaces of the foliation. Finally, we choose a diffeomorphism $\Phi : M \rightarrow \Sigma \times \mathfrak{R}$ such that each Cauchy surface of the foliation, $S$, is mapped onto a set of the form $\Sigma \times \{t\}$. We call such a diffeomorphism an *identification map*, since it gives us a way of identifying the leaves of the foliation with $\Sigma$. We use $\Phi$ to push forward $g$, so that $\mathcal{M}' = (\Sigma \times \mathfrak{R}, \Phi^* g)$ is isometric to $\mathcal{M}$; the surfaces $\Sigma \times \{t\}$ in $\mathcal{M}'$ are Cauchy surfaces isometric to the Cauchy surfaces of our preferred foliation of $\mathcal{M}$. Now let $q_{ab}(t)$ and $p^{ab}(t)$ characterize the geometry of the Cauchy surface $\Sigma \times \{t\}$ in $\mathcal{M}'$. As $t$ varies, $(q_{ab}(t), p^{ab}(t))$ sweeps out a curve in $N$ – the points on this curve representing a sequence of Cauchy surfaces. This curve is a dynamical trajectory of $(N, \sigma, H)$. Choosing a different foliation, time function or identification map gives us a new dynamical trajectory, which will be related to the first by a gauge transformation – i.e. one can map one dynamical trajectory on to the other via a transformation of phase space which preserves gauge orbits.

The trajectories that correspond to the models $\mathcal{M}$ and $\mathcal{M}'$ which appear in the hole argument are so related. This shows that our approach respects the general covariance of general relativity in the sense that it is indifferent to changes of foliation, time function, and identification map. Changing any of these simply carries us from one dynamical trajectory to a gauge-related one.[16]

Now suppose that we look at two points $x = (q, p)$ and $x' = (q', p')$ which lie in the same gauge orbit, and which can be joined by an integral curve of a vector field generated by the vector constraint. Then we find that there is a diffeomorphism $d : \Sigma \to \Sigma$ such that $d^*q = q'$ and $d^*p = p'$. That is, we can regard $x$ and $x'$ as agreeing on the geometrical structure of $\Sigma$, and disagreeing only as to how the underlying geometrical properties are shared out over the points of $\Sigma$ – $x$ and $x'$ may represent, for example, a geometry on $\Sigma$ which has a single point of maximum scalar curvature, but according to $x$ this point is $z \in \Sigma$, whereas according to $x'$ it is $z' \in \Sigma$. Thus, we can view the gauge transformations generated by the vector constraint as shuffling the geometrical roles played by the points of $\Sigma$.

Unfortunately, the gauge transformations generated by the scalar constraint are considerably more complex. *Very* roughly, they can be thought of as corresponding to time evolution – two points differ by a gauge transformation generated by the scalar constraint if they can be seen as representing *distinct* Cauchy surfaces in a given model. In a generic spacetime, distinct Cauchy surfaces can be expected to have very different geometries, so that points in $N$ which are related by a gauge transformation generated by the scalar constraint will not in general represent the same geometry. In general, of course, a given gauge transformation is generated by a combination of both sorts of constraint.

Next, suppose that we have two dynamical trajectories which correspond to the same model $(M, g)$. Suppose, further, that the trajectories differ by a gauge transformation generated by the vector constraint. Then, in terms of the construction above which establishes a correspondence between models and dynamical trajectories: we can use the same foliation by Cauchy surfaces and the same time function $\tau$ to generate both dynamical trajectories; the difference between the trajectories can be attributed solely to the freedom available in the choice of an identification map. If, on the other hand, the trajectories differ by a gauge transformation generated by the scalar constraint, then the difference can be traced to the freedom in the choice of foliation and time function on $(M, g)$.

Modulo technical difficulties to be discussed in Section 10.4, we can convert this gauge theory into a true Hamiltonian system by factoring out the action of the gauge transformations to construct the reduced phase space. It is illuminating to proceed in two steps: we first partially reduce the phase space by factoring by the action of the gauge transformations generated by the vector constraint; we then complete the reduction by removing the gauge freedom associated with the scalar constraint.

At the first stage, we identify any two points in $N$ which are related by a gauge transformation generated by the vector constraint. The partially reduced phase space which results can be constructed as follows. We return to the beginning of our construction of $(N, \sigma, H)$, and replace the configuration space $Q = \mathrm{Riem}(\Sigma)$ of metrics on $\Sigma$ by $Q_0 = \mathrm{Riem}(\Sigma)/\mathrm{Diff}(\Sigma)$, the set of equivalence classes of diffeomorphically related metrics on $\Sigma$. We call $Q_0$ *superspace*. We now construct $T^*Q_0$ and impose the scalar constraint, to construct the presymplectic geometry $(\tilde{N}, \tilde{\sigma})$. The gauge theory ($\tilde{N}, \tilde{\sigma}, H \equiv 0$) is the partially reduced phase space formulation of general relativity. By identifying diffeomorphically related 3-metrics from the start, we have eliminated the need for the vector constraint.

The gauge orbits remain infinite dimensional even after this partial reduction has been carried out. If we now identify points in $\tilde{N}$ which are related by a gauge transformation generated by the scalar constraint, then we end up with a Hamiltonian system ($\tilde{M}, \tilde{\omega}, H \equiv 0$), where points in the phase space correspond to equivalence classes of diffeomorphically related models of general relativity.

### 10.3.2 Interpretation

Classical substantivalists and relationalists about space agree with one another that space exists, and that it has some fixed geometrical structure. They are divided over the question of the nature of the existence of this peculiar entity. Substantivalists hold that it consists of parts which maintain their identity over time, and that these parts stand directly in geometrical relations to one another, while material objects stand in spatial relations only in virtue of the relations obtaining between those parts of space which they occupy. Relationalists deny the substantivalist claim that space has genidentical parts. They maintain that space is best thought of as the structure of possible spatial relations between bodies. Such relations are to be taken as primitive, rather than being reduced to relations holding among the points of an underlying substratum. There are a couple of vivid ways of putting the issue between the two factions. Substantivalists, but not relationalists, believe in genidentical points. This means that substantivalists, but not relationalists, can help themselves to a straightforward account of the nature of absolute motion – it is motion *relative* to the genidentical parts of space.[17] In addition, substantivalists will follow Clarke in affirming, while relationalists will follow Leibniz in denying, that two possible worlds could instantiate all of the same spatial relations, but differ in virtue of which point of space plays which role (I occupy *this* point rather than *that* one).

Relativistic physics, however, seems to demand that one think in terms of space-time rather than space. Thus, the traditional doctrines are often translated into the four-dimensional context. Substantivalists and relationalists will again agree that the world has some given geometrical structure. Substantivalists understand the existence of spacetime in terms of the existence of its pointlike parts, and gloss spatio-temporal relations between material events in terms of the spatio-temporal relations between points at which the events occur. Relationalists will deny that spacetime points enjoy this robust sort of existence, and will accept spatio-temporal relations between events as primitive. It is now somewhat more difficult to specify the nature of the disagreement between the two parties. It is no longer possible to cash out the disagreement in terms of the nature of absolute motion (absolute acceleration will be defined in terms of the four-dimensional geometrical structure that substantivalists and relationalists *agree* about). We can, however, still look to *possibilia* for a way of putting the issue. Some substantivalists, at least, will affirm, while all relationalists will deny, that there are distinct possible worlds in which the same geometries are instantiated, but which are nonetheless distinct in virtue of the fact that different roles are played by different spacetime points (in this world, the maximum curvature occurs at *this* point, while it occurs at *that* point in the other world). We will call substantivalists who go along with these sort of counterfactuals *straightforward* substantivalists. Not all substantivalists are straightforward: recent years have seen a proliferation of *sophisticated* substantivalists who ape relationalists' denial of the relevant counterfactuals (see Brighouse 1994, Butterfield 1989, Field 1985, Maudlin 1990). For the time being, however, we will bracket this option. We will address the virtues and vices of sophisticated substantivalism in Section 10.7.

It is easy to see that (straightforward) substantivalists are committed to giving a literal interpretation of general relativity. Consider two models, $\mathcal{M} = (M, g)$ and $\mathcal{M}' = (M, d^*g)$, which are related by a hole diffeomorphism, $d$. Fix a foliation, time function, and identification map, and use them to construct dynamical trajectories $x(t)$ and $x'(t)$ in the phase space of general relativity which correspond to $\mathcal{M}$ and $\mathcal{M}'$. Because $d$ is a hole diffeomorphism, we can assume that $x(t) = x'(t)$ for $t \leq 0$, but $x(1) \neq x'(1)$. Substantivalists will view $x(t)$ and $x'(t)$ as representing distinct physically possible histories: although they represent the same spatio-temporal geometry (lying as they do in the same gauge orbit), they represent different distributions of their shared set of geometrical properties over the points of spacetime (if $y$ is a point on the spacelike surface $t = 1$, then $x$ represents it as having *these* properties while $x'$ represents it as having *those*). Indeed whenever $x$ and $x'$ are distinct points in the phase space of general relativity, a substantivalist will view them as representing distinct physical situations: either they represent distinct possible geometries for a given spacelike hypersurface, or they represent the same pattern of geometric relations differently instantiated. This is just to say that substantivalists are committed to a literal construal of the gauge-theoretic formulation of general relativity. And, like any literal interpretation of a gauge theory, substantivalism implies that the theory is indeterministic: if $x(1)$ and $x'(1)$ correspond to distinct possible situations, then the state corresponding to $x_0 = x(0) = x'(0)$ has multiple physically possible futures. This is the content of the hole argument.

As in Section 10.2, the best way to avoid this sort of indeterminism is to adopt a gauge-invariant interpretation of the theory. We can do this by giving a literal interpretation of the reduced phase space formulation of general relativity. Recall from above that the points of the reduced phase space are just the equivalence classes of diffeomorphic models of general relativity. Thus, in order to avoid the indeterminism of the hole argument, we have to accept that diffeomorphic models always represent the same physically possible situation (this proposition is known as *Leibniz equivalence* in the literature on the hole argument). And this, of course, is just to deny that there could be two possible worlds with the same geometry which differ only in virtue of the way that this geometry is shared out over existent spacetime points. Thus, modulo the existence of an attractive form of sophisticated substantivalism, one must be a relationalist in order to give a deterministic interpretation of general relativity.

There is another, closely related, motive for adopting a gauge-invariant interpretation of general relativity. As was noted at the end of Section 10.2, the existence of gauge degrees of freedom in a theory seems to tell us that the theory contains excess variables. The natural response is to seek an interpretation in which all and only the variables which correspond to physical degrees of freedom are taken seriously. Typically, we will want to say that it is just those variables whose evolution is determined by the differential equations of the theory that should be taken seriously in this way. Recently, a number of philosophers have joined the majority of physicists in advocating such gauge-invariant interpretations of general relativity – although almost all philosophers opt for a form of sophisticated substantivalism, while many physicists adhere to a strict relationalism.

At this point a potential technical problem looms. Relatively little is presently known about the structure of the reduced phase space of general relativity. It is known that this space has singularities corresponding to models of general relativity with symmetries, and is smooth elsewhere (Marsden 1981). Interesting and extensive smooth open sets have been constructed (Fischer and Moncrief 1996; see footnote 25). But the concern is sometimes expressed that the structure of generic regions of this space may not be smooth (see p. 141 of Kuchař 1993a, p. 267 of Unruh 1991, p. 2600 of Unruh and Wald 1989). Equivalently, one can wonder whether there exists a full set of gauge-invariant quantities on the unreduced phase space of general relativity. In fact, very few such quantities are known (see Goldberg, Lewandowski, and Stornaiolo 1992 for a rare example). Furthermore, it is known that there are *no* local gauge-invariant quantities.[18]

Until some progress is made on these technical questions, a dark cloud hangs over the programme of providing gauge-invariant interpretations of general relativity. The problem is this. One knows that the reduced phase space of general relativity exists as a mathematical set with some topology (although this topology may not be well enough behaved to support any interesting global geometric structure). And one knows that the points of the reduced phase space can be characterized as equivalence classes of models of general relativity. Philosophers who have advocated gauge-invariant interpretations have been satisfied with this sort of approach, which we dub *extrinsic*, since the characterization of the points of the reduced phase space is in terms of the gauge orbits of the original phase space. Such an extrinsic approach may, indeed, yield some sort of interpretation of general relativity. But we feel that something is lacking from an interpretation which stops at this point. Ideally, one would like an interpretation of general relativity which was underwritten by some *intrinsic* characterization of the points of reduced phase space. Indeed, in order to formulate a gauge-invariant quantum theory, one would like to be able to find a set of co-ordinates on the reduced phase space – or, equivalently, a full set of gauge-invariant quantities on the original phase space. This would amount to isolating the true (i.e. gauge-invariant) degrees of freedom of the theory. Although this is *not* essential for Dirac quantization, it nonetheless seems to us that it is the approach to the theory which yields the deepest understanding, since it underwrites an explicit characterization of the classical and quantum degrees of freedom of the system.[19]

Thus, we conclude that the present state of ignorance concerning the structure of the reduced phase space of general relativity – and the lingering worry that this structure may be monstrous – should give pause to advocates of gauge-invariant interpretations of the theory. We will, however, bracket this technical objection to gauge-invariant interpretations, and move on to discuss the two other sorts of problem which plague such interpretations:

- It appears to be a consequence of any gauge-invariant interpretation of general relativity that change does not exist, since any such interpretation requires us to regard two points, $x$ and $x'$, of the phase space of general relativity which correspond to distinct Cauchy surfaces of the same model as representing the same state of affairs, since they are related by a gauge transformation generated by the scalar constraint. Equivalently, if the only physical quantities are gauge-invariant, then there is no such quantity which allows us to distinguish between two such Cauchy surfaces.
- Accepting a gauge-invariant interpretation of general relativity, and thus treating the general covariance of general relativity as analogous to the gauge invariance of electromagnetism, leads to nasty technical and interpretative problems when one attempts to quantize the theory. These problems are so intractable that some have called for a re-evaluation of the standard understanding of general covariance.

We will discuss these problems for advocates of gauge-invariant interpretations in Sections 10.4 and 10.5. In Section 10.6, we will survey some interpretations which lie outside of the gauge-invariant orthodoxy. All of these options will be seen to have serious shortcomings, as well as distinctive attractive features.

## 10.4 Gauge invariance and change

Is there room for time or change when general covariance is understood as a principle of gauge invariance? *Prima facie*, a gauge-invariant interpretation of general relativity is descriptively inadequate because it cannot accommodate real change.

> 'To maintain that the only observable quantities are those that commute with all the constraints [i.e. the gauge-invariant quantities] seems to imply that the Universe cannot change. For this reason, this standpoint on observables was dubbed the *frozen time formalism*. The frozen time formalism never successfully explained the evolution we see all around us'.
>
> (Kuchař 1992, p. 293)

> 'How can changes in time be described in terms of objects which are completely time independent? In particular, since the only physical, and thus measurable quantities are those which are time independent, how can we describe the rich set of time dependent observations we make of the world around us? . . . The time independent quantities in General Relativity alone are simply insufficient to describe time dependent relations we wish to describe with the theory'.
>
> (Unruh 1991, p. 266)

Kuchař and Unruh are putting their fingers on an important question about the nature of time. It will be helpful in what follows to be clear on the relationship between their question and questions about the nature of time which are currently at the centre of philosophical discussion. Kuchař and Unruh are *not* interested in: (i) the direction of time; (ii) the objectivity of the metric structure of time; (iii) the reducibility of temporal relations to causal relations; or (iv) the existence of a moving now or flow of time. Rather, they are interested in whether or not change itself exists. And, of course, to the extent that the existence of time and that of change are closely related, they are interested in the existence, or lack thereof, of time as well. Thus, it is tempting to see them as engaging the same problematic about time, change, and flux that so occupied the Ancients.

In the quotations above, Kuchař and Unruh are driving at the following point. If we accept that the only physically real quantities of general relativity are gauge-invariant, then it follows that for any given model there is no physically real quantity which takes on different values when evaluated on Cauchy surfaces corresponding to distinct times. Which is, they claim, just to say that there is no change when general relativity is understood in this fashion: there is no evolution in time of the values of the physically real quantities. *Prima facie*, people who hold such a view have a very simple view of the nature of change: it is illusory. For this reason, Kuchař associates the reading of the general covariance of general relativity as a principle of gauge invariance with the name of Parmenides (Kuchař 1993a, p. 139).

Both Kuchař and Unruh denounce this Parmenidean view. They maintain that it flies in the face of our experience of time and change, and are sceptical that any coherent conceptual framework for the articulation of a quantum theory of gravity can be built upon such a foundation. Many physicists working on quantum gravity seem to be swayed by these arguments. But if the Parmenidean view of change is to be rejected as descriptively inadequate, what sort of account should be erected in its place? This is clearly a philosophical problem. Now, the vast majority of contemporary philosophical discussion about the nature of change is concerned with the existence of a moving now. This literature, whatever its merits as metaphysics, seems to be entirely irrelevant to the physical problems with which we are here concerned – since it almost always presupposes a pre-relativistic world view, and turns upon a question (the viability of the tenseless view of time) which is likely to appear long-since settled to relativistically minded physicists.

Philosophers come closest to the physicists' questions when they attempt to motivate the idea of a moving now. In introducing philosophical theses related to this latter problem, Le Poidevin and MacBeath comment that:

> 'It is a commonplace that time, not space, is the dimension of change. There is a wholly uncontroversial sense in which this is true: genuine change involves temporal variation in the ordinary properties of things: a hot liquid cools, a tree blossoms, an iron gate rusts. Purely *spatial* variation, for example the distribution of colours in patterned rug, does not count as a genuine change. Uncontroversial as this is, it requires explanation. What is special about time?'
>
> (1993, p. 1)

In fact, it is not uncommon to introduce the moving now as a solution to this problem, before going on to consider whether it is a coherent notion (see, e.g. Mellor 1993, p. 163). Unfortunately, philosophers seem to have all too little to say about what distinguishes change from mere variation.

Unruh, however, has made a very interesting and influential suggestion along these lines – one which is clearly motivated by physical concerns, but which strikes us as being philosophically provocative. He calls his view Heraclitean, in honor of Heraclitus' characterization of time as a war of opposites. The fundamental insight is that 'Time is that which allows contradictory things to occur':

> 'At any *one* time, the statement that a cup is both green and red makes no sense;
> these are mutually contradictory attributes. At any one time, a single particle
> can have only one position. However, at different times a particle can
> have many different positions, as can the cup have many different
> colours'.

(Unruh 1988, pp. 254–55; see also p. 2602 of Unruh and Wald 1989)

Time *sets* the values of the other variables, in the sense that at any given time each object takes on exactly one property from any exhaustive and mutually exclusive set (such as position or colour), although the property assumed is allowed to vary as the time parameter varies. This is suggestive, but ultimately inadequate. After all, the patterned carpet is allowed to take on different colours in different regions of its spatial extension, just as the coloured cup assumes different colours in different parts of its temporal extension. Furthermore, Unruh's proposal is unsatisfactory at the classical level because it depends on a primitive notion of genidentity, which is unlikely to be attractive in the context of field theories.

But he goes on to make a suggestion about how to understand the Heraclitean aspect of time in the context of quantum theories which seems to provide a means to distinguish the spatial from the temporal. Let's suppose that we are given a two-dimensional spacetime continuum and a complex function $\psi$ on this continuum which we take to represent the wave function of some quantum particle. Can we distinguish the temporal dimension from the spatial dimensions? Well, let $\{x, y\}$ be an arbitrary set of co-ordinates. Suppose that we want to calculate the probability of finding the particle in a given region. In order for this to make sense, we will need the total probability to be normalized. But, if $\psi$ is really the wave function, then we expect $\int_{-\infty}^{\infty} \int_{-\infty}^{\infty} |\psi(x, y)|^2 \, dx \, dy$ to be infinite – at each time, the particle must be somewhere, so the integral over the temporal dimension diverges. The solution is obvious: we should be looking at $\int_{-\infty}^{\infty} |\psi(x, y_0)|^2 \, dx$ and $\int_{-\infty}^{\infty} |\psi(x_0, y)|^2 \, dy$, for fixed $x_0$ and $y_0$. If we find that the former is constant for all values of $y_0$, while the latter is extremely badly behaved, then this licenses us to conclude that $y$ is a time variable: integrating over surfaces of constant $y$ gives us normalized probability densities. In this situation, we can view $y$ but not $x$ as setting the conditions for the other variables in the following sense: fixing a value of the time parameter allows us to formulate a quantum theory in which we can interpret the square of the wave function as a probability for measurement outcomes for the other variable. This is the Heraclitean role of time in a quantum world.

Of course, it is not straightforward to implement this strategy in the case of quantum gravity. Our discussion above presupposed that we were simply handed the measures $dx$ and $dy$. But this is to tantamount to knowing the physically relevant inner product on our space of quantum states. And, in fact, to identify the correct inner product is to go a long way towards solving the problem of time in quantum gravity (see especially Kuchař 1993a on this question). Nonetheless, Unruh's suggestion provides a framework in which to talk about the notions of change and time in quantum theories. As such, it has been influential in shaping discussion of the problem of time in quantum gravity, and provides a useful point of departure for our own discussion.

In the next two sections we will sketch some of the most important Parmenidean and Heraclitean approaches to classical and quantum gravity, as well as some of the most telling objections to these proposals. We will begin by sketching a timeless approach to general relativity and quantum gravity in Section 10.5, before turning to the details of a couple of Heraclitean approaches in Section 10.6.

In what follows, it will be helpful to keep in mind the following picture of the dispute. All parties seem to agree that understanding what the general covariance of general relativity is telling us about change and time is a precondition for the formulation of a theory of quantum gravity. It is true, of course, that in the context of general relativity, we can always cash out talk about time and change in terms Cauchy surfaces in models of general relativity. But, it is maintained, for anyone interested in canonical quantization of general relativity the resources to speak about time and change in quantum gravity must be found in (or imposed upon) the structure intrinsic to the phase space of general relativity. Here one has two options, neither of which is entirely attractive: (i) to embrace the Parmenidean view and attempt to make sense of quantum and classical theories of gravity which are *prima facie* without change or time; or (ii) to turn away from gauge-invariant interpretations of general relativity, and thus to base one's theories of gravity upon some other interpretation of the significance of the general covariance of the classical theory.

## 10.5 Life without change

For the Parmenidean, the challenge provided by the general covariance of general relativity is to give an account of the theory in which time and change are not fundamental, but which (i) is consistent with our experience, and (ii) motivates a viable programme for quantization. We will begin with the first point before turning to the second.

### 10.5.1 The classical theory

Even if it is granted that change is not a fundamental reality, we are nevertheless owed an account of how we can understand the observations of experimental physics and everyday life – observations which would naively seem to involve recording the presence of different properties at different moments of time. Unruh attributes to Bryce DeWitt the suggestion that the accommodation is afforded by time-independent correlations between non-gauge-invariant quantities, a suggestion Unruh himself rejects:

> 'The problem is that all of our observations must be expressed in terms of the physically measurable quantities of the theory, namely those combinations of the dynamical variables which are independent of time. One cannot try to phrase the problem by saying that one measures gauge dependent variables, and then looks for time independent correlations between them, since the gauge dependent variables are not measurable quantities within the context of the theory'.
>
> 'For example, Bryce DeWitt has stated that one could express measurements in the form of correlations. As an example, one could define an instant of time by the

correlation between Bryce DeWitt talking to Bill Unruh in front of a large crowd of people, and some event in the outside world one wished to measure. To do so however, one would have to express the sentence "Bryce DeWitt talking to Bill Unruh in front of a large crowd of people" in terms of physical variables of the theory which is supposed to include Bryce DeWitt, Bill Unruh, and the crowd of people. However, in the type of theory we are interested in here, those physical variables are all time independent, they cannot distinguish between "Bryce DeWitt talking to Bill Unruh in front of a large crowd of people" and "Bryce DeWitt and Bill Unruh and the crowd having grown old and died and rotted in their graves." The complete future time development of any set of variables is described in this theory by exactly the same physical variables. The physical variables, those which commute with all the constraints, can distinguish only between different complete spacetimes, not between different places or times within any single spacetime ... The subtle assumption in a statement like the one ascribed to DeWitt, is that the individual parts of the correlation, e.g. DeWitt talking, are measurable, when they are not'.

(Unruh 1991, p. 267)

We think that there is a more charitable interpretation to DeWitt's proposal: take it not as a way of trying to smuggle real change through the back door, but as a way of explaining the illusion of change in a changeless world. The idea is that we measure hyphenated relative observables, such as clock-1-reads-$t_1$-when-and-where-clock-2-reads-$t_2$. Such relative observables can be gauge-invariant and hence measurable according to the theory. We then get the illusion of change because we think that we can dehyphenate these hyphenated relative observables and treat each of the component variables as a genuine observable.

Rovelli's proposal for constructing 'evolving constants' is the most sophisticated and cogent way of fleshing out this suggestion.[20] In order to avoid the complications of general relativity we illustrate the proposal by means of a toy Newtonian example, which is concocted in such a way as to resemble, in relevant features, general relativity as a constrained Hamiltonian system (see Section 10.8.4 for a general construction). Consider the Newtonian account of the motion of a free particle on a line. We model this using a Hamiltonian system $(T^*\Re, \omega, H)$ where $T^*\Re = \{(x, p_x)\}$ is the cotangent bundle of the configuration space $\Re$, $\omega$ is the canonical symplectic structure, and $H$ is the kinetic energy, $\frac{1}{2}p_x^2$. We now employ the following formal trick, known as *parameterization*. We enlarge the phase space, by adding the time, $t$, and its canonically conjugate momentum, $p_t$, and we impose the constraint $0 = p_t + \frac{1}{2}p_x^2 \equiv C$. We also take $H' \equiv C = 0$ as our Hamiltonian. We solve Hamilton's equations to find that our dynamical trajectories $(x(\tau), t(\tau); p_x(\tau), p_t(\tau))$ are determined by the equations $\dot{p}_t = 0$, $\dot{p}_x = -\dot{t}\,\partial H/\partial x = 0$, and $\dot{x} = \dot{t}\,\partial H/\partial p_x = \dot{t}p_x$, where $\dot{t}$ is arbitrary and the overdot indicates differentiation with respect to the arbitrary time parameter $\tau$. These equations are equivalent to our original equations of motion, $p_x = $ constant and $x = p_x t + x_0$. Thus, each gauge orbit of the parameterized system corresponds to a dynamical trajectory of the original Hamiltonian system.

Since $H' = 0$, we expect that the parameterized system will display some of the same peculiar features as general relativity. The vanishing of the Hamiltonian means that the dynamical trajectories lie in gauge orbits. This means that there are no gauge-invariant quantities which distinguish between two points lying on the same dynamical trajectory. Most strikingly, the position of the particle, $x$, fails to commute with the constraint, and hence is not gauge-invariant – and so is not measurable under the standard reading of gauge invariance. Thus, the parameterized system seems to describe a Parmenidean world in which there is no change – and, in particular, no motion.

This is paradoxical: after all, the parameterized system is empirically equivalent to the original Hamiltonian system, which can be thought of as describing an ordinary Newtonian world. How can we account for this? The obvious response is that we can deparameterize in a preferred way since $t$, which is supposed to represent the absolute time of Newtonian mechanics, is in principle observable (say, by reading an idealized clock). But suppose we did not know this, or that we wished to eschew absolute time. How might we describe change, or something enough like change to explain ordinary observations, in the parameterized system?

Choose a global time function on the augmented phase space: a function $T(x, t; p_x, p_t)$ whose level surfaces are oblique to the gauge orbits. Consider any phase function $F$, not necessarily a constant of motion. Define an associated one-parameter family of phase functions $\{F_\tau\}_{\tau \in \Re}$ by the following two requirements:

$(R_1) \quad \{F_\tau, C\} = 0$

$(R_2) \quad F_{\tau = T(x, t; p_x, p_t)} = F(x, t; p_x, p_t).$

Here $\{,\}$ is the Poisson bracket on the augmented phase space. The first requirement says that each $F_\tau$ is constant along the gauge orbits – so that the $F_\tau$, unlike $F$, are gauge-invariant. In the case at hand, that means that the $F_\tau$ are *constants of motion*. The second requirement says that the value of $F_\tau$ is equal to the value of the phase function $F$ when the phase point $(x, t; p_x, p_t)$ lies on the level surface $T(x, t; p_x, p_t) = \tau$. The $F_\tau$ are then the evolving constants which can be used to describe change. In our toy example, take $F(x, t; p_x, p_t) = x$. $F$ does not commute with the constraint, so it is not gauge-invariant. Take $T(x, t; p_x, p_t) = t$. Then using $R_2$, $F_\tau = x - p - x(t - \tau)$. The value of $F_\tau$ on an orbit with initial $x_0$ is $p_x \tau + x_0$. It is easy to verify that $R_1$ holds: the $F_\tau$'s are gauge-invariant. Being constants of the motion, the $F_\tau$ do not change. But the family $\{F_\tau\}$ can be said to 'evolve'. In our example, the law of evolution is $dF_\tau/d\tau = p_x$. Our tendency to group gauge-invariant quantities into families which can be viewed as 'evolving' is supposed to account for our experience of change.

Kuchař (1993a) interprets Rovelli's proposal as saying that change in a non-gauge-invariant quantity can be observed, at least indirectly, by observing the gauge-independent quantities $F_{\tau_1}$ and $F_{\tau_2}$ (say) and then inferring the change $\Delta x$ in $x$ from $t = \tau_1$ to $t = \tau_2$ to be $F_{\tau_1} - F_{\tau_2} = p_x(\tau_1 - \tau_2)$. Kuchař's objection is that we are not told how to observe $\tau$ – we can't do it by observing that the value of $t$ is $\tau$, for $t$ is not an observable in the theory (this echoes the objection of Unruh discussed above).

A possible response is that we *don't* have to observe $\tau$. The $F_\tau$ are constants of the motion so it doesn't matter when they are observed – in principle, all the $F_\tau$ could be observed at once. This doesn't make Kuchař happy either: 'If all $\tau$ is eternally present, all time is irredeemable' (1993a, p. 139). Perhaps another way to put the criticism is to observe that it is hard to see how one would know *which* $F_\tau$ one is measuring without measuring $\tau$, which brings us back to Unruh's objection.

These criticisms are misplaced if, as we suggested above, the evolving constants proposal is construed modestly as explaining the illusion of change. Each $F_\tau$ is to be taken as a hyphenated relative observable: the-position-$x$-of-the-particle-when-the-$t$-clock-reads-$\tau$. We think that there is real change because we (mistakenly) think that we can dehyphenate for various values of the 'time' to get differences in the particle position – but the resulting 'observable' would fail to be gauge-invariant. To be satisfying this line has to be extended to hook up with actual perceptions. Here one might worry that the Kuchař–Unruh challenge comes back to haunt us at the level of neurophysiology if, as the theory seems to demand, all explanations must ultimately be stated in terms of gauge-invariant quantities. But one has to be careful here: we certainly cannot expect the theory to recapitulate our full phenomenology of time – to do so would be to demand that the theory contain a moving now, a demand which physics left behind long ago. But it surely *is* reasonable to demand that there be a place in the theory for models of human beings. In particular, it is reasonable to demand that the theory explain the illusion of change – any gravitational theory which cannot save such basic phenomena as the expansion of the universe will be empirically inadequate.[21]

Kuchař himself seems to admit that the evolving constants framework does meet this latter challenge, and thus *does* provide a way to make sense of time and change in the context of general relativity (see his comments on pp. 138–40 of Ashtekar and Stachel 1991). But even if this is granted, there remain problems with the quantization of general relativity within the evolving constants framework. If these cannot be overcome, this will be a severe blow to the credibility of Parmenidean interpretations of general relativity.

### 10.5.2 Quantum gravity

One of the signal virtues of the Parmenidean approach is that it underwrites an approach to quantizing general relativity which is very clear in its broad outlines (although it is, like every other approach to quantum gravity, extremely difficult in its details). If one regards the general covariance of general relativity as being strictly analogous to the gauge invariance of electromagnetism, then one will treat the quantum constraints of quantum gravity just as one treats the constraints of the quantum theory of the electromagnetic field: one imposes quantum constraints $\hat{h}\psi = 0$ and $\hat{h}_a\psi = 0$ – corresponding to the scalar and vector constraints of the classical theory – on the space of physical states of quantum gravity.

Heuristically, we proceed as follows. We work in the Schrödinger representation, so that the quantum states, $\psi(q)$, are elements of $L^2(\text{Riem}(\Sigma), \mu)$, and we represent our canonical co-ordinates, $q_{ab}$ and $p_{ab}$, via $\hat{q}_{ab}(x)(\psi(q)) = q_{ab}\psi(q)$ and $\hat{p}_{ab}(x)(\psi(q)) = i(\partial/\partial q_{ab}(x))\psi(q)$. Then, again heuristically, writing the quantum vector constraint as $\hat{h}_a \equiv \hat{\nabla}_b \hat{p}_a^b$, we can show that imposing this constraint amounts to requiring that the quantum wave functions be invariant under three-dimensional diffeomorphisms. Formally, we can write $\hat{h}\psi = 0$ as

$$\sqrt{\det q}\left(\left(q_{ab}q_{cd} - \tfrac{1}{2}q_{ac}q_{bd}\right)\frac{\partial^2}{\partial q_{ac}\,\partial q_{bd}}\psi[q] - R(q)\,\psi(q)\right) = 0,$$

where $R$ is the scalar curvature of $q$. In this form, the quantum scalar constraint is known as the *Wheeler–DeWitt equation*. One then seeks a representation of an appropriate set of observables on the space of physical states, and looks for an appropriate inner product. Proponents of evolving constants hope to find an appropriate representation of the algebra of classical evolving constants as a set of linear operators on the space of physical states such that: (i) the quantum evolving constants are in fact constants of motion (i.e. they commute with the quantum Hamiltonian); and (ii) there is a unique inner product on the space of physical states which makes the quantum evolving constants self-adjoint (see Ashtekar and Tate 1994, Ashtekar 1995). Finally, one must construct a quantum Hamiltonian, $\hat{H}$. The classical Hamiltonian can be written as a sum of the classical constraints, so that it is identically zero on the constraint surface which forms the phase space of general relativity. Thus, it is natural to write the quantum Hamiltonian as a sum of the quantum constraints. But since these constraints annihilate the physical states, one concludes that $\hat{H}\psi = 0$.

This quantization programme faces some daunting technical problems (see Hájiček 1991, Section 6.4 of Isham 1991, and Section 15 of Kuchař 1992 for critical discussion). But there are also conceptual problems. The foremost is, of course, the problem of time: since the quantum Hamiltonian is zero, there is no equation which governs the dynamical evolution of the physical state. Thus there appears to be no change in quantum gravity. Now, we have seen above that the vanishing of the Hamiltonian is a direct consequence of requiring that the quantum constraints should annihilate the physical states. Parmenideans claim that this move is justified by analogy with the successful quantization of other gauge theories. If $q$ and $q'$ are related by a classical gauge transformation, then we expect that $\psi(q) = \psi(q')$. This principle is particularly plausible when $q$ and $q'$ are related by a gauge transformation generated by the vector constraint – demanding that $\hat{h}_a\psi = 0$ is equivalent to demanding that $\psi(q) = \psi(q')$ whenever $q' = d^*q$ for some diffeomorphism $d : \Sigma \rightarrow \Sigma$. And this is surely mandatory, since otherwise we could use quantum gravity to distinguish between the (classically) empirically indistinguishable spatial geometries $(\Sigma, q)$ and $(\Sigma, q')$.

In the case of the scalar constraint, no such direct geometric justification is available. Here, the Parmenidean must rely upon the general analogy between general relativity and other gauge theories, and upon the following consideration. The scalar constraint of general relativity implements time evolution, just as the constraint imposed on the parameterized Newtonian particle does. Now, we can apply our quantization algorithm to the parameterized particle. The configuration space of the particle is Newtonian spacetime, so the quantum states are wave functions on spacetime, subject to the constraint $(\hat{p}_t + \frac{1}{2}\hat{p}_x^2)\,\psi(x, t) = 0$. In the Schrödinger representation, in which $\hat{p}_t = -i\hbar\,\partial/\partial t$, the constraint becomes the familiar Schrödinger equation – modulo the fact that the wave functions of the parameterized particle are defined on spacetime rather than space.[22]

Thus, the quantum theory of the parameterized particle is intimately related to the quantum theory of the ordinary unparameterized particle. Now, the scalar constraint of general relativity is quadratic in momentum, whereas the constraint of the parameterized particle is linear in $p_t$, so the Wheeler–DeWitt equation is not even formally a Schrödinger equation – it cannot be solved for the time rate of change of the quantum state. Nonetheless, Parmenideans maintain, one may view the Wheeler–DeWitt equation as encoding all of the information about time and change that is relevant to quantum gravity, in analogy with the quantum constraint of the parameterized particle (since both of them, intuitively, are the quantum versions of classical constraints which generate time evolution).

But, if the Wheeler–DeWitt equation encodes this information, where is the key which will grant us access to it? This is where the evolving constants come in. One presumes that there is, for instance, a quantum evolving constant which corresponds to the classical evolving constant which measures the volume of the universe at different times. By asking for the expectation value of this quantum evolving constant, we can find evolution and change in the *prima facie* changeless world of quantum gravity.

It is at this point that the objections raised by Kuchař and Unruh return with redoubled force. Our discussion of the (classical) evolving constants of parameterized Newtonian particle proceeded smoothly only because we were working with a system in which Newtonian absolute time was merely hidden, and not absent from the beginning. But suppose that we are given a gauge system with a vanishing Hamiltonian, and that this system, like general relativity itself, does *not* arise via parameterization from a Hamiltonian system. Then there will be considerable arbitrariness in the selection of our evolving constants. In particular, we will not have any natural criterion to appeal to in place of $R_2$ above: we will not know which foliations of our phase space count as foliations by surfaces of constant time, and so our choice of evolving constants will be vastly underdetermined. In particular, we will have no way of guaranteeing that the foliation chosen corresponds to time rather than space – intuitively our family $\{F_\tau\}$ may correspond to the family {the-mass-of-the-object-at-the-point-$x_\tau$-of-space-at-time-$t_0$} rather than to {the-mass-of-the-rocket-at-time-$t = \tau$}. And here we are back to the problem discussed in Section 10.4: If one doesn't take time and change as fundamental realities, how is one to distinguish between mere spatial variation and true temporal change?

Rovelli himself takes a hard line on this question, and argues that, prior to quantization, any set of evolving constants is as good as any other. One expects, however, that different sets of evolving constants will lead to different quantizations, and that experiment will eventually allow one to determine which sets of evolving constants are viable. Indeed, this situation already arises in the context of quantum mechanics. Hartle (1996) discusses the quantum mechanics of a parameterized Newtonian description which results when non-standard time functions are employed. He finds that predictions that depart from those of standard non-relativistic quantum mechanics can result. This embarrassment can be overcome by restricting to 'good' time functions. Of course, it is not evident how such restrictions could be implemented in quantum gravity – so one expects to be faced with a highly ambiguous recipe for quantization. Whether this counts as a strength or a weakness will be a matter of taste.

Here we reach an impasse of a sort which is quite typical of debates concerning the conceptual foundations of classical and quantum gravity. On our reading, the heart of Kuchař and Unruh's objections to the Parmenidean view is to be found at this point. They both possess an intuition which runs directly counter to that of Rovelli. They see the distinction between change and variation as fundamental, and doubt that one will be able to formulate a fruitful approach to quantizing general relativity which is blind to this distinction. Thus they see Rovelli's willingness to accept *any* evolving constants as a sign of the conceptual bankruptcy of the Parmenidean approach. Rovelli, of course, rejects this interpretation. For him, tolerance of the radical underdetermination of the evolving constants is part of an attempt to shrug off outmoded classical intuitions about time and change. Indeed, one of the strengths of the Parmenidean approach has been its hints at the discrete structure of quantum spacetime.[23] Both sides agree that the proof will be in the pudding: vindication, if it comes at all, will come in the guise of a viable theory of quantum gravity. In the mean time, arguments about the proper way forward will continue to be cast in terms of disagreements concerning the nature of change – debates about content and method are inextricably intertwined.

### 10.6 *Vive le change*!

Not everyone accepts the Parmenidean approach. Some believe that the analogy between the general covariance of general relativity and principles of gauge invariance of theories like electromagnetism is profoundly misleading. These physicists are sceptical that the Parmenidean approach sketched in the previous section is either mathematically feasible (since they doubt that one will be able to find an appropriate inner product without appealing to a Heraclitean notion of time) or physically meaningful (since they doubt that one would be able to derive sensible physical predictions from a timeless theory). They believe that a Heraclitean time must be found within (or grafted on to) the conceptual structure of general relativity prior to quantization. In this section, we attempt to give the flavour of this approach.

Heracliteans comes in two varieties. They concur that Parmenideans profoundly misunderstand the nature of the general covariance of general relativity, but they disagree as to the correct account. On the one hand, there is a radical wing which forsakes a cornerstone of the traditional reading of general covariance: that in general relativity there is no preferred splitting of spacetime into space and time. On the other hand, there is a more conservative faction which attempts to hew to a traditional understanding of the general covariance of general relativity, while denying that it is a principle of gauge invariance.[24]

There are a number of varieties of radical Heracliteanism (see pp. 6–8 of Kuchař 1992 for an overview). The most straightforward is probably the doctrine that the mean extrinsic curvature is a good time variable for classical and quantum gravity (the mean extrinsic curvature at a point $x \in \Sigma$ of a Cauchy surface with geometry $(q, p)$ is $\tau = q_{ab} p^{ab} / \sqrt{q}$). The point of departure is the observation that there is a large open subset of the space of models of general relativity consisting of spacetimes which admit a unique foliation by surfaces of constant mean curvature (CMC surfaces).[25] If $(M, g)$ is a model which is CMC sliceable, then the mean extrinsic curvature, $\tau$, varies monotonically within the CMC foliation. This observation motivates the following programme (see Beig 1994, Fischer and Moncrief

1996). We restrict attention to that subset of the phase space of general relativity which corresponds to CMC sliceable models. We then solve the vector constraint, by moving to superspace. At this point, we have a gauge system in which the gauge orbits are infinite dimensional. We transform this into a gauge system with one-dimensional gauge orbits by stipulating that we are only interested in those points of phase space which represent CMC slices.[26] In effect, we have chosen a foliation for every model, and finessed the necessity to choose an identification map by working in superspace. The only remnant of the original general covariance of general relativity is the freedom to reparameterize the time parameter, $\tau$. Furthermore, the remaining constraint is linear in the momentum conjugate to $\tau$. Thus, general relativity is now written in the form of a parameterized system: by choosing a distinguished parameterization of $\tau$, one can construct a time-dependent Hamiltonian system whose parameterization is the CMC-reduced form of general relativity. The Hamiltonian, $H(\tau)$, measures the volume of the Cauchy surface of mean extrinsic curvature $\tau$. As noted by Isham (1991, p. 200), it is quite strange to have a theory of the entire universe in which the dynamics is driven by a time-dependent Hamiltonian – usually such Hamiltonians are employed to model the influence of the environment on the system. (This observation also applies to the internal time framework sketched below.)

One hopes that canonical quantization of this Hamiltonian system would lead to a quantum field theory of gravity, complete with a time variable and a (time-dependent) Hamiltonian which governs the evolution of the quantum state via an ordinary Schrödinger equation, and that the expectation value of the volume of the universe, $\langle \hat{H}(\tau) \rangle$, would vary with time. This quantization programme has been successfully carried out for $2 + 1$ general relativity, and is being actively developed in the full $3 + 1$ case (see Carlip 1998 for the $2 + 1$ case). One of the remarkable results obtained is that the CMC method of quantization is equivalent to some Parmenidean constructions of $2 + 1$ quantum gravity. But, of course, all models of $2 + 1$ general relativity are flat, so that the phase space of the classical theory is finite dimensional, and the quantum theory is a variety of quantum mechanics. One does not expect this sort of equivalence to arise in the $3 + 1$ case, where the classical phase space is infinite dimensional and the quantum theory is a quantum field theory.

Other radical Heraclitean proposals have similar structures. One much-discussed method is to postulate the existence of a form of matter which allows one to introduce a preferred foliation. For instance, one can postulate the existence of a cloud of dust, each mote of which is a clock. This fixes a reference frame and a time parameter.[27] One then uses this additional structure to reduce general relativity to a parameterized system, which, upon quantization, yields an ordinary Schrödinger equation. Breaking the general covariance of general relativity by introducing preferred frames allows one to introduce a time variable, $t$, at the classical level which is carried over to quantum gravity. This time variable is Heraclitean: the wave function describing the state of the gravitational field depends on $t$, and one is able to find an inner product on the space of instantaneous quantum states which is conserved in $t$. This allows one to make intelligible time-dependent predictions of measurement outcomes.

Such radical approaches view the general covariance of general relativity as an artifact of a particular formulation of the theory. Under this reading, it is true that general relativity can be given a Diff($M$)-invariant formulation. But, it is contended, this formulation is by no means the most perspicuous. By using our preferred co-ordinates to fix the gauge, we can bring to the fore the true physical content of the theory – just as the content of Newtonian physics is most clear when the theory is written in its traditional, non-generally covariant, form. Of course, this reading is vulnerable to the accusation that it betrays the spirit of general relativity:

> '... foliation fixing prevents one from asking what would happen if one attempted to measure the gravitational degrees of freedom on an arbitrary hypersurface. Such a solution ... amounts to conceding that one can quantize gravity only by giving up general relativity: to say that quantum gravity makes sense only when one fixes the foliation is essentially the same as saying that quantum gravity makes sense only in one coordinate system'.
>
> (Kuchař 1992, p. 228)

This criticism is extremely telling. To forsake the conventional reading of general covariance as ruling out the existence of preferred co-ordinate systems is to abandon one of the central tenets of modern physics. Unsurprisingly, radical Heracliteanism has few adherents – such approaches are explored because they are technically tractable, not because they are physically plausible.

Kuchař advocates a more conservative – and ambitious – brand of Heracliteanism. He articulates a subtle reading of general covariance which differs from both that of the Parmenideans and that of the radical Heracliteans: he denies that general covariance is a principle of gauge invariance, without countenancing the existence of a preferred foliation or a preferred set of co-ordinates (see Kuchař 1972 for the original proposal, and Kuchař 1992, 1993 for recent discussions). A good starting point for understanding his approach is to consider the dual role that time plays in a Newtonian world. On the one hand, we can construct a time function, $t(x)$, which assigns a time to each point in Newtonian spacetime. In this guise, time is a scalar function on spacetime. However, we can also think of time as a collection of instants. Because simultaneity is absolute in Newtonian physics, this collection can be thought of as a one-dimensional family, parameterized by $t$. Equivalently, the real numbers parameterize the ways in which one can embed an instant (surface of simultaneity) into spacetime. Of course, a time function on spacetime suffices to model this role of time as well: the permissible embeddings of instants are just the level surfaces of $t(x)$.

In the context of special relativity one doesn't have a preferred notion of simultaneity, and the two roles of time are no longer so tightly intertwined. One is still often interested in time functions, $t(x)$, on spacetime – especially the time functions associated with inertial observers. But in its guise as the space of instants, time can no longer be thought of as a one-parameter family, since the spirit of special relativity forbids us from identifying the possible embeddings of the instants with the level surfaces of the time function associated with any one inertial observer. In this context there is considerable ambiguity in the notion of an instant. For definiteness, let's fix upon surfaces of simultaneity relative to inertial observers. Then the family of instants will be four-dimensional: if we fix a fiducial instant, $\Sigma_0$, then an arbitrary instant, $\Sigma$, can be reached by applying a time translation and/or Lorentz boost to $\Sigma_0$. So we can think of time as being four-dimensional in Minkowski spacetime.[28]

Let's now consider a generic model, $\mathcal{M} = (M, g)$, of general relativity. As in the previous cases, it is easy to write down a time function, $t(x)$, on $\mathcal{M}$. One simply requires that its level surfaces be Cauchy surfaces. Let's fix such a time function – and the corresponding foliation of $M$ by Cauchy surfaces – and enquire after a coherent notion of 'instant' in general relativity. Here, in order to respect the traditional understanding of general covariance, we will want our set of instants to include all Cauchy surfaces of $\mathcal{M}$. Thus, time, *qua* the set of instants, becomes infinite dimensional. We are interested in examining the role that these two notions of time play in the phase space of general relativity.

To this end, we focus our attention on the gauge orbit in the phase space of general relativity which corresponds to $\mathcal{M}$. For a generic $\mathcal{M}$ admitting no symmetries we expect that a given point $(q, p)$ of the phase space represents a 3-geometry which occurs only once as a Cauchy surface of $\mathcal{M}$. That is, we expect that specifying the tensors $q$ and $p$ on $\Sigma$ is sufficient to determine a map $X : \Sigma \rightarrow M$ which tells us how $\Sigma$ must be embedded in $\mathcal{M}$ in order to induce the geometry $(q, p)$. Fixing an arbitrary co-ordinate system on $\Sigma$ and a co-ordinate system on $M$ of the form $\{x^\mu\} = \{t, x^a\}$, we find that specifying the twelve independent components of $q_{ab}(x)$ and $p_{ab}(x)$ on $\Sigma$ determines four real functions on $\Sigma$, $\{X^A(x)\} = \{T(x), X^a(x)\}$, which tell us how $\Sigma$ is embedded in $\mathcal{M}$.

We can think of these maps as functions on the phase space of general relativity: for each point $(q, p)$ of the phase space, $X^A(x)$ is a real number. Following Kuchař, we use the notation $X^A(x; q, p]$ to emphasize that each $X^A$ is a function on $\Sigma$ and a functional (in the physicist's sense) on phase space. This suggests that we could use the $X^A$ and their conjugate momenta, $P_B$, as co-ordinates on the phase space, in place of the $q_{ab}(x)$ and $p_{ab}(x)$. Now, of course, knowing the $q_{ab}(x)$ and $p_{ab}(x)$ for a given point of the phase space gives us more information than just the way that the instant is embedded in spacetime – it also tells us about the state of the gravitational field at that instant. Thus, the geometric variables, the $q_{ab}(x)$ and $p_{ab}(x)$, contain information beyond that which is contained in the embedding variables, $X^A(x; p, q]$ and $P_B(x; p, q]$. Indeed, $q_{ab}(x)$ and $p_{ab}(x)$ contain twelve independent components. So specifying the geometrical data gives us twelve functions on $\Sigma$, whereas specifying the embedding variables gives us only eight. One surmises that there must exist additional variables which represent the true physical degrees of freedom of the gravitational field relative to any given instant (i.e. relative to any fixed values of the embedding variables). Thus, we postulate that the dynamical state of the gravitational field at a given instant is represented by gravitational configuration variables, $\phi^r(x)$ $(r = 1, 2)$, on $\Sigma$, together with their momentum variables, $\pi_s(x)$ $(s = 1, 2)$.

So far, we have been restricting our attention to a single gauge orbit of the phase space of general relativity, and depending upon a particular set of co-ordinates for the corresponding model. More ambitiously, we could look for embedding variables, $X^A(x; q, p]$ and $P_B(x; q, p]$, defined globally on the phase space of general relativity. We then look for a canonical transformation of the phase space of general relativity of the form

$$\{q_{ab}(x), p_{ab}(x)\} \mapsto \{X^A(x), P_B(x); \phi^r(x), \pi_s(x)\}$$

(i.e. we are looking for a change of co-ordinates which preserves the presymplectic structure). Each of these new canonical variables associates a map from $\Sigma$ to the real numbers with each point of phase space. We require that the embedding variables satisfy the following two desiderata:

(1) Global time. Each gauge orbit of general relativity contains exactly one point corresponding to a given fixed value of the embedding variables.
(2) Spacetime interpretation. If $(q, p)$ and $(q', p')$ correspond to intersecting Cauchy surfaces of a given model, then we demand that $X^A(x; q, p] = X^A(x; q', p']$ for points $x \in \Sigma$ which lie in their intersection.

The first condition guarantees that the values assumed by the embedding variables at a given point of phase space do indeed single out a single instant in any given model of general relativity. The second condition guarantees the notion of time as a collection of instants is compatible with the notion of time as represented by a spacetime scalar: the time, $T = X^0(x; q, p]$ assigned to a given point $x$ of a relativistic spacetime is the same for all Cauchy surfaces $(q, p)$ passing through that point.[29] If we can find a canonical transformation satisfying these two desiderata, we then proceed to rewrite the constraints in terms of the new co-ordinates, where they will assume the form $C_A \equiv P_A(x) + h_A(x; X, \phi, \pi] = 0$.

If all this can be achieved, then we have rewritten general relativity in the *internal time formulation* (time is said to be internal in this formalism because it depends only on phase space variables). It would allow us to reconcile the two roles of time. The internal time on phase space admits an interpretation as a spacetime scalar for any particular model. But general covariance is not broken: there are no preferred foliations or co-ordinate systems.[30] One can, if one likes, pay special attention to the level surfaces of the time function which the internal time induces on models. But the formalism itself does not privilege these level surfaces: the constraints can be viewed as governing the evolution of the gravitational degrees of freedom between arbitrary instants.

One could go on to apply the Dirac quantization algorithm to the internal time formulation of general relativity. Here, the configuration variables are the embedding variables, $X^A$, and the gravitational variables, $\phi^r$. Thus, the quantum states will be wave functions over the classical configurations of $X$ and $\phi$, of the form $\psi(X, \phi)$. We will want to impose the constraints, $\hat{C}_A \psi[X, \phi] = 0$. Because the classical constraints are linear in the momentum, the quantum constraints become Schrödinger equations:

$$-i \frac{\partial \psi(X, \phi)}{\partial X^A(x)} = \hat{h}_A(x; X, \hat{\phi}, \hat{\pi}] \psi(X, \phi),$$

which govern the change in all of the configuration variables under small variations in the embedding variables. Let us denote the space of wave functions satisfying these constraints by $V_0$. At this point, one could proceed as in the Parmenidean programme of Section 10.5: complete the quantum theory by finding quantum evolving constants, and find an inner product on $V_0$ which renders them self-adjoint.

Kuchař, however, rejects the Parmenidean reading of the significance of the quantum constraints and as a result, he denies that the observables of quantum gravity are self-adjoint operators on $V_0$ which commute with the constraints. His objections concerning the quantum constraints can be traced back to a subtle difference between his reading of the significance of the general covariance of general relativity, and that of the Parmenideans. As we saw above, his programme for quantization takes as its point of departure a formulation of general relativity which fully respects the general covariance of the theory. Kuchař does not, however, subscribe to the Parmenidean dogma that the constraints of general relativity should be understood as the generators of gauge transformations. Rather, he draws a sharp distinction between the role of the vector constraint, and that of the scalar constraint. In particular, he holds that the observable quantities of general relativity must commute with the vector constraint, but that they need not commute with the scalar constraint.

The rationale is as follows. In the case of the vector constraint, we can say that '[t]wo metric fields, $q_{ab}(x)$ and $q'_{ab}(x)$, that differ only by the action of Diff($\Sigma$), i.e. which lie on the same orbit of $h_a(x)$, are physically indistinguishable. This is due to the fact that we have no direct way of observing the points $x \in \Sigma$'.[31] The difference between two geometries, $q_{ab}(x)$ and $q'_{ab}(x)$, related by a transformation generated by the vector constraint is unobservable: it is the difference between identical spatial geometries, which differ only in virtue of *which* point of $\Sigma$ plays *which* geometrical role. The role of the scalar constraint, $h$, is very different:

> '... it generates the dynamical change of the data from one hypersurface to another. The hypersurface itself is not directly observable, just as the points $x \in \Sigma$ are not directly observable. However, the collection of the canonical data $(q_{ab}(x), p_{ab}(x))$ on the first hypersurface is clearly distinguishable from the collection $(q'_{ab}(x), p'_{ab}(x))$ of the evolved data on the second hypersurface. If we could not distinguish between those two sets of data, we would never be able to observe dynamical evolution'.
>
> (Kuchař 1993, p. 137)

Or, again, '[t]wo points on the same orbit of [the scalar constraint] are two events in the dynamical evolution of the system. Such events are physically distinguishable rather than being descriptions of the same physical state' (Kuchař 1992, p. 293). Thus, Kuchař believes that there are physically real quantities which do not commute with the scalar constraint of general relativity.

In the internal time formulation, this point will take the following form. $T = X^0(x; q, p]$ has a different status from the $X^a(x; q, p]$ for $a = 1, 2, 3$. The former can be thought of as specifying the instant corresponding to $(q, p)$, while the latter specify how $\Sigma$ is mapped on to this instant. That is, $T$ specifies a Cauchy surface, while the $X^a$ tell us how $\Sigma$ is mapped on to this Cauchy surface. Thus, the constraint $C_T \equiv P_T(x) + h_T(x; X, \phi, \pi] = 0$ should be thought of as governing time evolution, while the $C_a \equiv P_a(x) + h_a(x; X, \phi, \pi] = 0$ generate gauge transformations which correspond to altering the way that $\Sigma$ is mapped on to a given Cauchy surface.

Thus, according to Kuchař's analysis, the quantum constraints should not be treated uniformly, as they are within the Parmenidean approach. The quantum constraints $\hat{C}_a \psi = 0$ should be imposed as in the standard approach. This will, as usual, ensure that the theory is indifferent to diffeomorphisms acting on $\Sigma$. But the quantum constraint $\hat{C}_T$ demands a different approach. Kuchař recommends the following procedure. Begin by arbitrarily fixing values $T(x) = X^0(x)$ for all $x \in \Sigma$.

This specifies an instant in general relativity. Fixing the state of the gravitational field now amounts to fixing the values of $\phi^r(x)$ and $\pi_s(x)$ on $\Sigma$, and the classical observables at this instant are just functions of these field variables. It is natural to think of an instantaneous state of the gravitational fields as a wave function $\Psi(\phi^1, \phi^2)$ over field configurations over $\Sigma$ (with configurations identified if they are related by a diffeomorphism of $\Sigma$). An observable will then be any (diffeomorphism-invariant) function of $\hat{\phi}^r(x)$ and $\hat{\pi}_s(x)$. In order to bring dynamics into the picture, we impose the constraint $\hat{C}_T\Psi(X, \phi) = 0$. This gives us the Schrödinger equation,

$$-i\frac{\partial\Psi(T, \phi)}{\partial T(x)} = \hat{h}_T(x; X, \hat{\phi}, \hat{\pi}]\Psi(T, \phi),$$

which tells us how the quantum states change under infinitesimal changes of our instant. Now one attempts to find an inner product on the space of instantaneous states which is preserved under the evolution induced by the constraints. If one is successful, then one has a quantum theory of gravity: a Hilbert space, observables, and dynamics. Notice that although the states are gauge-invariant (since they satisfy the quantum constraints), the observables need not be: in general, one expects the expectation values of a function of $\phi^r(x)$ and $\pi_s(x)$ to vary from embedding to embedding (i.e. from instant to instant). Hence, Kuchař's proposal leads to a theory of quantum gravity in which the infinite dimensional internal time plays the role of a Heraclitean time variable.

This quantization procedure has been successfully applied to a number of theories which arise from general relativity by killing infinitely many degrees of freedom (see Section 6 of Kuchař 1992). Before it can be applied to full general relativity, however, a number of severe technical difficulties must be overcome – including the fact that it appears to be impossible to satisfy Global Time for full general relativity, and the fact that no one has yet been able to write down in closed form an internal time variable which satisfies Spacetime Interpretation.[32] Nonetheless, work continues on the programme, in the hope that it is possible to overcome these difficulties (perhaps by modifying the original programme).

There are also a number of potential difficulties in interpreting the formalism. The great advantage of the internal time proposal is that it casts quantum gravity into a familiar form: one has a quantum field theory whose states are wave functions over the classical configuration space, and a Hamiltonian which determines the temporal evolution of these states. The chief novelty is that time is now an infinite dimensional parameter, since there are as many ways of specifying an instant as there are Cauchy surfaces in a model.[33] Thus, it seems that the interpretation of such a quantum theory of gravity should be no more (or less) difficult a task than the interpretation of a standard quantum field theory. But this is not quite the case, for three reasons. (i) The fact that the observables are not required to commute with the constraints complicates the measurement problem. If $\hat{O}$ is an observable which does not commute with the constraint $\hat{C}$, then we can find a state, $\Psi$, of quantum gravity such that $\hat{C}\hat{O}\Psi \neq \hat{O}\hat{C}\Psi = 0$. Thus, $\hat{O}\Psi$ is *not* a state of quantum gravity (if it were, it would be annihilated by the constraint $\hat{C}$). So, naively carrying over the formalism of quantum mechanics, it appears that measurement can throw states out of the space of physically possible states. (ii) Since general relativity is a theory of the structure of spacetime, one expects to be able to recover spatio-temporal information from states of quantum gravity, at least approximately. But this appears to be an extremely difficult problem: one expects that the relationship between the geometric data $(q, p)$ and the gravitational degrees of freedom $(\phi, \pi)$ is highly non-local at the classical level. The inversion of this relationship at the quantum level presents a formidable problem. (iii) One also has to wonder how to make sense of quantum states which are defined as wave functions on classical instants – since these instants originally derived their significance from the classical structure of spacetime.

Each of these problems is potentially very serious. By attempting to cast quantum gravity into a familiar quantum field theoretic form, the advocates of internal time may be creating an unintelligible formalism, rather than one whose interpretation is straightforward. Such a turn of affairs would come as no surprise to Parmenideans. On their view, Kuchař's proposal is an attempt to carry classical notions of time over to quantum gravity. From the Parmenidean perspective, it *might* be possible to formulate a consistent theory along these lines, but one should not expect it to be a full theory of *quantum gravity* – since, by all rights, quantum gravity should be a quantum theory of space and time, as well as a quantum field theory of gravity. (As noted above, Parmenideans claim, with some justice, that their discrete-spectra area and volume operators are the first hints of the quantum nature of space and time at the Planck length; see footnote 23.) Here we again reach an impasse: Parmenideans and Heracliteans have divergent intuitions about the nature of time and change, and these intuitions condition their taste in approaches to quantizing gravity.

### 10.7 The status of spacetime

In the preceding sections, we sketched three proposals for quantizing general relativity: evolving constants, CMC gauge fixing, and internal time. These proposals are underwritten by three very different attitudes towards the general covariance of general relativity, and lead to three very different approaches to quantum gravity. Most notably, differences of opinion about general covariance are directly linked to differences of opinion about the existence and nature of change at both the classical and the quantum level. This divergence of opinion cannot be dismissed as *merely* philosophical: it has important ramifications for questions about which quantities are physically real and/or observable in classical and quantum gravity. Indeed, one has every reason to expect that these proposals, if successfully executed, would lead to three inequivalent theories of quantum gravity, which would make very different predictions about the quantum behaviour of the gravitational fields.

Before bringing this discussion to a close, we would like to return to the question of the status of the spacetime of general relativity. We proceed by constructing the most plausible interpretation of general relativity which would underwrite each of our quantization procedures.[34] The underlying presumption is that if a given proposal, and no other, were to lead to a successful quantum theory of gravity, that would be a reason to prefer the corresponding interpretation of general relativity over its rivals.

### 10.7.1 Rovelli's evolving constants

The motivation for this programme is the conviction that general covariance should be understood as a principle of gauge invariance. Thus, one is led to deny that there are any physically real quantities in general relativity which fail to commute with the constraints. As we argued in Section 10.3, this drives one towards a relationalist understanding of general relativity – or, perhaps, towards a sophisticated form of substantivalism (more on this below). If spacetime points enjoy existence, then it seems reasonable that a quantity like 'the curvature at $x$' should be physically real. But such quantities do not commute with the constraints, and hence cannot be physically real.[35] Therefore: spacetime points do not exist. Rovelli himself enthusiastically embraces the relationalism which follows from this line of thought.[36]

### 10.7.2 Constant mean curvature as time

Under this proposal, any admissible model of general relativity comes equipped with a preferred foliation by Cauchy surfaces, as well as a preferred parameterization of the time variable which labels these Cauchy surfaces. The CMC time is absolute in some respects, but not others. There is a preferred notion of simultaneity, and a preferred parameterization of time. But this parameterization is determined by the dynamics of the theory rather than being imposed from outside. Thus, the time which results in this case certainly is not the absolute time of Newton. Using this time variable, we can write general relativity as a Hamiltonian system whose configuration space is a subset of superspace (the space of equivalence classes of metrics on $\Sigma$). Thus, general relativity becomes a theory of the evolution in time of the geometry of space. Here, space is best conceived of in relationalist terms: because we take $\mathrm{Riem}(\Sigma)/\mathrm{Diff}(\Sigma)$, rather than $\mathrm{Riem}(\Sigma)$, as our configuration space, we cannot imagine two identical geometries, differently instantiated.

### 10.7.3 Internal time

The core of Kuchař's reading of general covariance is that, properly understood, the observables of general relativity should commute with the vector constraint, but not with the scalar constraint – the qualification being essential since, as noted above, Kuchař explicitly allows that the evolving constants proposal provides a coherent framework for understanding observables in the classical theory. The following interpretative stance underlies this approach.[37] In Section 10.3, we argued that straightforward substantivalists are committed to the doctrine that each point of the phase space of general relativity represents a distinct physically possible state. This implies that there are physically real quantities which do not commute with any of the constraints of general relativity: presumably, for any two points of phase space which represent distinct physically possible states there must exist a physically real quantity which takes on different values when evaluated at these two points; we could take these quantities to be of the form 'the curvature at point $x$ of spacetime'. Now let $(q, p)$, $(q', p')$, and $(q'', p'')$ be three points which lie in the same gauge orbit of the phase space of general relativity, and suppose that $(q, p)$ is related to $(q', p')$ by a gauge transformation generated by the vector constraint, and is related to $(q'', p'')$ by a gauge transformation generated by the scalar constraint. Then $(q, p)$

and $(q', p')$ represent the same geometry of $\Sigma$ – they differ merely as to how this geometry is instantiated by the points of $\Sigma$ – while $(q, p)$ and $(q'', p'')$ represent distinct geometries. Thus, although $(q, p)$ and $(q', p')$ represent distinct states of affairs for the substantivalists, they represent states of affairs which are empirically indistinguishable. Although there are, according to substantivalists, physically real quantities which distinguish between $(q, p)$ and $(q', p')$, these quantities are not observables in any literal sense. On the other hand, the states represented by $(q, p)$ and $(q'', p'')$ *are* distinguishable – otherwise we could not observe change. Thus, the physically real quantities which distinguish between $(q, p)$ and $(q'', p'')$ should be empirically accessible. If we now grant that quantum observables correspond to classical quantities which are not only physically real, but also empirically observable (i.e. they do not distinguish between empirically indistinguishable states of affairs), then we see that substantivalists can provide a coherent motivation for the internal time approach.

Do the points of the spacetime of general relativity, then, exist, or not? Given that general relativity is almost certainly false – since it appears to be impossible to marry a quantum account of the other three forces with a classical account of gravity – the only sense that we can make of this question is whether, given our total physics, the best interpretation of general relativity postulates the existence of spacetime. This question will remain open until the nature of quantum gravity is clarified: if distinct interpretations of general relativity mandate distinct quantizations of gravity, then the empirical success of one or another theory of quantum gravity will have repercussions for our understanding of the spacetime of general relativity.

Even at the present stage, however, we can say something about the lessons of the hole argument for our understanding of classical spacetime. (1) Despite widespread scepticism among philosophers, physicists are correct in seeing the hole argument as pointing up a knot of problems concerning: the existence of spacetime points; the difficult notion of 'observable' in classical and quantum gravity; and the nature of time and change in physical theory.[38] (2) There is no easy solution to the hole argument. We have seen that both traditional relationalism and traditional substantivalism are associated with some extremely difficult technical and conceptual problems when one turns one's attention to quantum gravity – and that these difficulties arise directly out of the doctrines' respective pronouncements on the nature of general covariance. Ultimately, one of the other of these positions may triumph. But it will be because physical and mathematical ingenuity show how the attendant problems can be overcome, not because either position can be ruled out on strictly philosophical grounds internal to general relativity. (3) That being said, we maintain that there is one sort of response to the hole argument which *is* clearly undesirable: the sort of sophisticated substantivalism which mimics relationalism's denial of the Leibniz–Clarke counterfactuals. It would require considerable ingenuity to construct an (intrinsic) gauge-invariant substantivalist interpretation of general relativity. And if one were to accomplish this, one's reward would be to occupy a conceptual space already occupied by relationalism. Meanwhile, one would forego the most exciting aspect of substantivalism: its link to approaches to quantum gravity, such as the internal time approach. To the extent that such links depend upon the traditional substantivalists' commitment to the existence of physically real quantities which do not commute with the constraints, such approaches are clearly unavailable to relationalists. Seen in this light, sophisticated substantivalism, far from being the saviour of substantivalism, is in fact a pallid imitation of relationalism, fit only for those substantivalists who are unwilling to let their beliefs about the existence of space and time face the challenges posed by contemporary physics.

We conclude that there is indeed a tight connection between the interpretative questions of classical and quantum gravity. There is a correspondence between interpretations of the general covariance of general relativity and approaches to – and interpretations of – quantum gravity. This correspondence turns upon the general covariance of the classical theory and is mediated via the processes of quantization and the taking of classical limits. One demands that one's interpretation of general relativity should underwrite an approach to quantization which leads to a viable theory of quantum gravity, and that one's understanding of quantum gravity should lead to a way of viewing general relativity as an appropriate classical limit. This provides a cardinal reason to take the interpretative problems surrounding the general covariance of general relativity seriously – at the very least, one wants to know which interpretative approaches to general relativity mandate, open up, or close off which approaches to quantum gravity. More ambitiously, one can hope that clarity concerning the general covariance of the classical theory will provide insights which prove helpful in the quest for a quantum theory of gravity.

## 10.8 Appendix

In this Appendix we provide a few details about the definitions and constructions mentioned in the text. It falls into four sections, corresponding to material supporting Sections 10.2.1, 10.2.2, 10.2.3, and 10.5.1, respectively.

### 10.8.1 Hamiltonian systems

Our phase spaces will always be manifolds. These may be either finite dimensional or infinite dimensional. In the latter case, we require that our space be locally homeomorphic to a Banach space rather than to $\Re^n$. For details and for the infinite dimensional versions of the material discussed below, see Choquet-Bruhat et al. (1982) or Schmid (1987).

*Definition*: Non-degenerate forms. A two form, $\omega$, on a manifold, $M$, *non-degenerate* if for each $x \in M$ the map $v \in T_xM \mapsto \omega_x(v,\cdot) \in T_x^*M$ is one-to-one. If this map fails to be one-to-one, then there will be non-trivial $v \in T_xM$ with $\omega_x(v,\cdot) = 0$. These are called the *null vectors* of $\omega$.

*Definition*: Symplectic form. A *symplectic form* on a manifold, $M$, is a closed, non-degenerate, two form, $\omega$, on $M$.

*Definition*: Hamiltonian vector field. The *Hamiltonian vector field*, $X_f$, of $f$ in $(M, \omega)$ is the solution of the equation $\omega(X_f, \cdot) = dH$. When $H$ is the Hamiltonian, we call the integral curves of $X_H$ the *dynamical trajectories* of the system.

*Definition*: Poisson brackets. $\{f, g\} = \omega(X_f, X_g) = X_f(g)$.

*Construction*: Canonical co-ordinates. When $M$ is finite dimensional, we can find local co-ordinates $(q^1, \ldots, q^n; p_1, \ldots, p_n)$ such that $\omega$ can be written as

$\omega = dq^i \wedge dp_i.$[39] Equivalently, in such co-ordinates we have:

$$\omega = \begin{vmatrix} 0 & I \\ -I & 0 \end{vmatrix},$$

where $I$ is the $n \times n$ identity matrix. Co-ordinates of this kind are known as *canonical co-ordinates*; we speak of the $q^i$ as being *canonically conjugate* to the $p_i$. In canonical co-ordinates the equations for our dynamical trajectories assume their familiar form: $\dot{q} = \{q, H\}$ and $\dot{p} = \{p, H\}$. Notice that conservation of energy is a trivial consequence of the formalism: $\{H, H\} = \omega(X_H, X_H) = 0$ since $\omega$ is antisymmetric; so $H$ is a *constant of motion* (i.e. $H$ is constant along each dynamical trajectory).

*Construction*: Cotangent symplectic structure. Let $(q^1, \ldots, q^n)$ be a set of local co-ordinates on $Q$. We then construct a co-ordinate system on $T^*Q$ of the form $(q^1, \ldots, q^n; p_1, \ldots, p_n)$, where the $p_i$ are just the components of covectors relative to our co-ordinate system. We can now construct the canonical symplectic form, $\omega = dq^i \wedge dp_i$. That is: the $q^i$ and $p_i$ are canonically conjugate co-ordinates. This construction is independent of the original co-ordinate system on $Q$, $(q^1, \ldots, q^n)$, and can be extended to construct a unique symplectic form for all of $T^*Q$. Thus, the cotangent bundle structure singles out a preferred symplectic structure on $T^*Q$. This construction can be generalized to the infinite dimensional case.

### 10.8.2 Gauge systems

*Definition*: Presymplectic form. A *presymplectic form* on a manifold, $N$, is a closed two form, $\sigma$, with the property that its space of null vectors has the same dimensionality at each point in $N$.

*Definition*: Gauge orbit. Two points lie in the same *gauge orbit* if they can be connected by a curve, all of whose tangent vectors are null vectors of $\sigma$.

*Construction*: Gauge orbits. The gauge orbits are constructed by integrating the null distribution of $\sigma$. That they are manifolds follows from Frobenius' theorem together with the following fact: $[X_f, X_g] = X_{[f,g]}$, where $[\cdot, \cdot]$ is the Lie bracket, so that the map $f \mapsto X_f$ is a Lie algebra homomorphism of $C^\infty(N)$ into $\Xi(N)$, the algebra of vector fields on $N$. Since the dimensionality of the null space is constant on $N$, our phase space is foliated by gauge orbits of a fixed dimensionality.

*Definition*: Dynamical trajectories. Again, we look at the integral curves of vector fields, $X_f$, which solve $\sigma(X_f, \cdot) = dH$.

*Discussion*: Dynamical trajectories on constraint surfaces. If we are thinking of $(N, \sigma, H)$ as being imbedded in $(M, \omega)$, it is natural to wonder about the relationship between the dynamical trajectories of $(N, \sigma, H)$ and the restriction to $N$ of the Hamiltonian vector fields of $(M, \omega)$. We call $h \in C^\infty(M)$ an *extension* of $H$ to $(M, \omega)$ if: (i) $h|_N = H$; (ii) $\{h, c\} \cong 0$ for all constraints $c$. The latter condition means that flow generated by $h$ carries points on $N$ to points on $N$, since the Hamiltonian vector field of $h$ is everywhere tangent to $N$. If $h$ is an extension of $H$, and $X_h$ is the Hamiltonian vector field of $h$ in $(M, \omega)$, then $X_h|_N$ is a Hamiltonian vector field of $H$ in $(N, \sigma)$. Conversely, every Hamiltonian vector field of $H$ in $(N, \sigma)$ arises in this manner, for some extension $h$ of $H$. It is not difficult to prove that any two extensions, $h$ and $h'$ of $H$ differ by a linear combination of first-class constraints. It follows that the transformation $h \mapsto h + u^a \gamma_a$ carries us from one set of dynamical trajectories of $(N, \sigma, H)$ to another, where the $u^a$ are arbitrary functions on $M$; conversely, every pair of sets of dynamical trajectories are so related.

Whereas in the Hamiltonian case Hamilton's equations $\dot{q} = \{q, h\}$ and $\dot{p} = \{p, h\}$ determine a unique dynamical trajectory of the form $(q^i(t); p_i(t))$ through each $x \in M$, we see that in the case of a constrained Hamiltonian system, Hamilton's equations determine a different set of dynamical trajectories for each $h$ which extends $H$. Given our freedom to replace $h$ by $h' = h + u^a \gamma_a$, we can write Hamilton's equations as $\dot{q} = \{q, h\} + u^a \{q, \gamma_a\}$ and $\dot{p} = \{p, h\} + u^a \{p, \gamma_a\}$. Thus, the solutions of Hamilton's equations which determine the dynamical trajectories of $(N, \sigma, H)$ contain as many arbitrary functions of time as there are first-class constraints. (Here, for convenience, we have chosen a set of canonical co-ordinates on $(M, \omega)$.)

### 10.8.3 Reduced phase spaces

*Construction*: Reduced phase space. The points of the *reduced phase space*, $\bar{M}$, are the gauge orbits $[x]$ of $(N, \sigma)$, equipped with the projection topology induced by the projection $\pi : N \to \bar{M}$. The symplectic form $\bar{\omega}$ is given by $\pi^* \sigma$. Since $H$ is gauge-invariant, $\bar{M}$ is well defined by $\bar{M}([x]) = H(x)$. The set of dynamical trajectories of $(N, \sigma, H)$ which pass through $x \in N$ projects down to the single dynamical trajectory of $(\bar{M}, \bar{\omega}, \bar{H})$ which passes through $[x]$.

*Example*: Bad topology. Here is one way in which this problem can arise. One can construct a constrained Hamiltonian system by starting with a Hamiltonian system $(M, \omega, H)$ and imposing the constraint $H = c$. That is, one looks at a surface of constant energy. This is a presymplectic manifold since it has an odd number of dimensions. The gauge orbits of the resulting presymplectic geometry are just the dynamical trajectories of the original Hamiltonian system. Imposing the Hamiltonian $h = 0$ leads to a gauge theory with these gauge orbits as its dynamical trajectories. We can go on to construct the reduced phase space. We simply identify all the points which lie on the same trajectory, and impose the projection topology on the resulting space of dynamical trajectories. What is this reduced phase space like? This depends on the details of the system we started with. If it is integrable, we can find constants of motion (= gauge-invariant quantities) which project down to co-ordinates on the reduced phase space – the latter will, therefore, be a manifold. If, however, our original Hamiltonian system was chaotic, the phase space will be a mess. If our system is ergodic then we will be unable to find constants of motion other than the Hamiltonian, and each trajectory will wander over the entire energy surface. Thus, we will be unable to find a sufficient number of gauge-invariant quantities to co-ordinatize the reduced phase space. Indeed, the topology of the reduced phase space will not even be Hausdorff: since each trajectory of the gauge system approaches every other arbitrarily closely, it will be impossible to separate points of the reduced phase space by open sets.

### 10.8.4 Parametrized systems

*Construction*: Parameterization. Let $(M, \omega, h)$ be a Hamiltonian system. We construct $M' = \Re^2 \times M$ by adding to $M$ the canonically conjugate variables $t$ and $u$. Let the symplectic form on $M'$ be given by $\omega' = \omega - du \wedge dt$. Let $H = h + u$ and let

$N$ be the submanifold of $M'$ determined by the constraint $H \equiv 0$ (we extend $h$ to $M'$ in the obvious way, by making it independent of $t$ and $u$). Then the constrained Hamiltonian system $(N, \sigma, H)$, with $\sigma = \omega'|_N$ and the Hamiltonian given by $H = 0$, is called the *parameterization* of $(M, \omega, h)$. We can think of $(N, \sigma, H)$ as the result of including time among the position variables of the system, with the energy $h$ as its canonically conjugate momentum (since $h = -u$ on $N$). Notice that $(N, \sigma)$ is presymplectic (in the finite dimensional case this is obvious since dim $N$ is odd). The gauge orbits are one-dimensional and coincide with the dynamical trajectories since $H \equiv 0$ (so that the solutions of $X_H$ are just the null vector fields of $\sigma$). Each dynamical trajectory on $(M, \omega, h)$ corresponds to a gauge orbit on $(N, \sigma, H)$. Pick a time $t$ and a point $x \in M$, and look at the dynamical trajectory $(q(t), p(t))$ on $M$. Then the dynamical trajectories through $(t, x) \in N$ will be of the form $(\tau, (q(t), p(t)))$, where $\tau(t)$ is some re-parameterization of time. The gauge orbit in $N$ which corresponds to the trajectory $(q(\tau), p(\tau))$ in $M$ will include all the points in $N$ which are images of the maps $(\tau, (q(t), p(t)))$, for all parameters $\tau$. Thus, the loss of the preferred parameterization of time is the price of including time among the canonical variables. The reduced phase space of a parameterized system is just the original Hamiltonian system.

## Notes

We would like to thank Karel Kuchař and Carlo Rovelli for invaluable tutelage, and Craig Callender, Nick Huggett, and Steve Weinstein for helpful comments.

1. Here and throughout we restrict our attention to the canonical approach to quantum gravity.
2. See Norton (1993) for a discussion of this and other episodes in the long debate over general covariance.
3. See Belot and Earman (1999) for a discussion of some related contrasts between the pessimistic attitudes of (many) philosophers and the optimistic attitudes of (some) physicists with respect to the relevance of interpretative work on general relativity to ongoing research on quantum gravity.
4. The main text presumes that the reader is familiar with the formalism of non-relativistic quantum mechanics and with enough differential geometry to be able to read the standard textbook presentations of general relativity. The most important technical details are collected together in an Appendix. Although we hope that our presentation is not misleading, it does of course leave out many details.
5. The original development of the formalism, Dirac (1964), remains the best place to learn about gauge systems. Chapters 1–3 of Henneaux and Teitelboim (1992) include many invaluable examples, as well as an introduction to the modern geometric point of view.
6. A regular submanifold is one which is given locally by stipulating that some subset of a set of co-ordinates on $M$ take on a given constant value.
7. Here we gloss over the subtleties surrounding the Dirac conjecture. See pp. 16–20 of Henneaux and Teitelboim (1992).
8. Note that if we imposed both $p_1 = 0$ and $q^1 = 0$, then the constraints would be second class, and $(N, \sigma)$ would be a symplectic geometry.
9. See Redhead (1975) for the notion of surplus structure, and its relevance to the interpretative enterprise.
10. When space is multiply connected, the correspondence between gauge orbits and magnetic fields is many-to-one. This means that there is additional structure that is not captured by $B$ – structure which is empirically accessible via the Aharonov–Bohm effect. See Belot (1998) for an account of how this complication forces the would-be interpreter of electromagnetism to

choose between non-locality and indeterminism. The moral is that it is not always easy to find a gauge-invariant interpretation of a given gauge theory.

11. Again, Dirac (1964) remains one of the best sources. See also Chapter 13 of Henneaux and Teitelboim (1992). The following description, although accurate enough for present purposes, glosses over a large number of technicalities. One would hope, of course, that the two techniques of quantization would lead to the same results in cases where they can both be carried out. Unfortunately, this is not always the case; see Plyushchay and Razumov (1996). See Landsman (1995, 1998) for an alternative to Dirac quantization.

12. Note that operator ordering problems introduce considerable ambiguity at this stage. We ignore these below. In general, it is safe to assume that quantum gravity is beset by all of the problems of ordinary quantum field theories – operator ordering problems, divergences, anomalies, problems of renormalization and regularization – and then some.

13. We limit discussion to globally hyperbolic vacuum solutions of general relativity with compact Cauchy surfaces. As a rule of thumb, one can think of the content of these restrictions as follows. The restriction to globally hyperbolic spacetimes is substantive – much of what follows is simply false, or poorly understood in the non-globally hyperbolic case. The restriction to vacuum solutions, on the other hand, is largely for convenience's sake. Much of what will be said is true when matter fields are taken into consideration – although the formalism involved is often more unwieldy if matter is included. The restriction to spatially compact spacetimes lies somewhere between these two extremes. There are some interesting and important differences between the compact and the asymptotically flat cases. But, for the most part, taking these differences into account would involve adding many qualifications to our technical treatment, without substantially altering the interpretative theses defended below.

14. There are two ways to proceed here. We follow that more familiar route, and characterize the geometry of $S$ using a metric tensor (geometrodynamics). It is also possible to work in terms of connections (connection dynamics). This approach, pioneered by Ashtekar, is in many ways more tractable and has led to many significant results in recent years. The best intuitive introductions to connection dynamics are contained in Baez and Munian (1994) and Kuchař (1993a); see Ashtekar (1995) for a more detailed presentation. We believe that at the level of detail of the present chapter, nothing is lost by focusing on geometrodynamics to the exclusion of connection dynamics.

15. See Appendix E of Wald (1984) for a more complete treatment. Note that most formulations of general relativity as a gauge theory make use of the lapse and shift as Lagrange multipliers. In order to avoid this complication, we have followed Beig (1994) in adopting a more geometric approach in which the lapse and shift are eliminated.

16. Actually, this glosses over an interesting detail: it could be argued that our formalism *fails* to be diffeomorphism invariant, since our phase space only contains *spacelike* geometries. This is closely related to the fact, emphasized to us by Steve Weinstein, that it is far from trivial to see how the group of four-dimensional diffeomorphisms acts on the phase space of 3-geometries. See Kuchař (1986), Isham (1991), and Weinstein (1998) for illuminating discussions of this problem in the classical and quantum theories. The results announced in Gotay, Isenberg, and Marsden (1998) promise to shed a great deal of light on this problem.

17. There are, however, some more sophisticated ploys which relationalists can adopt to make sense of inertial effects. See Barbour (1982), Belot (1999), and Lynden-Bell (1995).

18. See Torre (1993). Here a quantity is local if it is an integral over $\Sigma$ of the canonical variables, $p$ and $q$, and a finite number of their derivatives. The situation is slightly more encouraging if we work with asymptotically flat spacetime, rather than spatially compact spacetimes. In that case, their are a finite number of known local gauge-invariant quantities, such as the ADM momenta.

19. This is, we believe, part of the explanation of the current vogue for non-local interpretations of electromagnetism and its non-abelian cousins; see Baez and Munian (1994) for a nice introduction to these issues. It also justifies the demand, expressed in Earman (1989), that relationalists should produce formulations of physical theories which can be expressed in relationally pure vocabulary.

20. See Rovelli (1991a,b,c). For criticism and discussion, see Cosgrove (1996), Hájiček (1991), Hartle (1996), Section 6.4 of Isham (1993), and Section 15 of Kuchař (1992).

21. One must also face the challenge posed by non-locality: in (spatially compact) general relativity, each gauge-invariant quantity – and hence each member of some family which we want to view as an evolving constant – is a non-local quantity, while we are accustomed to believe that the quantities that we measure are local.

22. Heracliteans will object that this is a crucial caveat: by their lights, the choice of a correct inner product for wave functions on spacetime is equivalent to the choice of a Heraclitean time variable. Parmenideans hope to finesse this objection by showing that their approach singles out a unique candidate for the correct inner product for quantum gravity. See Ashtekar and Tate (1994).

23. Recent work in Ashtekar's connection–dynamical formulation of general relativity, has produced area and volume operators for quantum gravity which have *discrete* spectra (see Ashtekar 1995, 1998 for a survey of these results). This is indeed exciting. But the reader is urged to take these results with a grain of salt: (i) the operators in question do not commute with all of the constraints, and hence are *not* observables within the Parmenidean framework; (ii) although these operators have discrete spectra, the family of such operators is parameterized by the family of volumes and areas in $\Sigma$, so that an underlying continuum remains. Furthermore, there appears to be some difficulty in defining a physically reasonable version of the quantum scalar constraint for the Ashtekar variables.

24. The proposal developed by Unruh and Wald (Unruh 1988, Unruh and Wald 1989) does not quite fit into this classification, since their unimodular time introduces a preferred volume element rather than a preferred slicing. Nonetheless, it is very similar to radical proposals. See Section 4 of Kuchař (1993) or Section 4.4 of Isham (1993) for discussion.

25. The extent of this open set is an open question. See Isenberg and Moncrief (1996) for a recent discussion. Fischer and Moncrief (1996) show that there are three manifold topologies such that if we look at the phase space of CMC sliceable solutions with Cauchy surfaces of such a topology, we find that the reduced phase space is a manifold with *no* singularities.

26. This is a variety of gauge fixing. In completely fixing the gauge, one kills off gauge freedom by adding further constraints in such a way that the expanded set of constraints are all second class and the new system is strictly Hamiltonian rather than being gauge; see footnote 8. In electromagnetism, one can impose the Lorentz gauge condition, $\partial_a A_a = 0$. This completely fixes the gauge, in the sense that the resulting second-class constraint surface intersects each gauge orbit exactly once. The CMC gauge choice only partially fixes the gauge: the resulting constraint surface still has a one-dimensional intersection with each of the original gauge orbits.

27. These will depend upon the Lagrangian which governs the dynamics of the matter: postulating four non-interacting massless scalar fields privileges a system of harmonic co-ordinates; introducing a cloud of non-rotating and heat-conducting dust leads to Gaussian co-ordinates. See Kuchař (1993a).

28. Different notions of 'instant' in Minkowski spacetime produce families of instants of different dimensionalities. See Hájiček (1994) for an analysis and comparison of the distinct varieties of relativistic quantum mechanics which correspond to different notions of instant in Minkowski spacetime.

29. It may help to consider how the CMC time fits into this scheme. If we attempt to define $T(x; q, p]$ as the extrinsic curvature of $(q, p)$ at $x$, then we run afoul of the requirement Spacetime Interpretation: if we look at two Cauchy surfaces passing through the same point of spacetime, we expect them to have different extrinsic curvatures at that point. We could, in accord with the CMC proposal discussed above, attempt to define $T(x; q, p]$ to be the value of the extrinsic curvature of the CMC slice through $x$. This satisfies Spacetime Interpretation. But it requires solving the Einstein equations in order to define the time variable on phase space. See Section 6 of Kuchař (1992) or Section 4.2.4 of Isham (1993) for discussion.

30. Indeed, Kuchař (1986) argues that the Diff($M$)-invariance of general relativity is hidden in the ordinary Hamiltonian formulation of general relativity, but is manifest in the internal time formulation.

31. Kuchař (1993), p. 136. Here and below, we have slightly altered Kuchař's notation to conform to our own.

32. See Schön and Hájiček (1990) and Torre (1992) for the former, and Sections 1, 2, and 6 of Kuchař (1992) and Sections 3.4, 4.4, and 4.2 of Isham (1993) for the latter.

254

33. But see Hájiček and Isham (1996, 1996a) for formulations of classical and quantum field theories in terms of embedding variables. Also see Torre and Varadarajan (1998) for a problem with unitarity in this context.

34. We note that among the many proposals for quantum gravity which we have not touched upon, there are a number which are particularly rich in connections to the substantival–relational debate. See especially the discussions of Barbour (1994, 1994a) and Smolin (1991).

35. But recall that the quantities which *do* commute with the constraints are non-local. So today's relationalists find themselves in a somewhat uncomfortable position: there appears to be a mismatch between their relationalist ontology and the non-local ideology which is forced upon them. Traditional substantivalists and relationalists had no such problems – there was a perfect match between their ontology and ideology (spacetime points and the relations between them *or* bodies and the relations between them).

36. See Rovelli (1997) and Chapter 4 of this book. See also Rovelli (1991d), where he discusses in some detail the relationship between the hole argument and the view that general covariance is a principle of gauge invariance.

37. We believe that Kuchař himself is committed to this interpretative stance. We do not argue in favour of this claim here. But note the similarity between the reasons adduced below, and the considerations which Kuchař uses to motivate the internal time approach.

38. Some physicists might balk at these claims. But, we believe, almost all would agree with the following statement: 'the question as to what should be the correct notion of observables in canonical G.R., which is clearly important for any quantum theory of gravity, is not fully understood even on the classical level' (Beig 1994, p. 77). We maintain that once the claim about observables is granted, the others follow.

39. Many infinite dimensional symplectic geometries admit (suitably generalized) canonical co-ordinates.

*Note added in proof:*
It now seems to us that the worry of Kuchař and Unruh about the structure of the reduced phase space of general relativity should be largely allayed by the results of James Isenberg and Jerrold Marsden (1982).

11      # The origin of the spacetime metric: Bell's 'Lorentzian pedagogy' and its significance in general relativity

Harvey R. Brown and Oliver Pooley

## 11.1 Introduction

In 1976, J.S. Bell published a paper on 'How to teach special relativity' (Bell 1976). The paper was reprinted a decade later in his well-known book *Speakable and unspeakable in quantum mechanics* – the only essay to stray significantly from the theme of the title of the book. In the paper Bell was at pains to defend a dynamical treatment of length contraction and time dilation, following 'very much the approach of H.A. Lorentz' (Bell 1987, p. 77).

Just how closely Bell stuck to Lorentz's thinking in this connection is debatable. We shall return to this question shortly. In the meantime, we briefly rehearse the central points of Bell's rather unorthodox argument.

Bell considered a single atom modelled by an electron circling a more massive nucleus, ignoring the back-effect of the field of the electron on the nucleus. The question he posed was: what is the prediction in Maxwell's electrodynamics (taken to be valid in the frame relative to which the nucleus is initially at rest) as to the effect on the electron orbit when the nucleus is set (gently) in motion in the plane of the orbit? Using only Maxwell's field equations, the Lorentz force law and the relativistic formula linking the electron's momentum and its velocity – which Bell attributed to Lorentz – he concluded that the orbit undergoes the familiar longitudinal ('Fitzgerald' (*sic*)) contraction, and its period changes by the familiar ('Larmor') dilation. Bell went on to demonstrate that there is a system of primed variables such that the description of the moving atom with respect to them coincides with that of the stationary atom relative to the original variables, and the associated transformations of co-ordinates is precisely the familiar Lorentz transformation.

Bell carefully qualified the significance of this result. He stressed that the external forces involved in boosting a piece of matter must be suitably constrained in order that the usual relativistic kinematical effects such as length contraction be observed (see Section 11.5). More importantly, Bell acknowledged that Maxwell–Lorentz theory is incapable of accounting for the stability of solid matter, starting with that of the very electronic orbit in his atomic model; nor can it deal with cohesion of the nucleus. (He might also have included here the cohesion of the electron itself.) How Bell addressed this shortcoming of his model is important, and will be discussed in Section 11.3. In the meantime, we note that the positive point Bell wanted to make was about the wider nature of the Lorentzian approach: that it differed from that of Einstein in 1905 in both *philosophy* and *style*.

The difference in philosophy is well known, and Bell did not dwell on it. It is simply that Lorentz believed in a preferred frame of reference – the rest-frame of the ether – and Einstein did not, regarding the notion as superfluous. The interesting distinction, rather, was that of style. Bell argues first that 'we need not accept Lorentz's philosophy to accept a Lorentzian pedagogy. Its special merit is to drive home the lesson that the laws of physics in any *one* reference frame account for all physical phenomena, including the observations of moving observers'. He went on to stress that Einstein postulates what Lorentz is attempting to prove (the relativity principle). Bell has no 'reservation whatever about the power and precision of Einstein's approach'; his point is that 'the longer road [of FitzGerald, Lorentz, and Poincaré] sometimes gives more familiarity with the country' (Bell 1987, p. 77).

The point, then, is not the existence or otherwise of a preferred frame – and we have *no* wish to defend such an entity in this chapter. It is how best to understand, and teach, the origins of the relativistic 'kinematical' effects. Near the end of his life, Bell reiterated the point with more insistence:

> 'If you are, for example, quite convinced of the second law of thermodynamics, of the increase of entropy, there are many things that you can get directly from the second law which are very difficult to get directly from a detailed study of the kinetic theory of gases, but you have no excuse for not looking at the kinetic theory of gases to see how the increase of entropy actually comes about. In the same way, although Einstein's theory of special relativity would lead you to expect the FitzGerald contraction, you are not excused from seeing how the detailed dynamics of the system also leads to the FitzGerald contraction'.
>
> (Bell 1992, p. 34)

There is something almost uncanny in this exhortation. Bell did not seem to be aware that just this distinction between thermodynamics and the kinetic theory of gases was foremost in Einstein's mind when he developed his fall-back strategy for the 1905 relativity paper (see Section 11.2).

It is the principal object of this chapter to analyse the significance of what Bell calls the 'Lorentzian pedagogy' in both special relativity and general relativity. Its merit is to remind us that in so far as rigid rods and clocks can be used to survey the metrical structure of spacetime (and the extent to which they do will vary from theory to theory), their status as *structured* bodies – as 'moving atomic configurations' in Einstein's words – must not be overlooked. The significance of the dynamical nature of rods and clocks, and the more general theme of the entanglement between kinematics and dynamics, are issues which in our opinion deserve more attention in present-day discussions of the physical meaning of spacetime structure.

## 11.2 Chalk and cheese: Einstein on the status of special relativity theory

Comparing the explanation in special relativity (SR) of the *non-null* outcome of the celebrated 1851 Fizeau interferometry experiment – a direct corroboration of the Fresnel drag coefficient – with the earlier treatment given by Lorentz can seem like comparing chalk and cheese.

From the perspective of SR the drag coefficient is essentially a simple consequence (to first order) of the relativistic velocity transformation law, itself a direct consequence of the Lorentz transformations. The explanation appears to be entirely kinematical. Lorentz, on the other hand, had provided a detailed dynamical account – based on his theory of the electron – of the microstructure of the moving transparent medium (water in the case of the Fizeau experiment) and its interaction with the light passing through it (Lorentz 1892). In his 1917 text *Relativity*, Einstein noted with satisfaction that the explanation of Fizeau's experiment in SR is achieved 'without the necessity of drawing on hypotheses as to the physical nature of the liquid' (Einstein 1961, p. 57).

Yet Lorentz had achieved something remarkable, and Einstein knew it. In deriving the drag coefficient from principles contained within his theory of the electron, Lorentz was able to reconcile the null results of *first-order* ether-wind experiments (all of which incorporated moving transparent media) with the claimed existence of the luminiferous ether itself. There were few, if any, complaints that such a surprising reconciliation was obtained on the basis of *ad hoc* reasoning on Lorentz's part. But the case of the *second-order* ether-wind experiments was of course different, and it is worth noting Einstein's take on these, again as expressed in *Relativity*.

In order to account for the null result of the 1887 Michelson–Morley (M-M) experiment, Lorentz and FitzGerald assumed, says Einstein, 'that the motion of the body [the Michelson interferometer] relative to the aether produces a contraction of the body in the direction of motion' (Einstein 1961, p. 59). Einstein's claim is not quite right. In fact, both Lorentz and FitzGerald had correctly and independently realized that it was sufficient to postulate any one of a certain family of motion-induced distortions: the familiar longitudinal contraction is merely a special case and not uniquely picked out by the M-M null result.[1] But this common historical error (repeated by Bell) should not detain us, for the real issue lies elsewhere. In SR, Einstein stresses, 'the contraction of moving bodies follows from the two fundamental principles of the theory [the relativity principle and the light postulate], without the introduction of particular hypotheses' (Einstein 1961, p. 59).

The 'particular hypotheses' of FitzGerald and Lorentz went beyond the phenomenological claim concerning the distortion of rigid bodies caused by motion through the ether. Both these physicists, again independently, attempted to justify this startling claim by surmising, not unreasonably, that the molecular forces in rigid bodies, and in particular in the stone block on which the Michelson interferometer was mounted, are affected by the ether-wind in a manner similar to that in which electromagnetic interactions are so affected. Unlike their contemporary Larmor, neither FitzGerald nor Lorentz was prepared to commit himself to the claim that the molecular forces *are* electromagnetic in origin. In this sense, their courageous solution to the conundrum posed by the M-M experiment did involve appeal to hypotheses outside what Einstein referred to as the 'Maxwell–Lorentz theory'.

Indeed, it was precisely their concern with rigid bodies that would have made FitzGerald and Lorentz less than wholly persuaded by Bell's construction above, as it stands. It is not just that Bell's atomic model relies on post-1905 developments in physics. The point is rather that Bell does not discuss the forces that glue the atoms together – the analogue of the 'molecular forces' – to form a rigid body like Michelson's stone.[2] (Bell of course must have known that they were also electromagnetic in origin while, as we have seen, FitzGerald and Lorentz were uncertain and unwilling to commit themselves on this point.)

Returning to Einstein, in *Relativity* he also mentions the case of predictions concerning the deflection of high-velocity electrons (cathode- and beta-rays) in electromagnetic fields (Einstein 1961, p. 56). Lorentz's own predictions, which coincided with Einstein's, were obtained by assuming *inter alia* that the electron itself deforms when in motion relative to the ether. It is worth recalling that predictions conflicting with those of Lorentz and Einstein had been made by several workers; those of M. Abraham (an acknowledged authority on Maxwellian electrodynamics) being based on the hypothesis of the non-deformable electron.[3] Einstein's point was that whereas Lorentz's hypothesis is 'not justifiable by any electrodynamical facts' and hence seems extraneous, the predictions in SR are obtained 'without requiring any special hypothesis *whatsoever* as to the structure and behaviour of the electron' (Einstein 1961, p. 56, our emphasis).

Whatsoever? Not quite, as we see in the next section. But for the moment let us accept the thrust of Einstein's point. The explanations of certain effects given by Maxwell–Lorentz theory and SR differ in both style and degree of success: in some cases the Maxwell–Lorentz theory actually seems incomplete. Yet there was a stronger reason for the difference in style, and for the peculiarities of the approach that Einstein adopted in 1905 (peculiarities which should be borne in mind when evaluating claims – which resurface from time to time – that Poincaré was the true father of SR). Part of the further story emerged in 1919, in a remarkable article Einstein wrote for the London *Times* (Einstein 1919), when he characterized SR as an example of a 'principle theory', methodologically akin to thermodynamics, as opposed to a 'constructive theory', akin to the kinetic theory of gases. Like all good distinctions, this one is hardly absolute, but it is enlightening. It is worth dwelling on it momentarily.

In 1905, Einstein was faced with a state of confusion in theoretical physics largely caused by Planck's 1900 solution to the vexing problem of blackbody radiation. Not that the real implications of Planck's quantum revolution were widely appreciated by 1905, even by Planck; but that year saw Einstein himself publish a paper with the revolutionary suggestion that free radiation itself had a quantized, or granular structure (Einstein 1905a). What his light-quantum proposal undoubtedly implied in Einstein's mind was that the Maxwell–Lorentz theory was probably only of approximate, or statistical validity. Now within that theory, Lorentz, with the help of Poincaré, had effectively derived the Lorentz (co-ordinate) transformations as the relevant subgroup of the linear covariance group of Maxwell's equations, consistent moreover with the FitzGerald–Lorentz deformation hypothesis for rigid bodies. But if Maxwell's field equations were not to be considered fundamental, and if, furthermore, the nature of the various forces of cohesion within rigid bodies and clocks was obscure, how was one to provide a rigorous derivation of these co-ordinate transformations, which would determine the behaviour of moving rods and clocks? Such a derivation was essential if one wanted, as Einstein did, to tackle the difficult problem of solving Maxwell's equations in the case of moving charge sources.

It is important to recognize that Einstein's solution to this conundrum was the result of despair, as he admits in his *Autobiographical Notes* (Einstein 1969). Einstein could not see any secure foundation for such a derivation on the basis of 'constructive efforts based on known facts' (Einstein 1969, p. 53). In the face of this *impasse*, Einstein latched on to the example of thermodynamics. If for some reason one is bereft of the means of mechanically modelling the internal structure of the gas in a single-piston heat engine, say, one can always fall back on the laws of thermodynamics to shed light on the performance of that engine – laws which stipulate nothing about the structure of the gas, or rather which hold whatever that structure might be. The laws or principles of thermodynamics are phenomenological, the first two laws being susceptible to formulation in terms of the impossibility of certain types of perpetual motion machines. Could similar, well-established phenomenological laws be found, Einstein asked, which would constrain the behaviour of moving rods and clocks without the need to know in detail what their internal dynamical structure is?

In a sense, Galileo's famous thought-experiment involving a ship in uniform motion is an impossibility claim akin to the perpetual-motion dictates of thermodynamics: no effect of the ship's motion is detectable in experiments being performed in the ship's cabin. The Galileo–Newton relativity principle was probably originally proposed without any intention of restricting it to non-electromagnetic or non-optical experiments (see Brown and Sypel 1995). In the light of the null ether-wind experiments of the late nineteenth century, Einstein, like Poincaré, adopted the principle in a form which simply restored it to its original universal status. In Einstein's words:

> 'The universal principle of the special theory of relativity [the relativity principle]
> . . . is a restricting principle for natural laws, comparable to the restricting principle
> of the non-existence of the *perpetuum mobile* which underlies thermodynamics'.
> (Einstein 1969, p. 57)

Turning to Einstein's second postulate, how apt, if at first sight paradoxical, was its description by Pauli as the 'true essence of the aether point of view' (Pauli 1981, p. 5). Einstein's light postulate – the claim that relative to a certain 'resting' co-ordinate system, the two-way light-speed is constant (isotropic and independent of the speed of the source) – captures that phenomenological aspect of all ether theories of electromagnetism which Einstein was convinced would survive the maelstrom of changes in physics that Planck had started. Combined now with the relativity principle, it entailed the invariance of the two-way light-speed. This was not the only application of the relativity principle in Einstein's 1905 derivation of the Lorentz transformations (1905), as we discuss in the next section.[4]

Einstein had now got what he wanted in the Kinematical Part of his 1905 paper, without committing himself therein to the strict validity of Maxwell's equations and without speculation as to the detailed nature of the cohesion forces within material bodies such as rods and clocks. But there was a price to be paid. In comparing 'principle theories' such as thermodynamics with 'constructive theories' such as the kinetic theory of gases in his 1919 *Times* article, Einstein was quite explicit both that special relativity is a principle theory, and that principle theories lose out to constructive theories in terms of explanatory power:

> '. . . when we say we have succeeded in understanding a group of natural processes, we invariably mean that a constructive theory has been found which covers the processes in question'.
>
> (Einstein 1982, p. 228)

This was essentially the point Bell was to make half a century later.[5]

## 11.3 The significance of the Lorentzian pedagogy

We saw in Section 11.1 that Bell was aware in his 1976 essay of the limitations of the Maxwell–Lorentz theory in accounting for stable forms of material structure. He realized that a *complete* analysis of length contraction, say, in the spirit of the Lorentzian pedagogy would also require reference to forces other than of electromagnetic origin, and that the whole treatment would have to be couched in a quantum framework. But it is noteworthy that Bell did not seem to believe that articulation of a *complete* dynamical treatment of this kind was a necessary part of the Lorentzian pedagogy. In order to predict, on dynamical grounds, length contraction for moving rods and time dilation for moving clocks, Bell recognized that one need not know exactly how many distinct forces are at work, nor have access to the detailed dynamics of all of these interactions or the detailed microstructure of individual rods and clocks. It is enough, said Bell, to assume Lorentz covariance of the complete dynamics – known or otherwise – involved in the cohesion of matter. We call this the *truncated* Lorentzian pedagogy.

It is at this important point in Bell's essay that one sees something like a re-run of the thinking that the young Pauli brought to bear on the significance of relativistic kinematics in his acclaimed 1921 review article on relativity theory (Pauli 1981). Pauli was struck by the 'great value' of the apparent fact that in 1905 Einstein, unlike Lorentz, had given an account of his kinematics which was free of assumptions about the constitution of matter. He wrote:

> 'Should one, then, completely abandon any attempt to explain the Lorentz contraction atomistically? We think that the answer to this question should be No. The contraction of a measuring rod is not an elementary but a very complicated process. It would not take place except for the covariance with respect to the Lorentz group of the basic equations of electron theory, as well as of those laws, as yet unknown to us, which determine the cohesion of the electron itself'.
>
> (Pauli 1981, p. 15)

Both Pauli and Bell seem then to contrast the dynamical underpinning of relativistic kinematics with Einstein's 1905 argument. But it seems to us that once

the Lorentzian pedagogy relinquishes detailed specification of the dynamical interactions involved – in other words once it takes on the truncated form – the
difference between it and Einstein's approach, although significant, can easily be
overstated. Indeed, we regard it as plain wrong to construe Einstein's 1905 'kinematical' derivation of the Lorentz transformations as free of assumptions about the
constitution of matter, despite the distance between SR and the Maxwell–Lorentz
theory that Einstein urges (see above) in his *Relativity*.

This is best seen in the second application of the relativity principle in Einstein's
argument. The first application, it will be recalled, establishes the invariance of the
two-way light-speed, given the light postulate. Adopting the Einstein convention for
synchronizing clocks in both the moving and rest frames, this entails that the linear
co-ordinate transformations take on the form of the Lorentz transformations up to
a non-trivial scale or conformal factor. Einstein is now faced with the problem of
reducing this factor to unity (a problem which bedevilled Lorentz virtually throughout the development of his theory of the electron). Einstein achieves this (as did
Poincaré independently) by a second appeal to the relativity principle, in order to
guarantee the group property of the transformations – in particular to ground the
claim that the form of the transformation does not depend on the choice of frames.
This, together with an appeal to the principle of spatial isotropy, does the trick. The
details need not concern us; the interesting question is how this second application
of the relativity principle should be understood.

The co-ordinate transformations encode the behaviour of moving ideal rulers
and clocks under the crucial and universally accepted convention that these devices
retain their rest lengths and periods respectively under boosts. Suppose now that
the co-ordinate transformations between frames $S$ and $S'$ are different in form from
their inverses. We expect in this case either the length contraction factor or the
time dilation factor (if any), or both, to differ when measured relative to $S$ and
when measured relative to $S'$. And this would imply a violation of the relativity
principle. Specifically, it would be inconsistent with the claim that the dynamics of all the fundamental non-gravitational interactions which play a role in the
cohesion of these bodies satisfy the relativity principle. Thus, the *dynamical* relativity principle constrains the form of the *kinematical* transformations, because
such kinematics encodes the universal dynamical behaviour of rods and clocks in
motion.

It was clearly of importance to Bell that the Lorentzian pedagogy relied on physics
specified relative to a *single* inertial frame in order to account for the 'observations
of moving observers', and, in particular, the very validity of the relativity principle
itself. But ultimately that physics amounted to the claim that the complete theory
of the construction of matter is Lorentz covariant, of which the relativity principle
*inter alia* is a consequence. Einstein on the other hand started with the relativity
principle and the light postulate, and derived (using the isotropy of space) Lorentz
covariance. In comparing these two approaches, two points must not be lost sight of.
The first is that Einstein's argument is dynamical, since kinematics and dynamics in
this context cannot be disentangled.[6] The second point is that his 'principle theory'
approach to relativistic kinematics ruled out the truncated Lorentzian pedagogy as
a possible starting point for Einstein.

## 11.4 Einstein's unease about rods and clocks in special relativity

The extent to which Einstein understood the full dynamical implications of his 1905 derivation of the Lorentz transformations is perhaps unclear. Specifically, it is not clear that he recognized the role that rods and clocks can be seen to play in the derivation as *structured* bodies. What is clearer is that he harboured, or developed, a sense of unease about the status of these bodies in his initial formulation of SR. Einstein made use of these devices in the first instance to operationalize the spatial and temporal intervals, respectively, associated with inertial frames, but he never explained where they come from. In an essay entitled 'Geometrie und Erfahrung' (Einstein 1921), Einstein wrote:

> 'It is ... clear that the solid body and the clock do not in the conceptual edifice of physics play the part of irreducible elements, but that of composite structures, which must not play any independent part in theoretical physics. But it is my conviction that in the present stage of development of theoretical physics these concepts must still be employed as independent concepts; for we are still far from possessing such certain knowledge of the theoretical principles of atomic structure as to be able to construct solid bodies and clocks theoretically from elementary concepts'.
>
> (Einstein 1982)

Einstein's unease is more clearly expressed in a similar passage in his 1949 *Autobiographical Notes*:

> 'One is struck [by the fact] that the theory [of special relativity] ... introduces two kinds of physical things, i.e. (1) measuring rods and clocks, (2) all other things, e.g. the electromagnetic field, the material point, etc. This, in a certain sense, is inconsistent; strictly speaking measuring rods and clocks would have to be represented as solutions of the basic equations (objects consisting of moving atomic configurations), not, as it were, as theoretically self-sufficient entities. However, the procedure justifies itself because it was clear from the very beginning that the postulates of the theory are not strong enough to deduce from them sufficiently complete equations ... in order to base upon such a foundation a theory of measuring rods and clocks. .... But one must not legalize the mentioned sin so far as to imagine that intervals are physical entities of a special type, intrinsically different from other variables ('reducing physics to geometry', etc.)'.
>
> (Einstein 1969, pp. 59, 61)

It might seem that the justification Einstein provides for the self-confessed 'sin' of treating rods and clocks as 'irreducible', or 'self-sufficient' in 1905 is different in the two passages. In the 1921 essay, Einstein is saying that the constructive physics of atomic aggregation is still too ill-defined to allow for the modelling of such entities, whereas in the 1949 passage the point is that his 1905 postulates were insufficient in the first place to constrain the theory of matter in the required way – which is little more than a restatement of the problem. But as we have seen, it was precisely the uncertainties surrounding the basic constructive principles of matter and radiation that led Einstein in 1905 to base his theory on simple, phenomenological postulates.

Now there are two ways one might interpret these passages by Einstein. One might take him to be expressing concern that his 1905 derivation fails to recognize that rods and clocks are complex, structured entities. We argued in the previous section that there is no such failure. While the derivation is independent of the details of the laws which describe their internal structure, it is completely consistent with the true status of rods and clocks as complex solutions of (perhaps unknown) dynamical equations. In fact, the derivation implicitly treats them as such when the second appeal to the relativity principle is made.

Alternatively, one might take Einstein to be concerned about the fact that his postulates could not account for the availability of rods and clocks in the world in the first place. If this is his concern, then it is worth noting that the possibility of the existence of rods and clocks likewise does not follow from the mere assumption that all the fundamental laws are Lorentz covariant. It is only a full-blown quantum theory of matter, capable of dealing with the formation of stable macroscopic bodies that will fill the gap.

The significance of this point for the truncated Lorentzian pedagogy and the role of the constructive theory of rods and clocks in Einstein's thought are themes we will return to in Section 11.6. In the meantime, two points are worth making. First, it is perhaps odd that in his *Autobiographical Notes*, Einstein makes no mention of the advances that had occurred since 1905 in the quantum theory of matter and radiation. The understanding of the composition of bodies capable of being used as rods and clocks was far less opaque in 1949 than it was in 1905. Second, there is a hint at the end of the second passage above that towards the end of his life, Einstein did not view geometrical notions as fundamental in the special theory. An attempt to justify this scepticism, at least in relation to four-dimensional geometry, is given in Section 11.6.

### 11.5 A digression on rods and clocks in Weyl's 1918 unified field theory

In discussing the significance of the M-M experiment in his text *Raum-Zeit-Materie* (Weyl 1918a), Hermann Weyl stressed that the null result is a consequence of the fact that 'the interactions of the cohesive forces of matter as well as the transmission of light' are consistent with the requirement of Lorentz covariance (Weyl 1952, p. 173). Weyl's emphasis on the role of 'the mechanics of rigid bodies' in this context indicates a clear understanding of the dynamical underpinnings of relativistic kinematics. But Weyl's awareness that rigid rods and clocks are structured dynamical entities led him to the view that it is wrong to define the 'metric field' in SR on the basis of their behaviour.

Weyl's concern had to do with the problem of accelerated motion, or with deviations from what he called 'quasi-stationary' motion. Weyl's opinion in *Raum-Zeit-Materie* seems to have been that if a clock, say, is undergoing non-inertial motion, then it is unclear in SR whether the proper time read off by the clock is directly related to the length of its world-line determined by the Minkowski metric. For Weyl, clarification of this issue can only emerge 'when we have built up a **dynamics** based on physical and mechanical laws' (Weyl 1952, p. 177). This theme was to re-emerge in Weyl's responses to Einstein's criticisms of his 1918 attempt at a unified field theory. Before turning to this development, it is worth looking at Weyl's comments on SR.

In a sense Weyl was right. The claim that the length of a specified segment of an arbitrary time-like curve in Minkowski spacetime – obtained by integrating the

Minkowski line element $ds$ along the segment – is related to proper time rests on the assumption (now commonly dubbed the 'clock hypothesis') that the performance of the clock in question is unaffected by the acceleration it may be undergoing. It is widely appreciated that this assumption is not a consequence of Einstein's 1905 postulates. Its justification rests on the contingent dynamical requirement that the external forces accelerating the clock are small in relation to the internal 'restoring' forces at work inside the clock. (Similar considerations also hold, of course, in the case of rigid bodies.)

Today we are more sanguine about the clock hypothesis than Weyl seems to have been in *Raum-Zeit-Materie*. There is experimental confirmation of the hypothesis for nuclear clocks, for instance, with accelerations of the order $10^8$ cm s$^{-2}$. But the question remains as to whether the behaviour of rods and clocks captures the full significance of the 'metric field' of SR. Suppose accelerations exist such that for no known clock is the hypothesis valid (assuming the availability of the external forces in question!). Mathematically, one can still determine – using the prescription above – the length of the time-like worldline of any clock undergoing such acceleration if it does not disintegrate completely. From the perspective of the Lorentzian pedagogy, should one say that such a number has no physical meaning in SR? We return to this issue in the next section.

Weyl's separate publication of a stunning, though doomed unification of gravitational and electromagnetic forces (Weyl 1918) raised a number of intriguing questions about the meaning of spacetime structure which arguably deserve more attention than they have received to date (see, however, Ryckman 1994). Space prevents us from giving more than a sketch of the theory and its ramifications; our emphasis will be on the role of the Lorentzian pedagogy in evaluating the theory.

Weyl started from the claim that the pseudo-Riemannian spacetime geometry of Einstein's general relativity is not sufficiently local in that it allows the comparison of the lengths of distant vectors. Instead, Weyl insisted that the choice of unit of (spacetime) length at *each* point is arbitrary: only the ratios of the lengths of vectors *at the same point* and the angles between them can be physically meaningful. Such information is invariant under a *gauge transformation* of the metric field: $g_{ij} \rightarrow g'_{ij} = e^{2\lambda(x)} g_{ij}$ and constitutes a conformal geometry.

In addition to this conformal structure, Weyl postulated that spacetime is equipped with an affine connection that preserves the conformal structure under infinitesimal parallel transport. In other words, the infinitesimal parallel transport of all vectors at $p$ to $p'$ is to produce a similar image at $p'$ of the vector space at $p$.[7] For a given choice of gauge, the constant of proportionality of this similarity mapping will be fixed. Weyl assumed that it differed infinitesimally from 1 and thereby proceeded to show that the coefficients of the affine connection depended on a one-form field $\phi_i$ in addition to the metric coefficients $g_{ij}$ in such a way that the change in any length $l$ under parallel transport from $p$ (co-ordinates $\{x^i\}$) to $p'$ (co-ordinates $\{x^i + dx^i\}$) is given by:

$$dl = l\phi_i \, dx^i. \tag{11.1}$$

Under the gauge transformation $g_{ij} \rightarrow g'_{ij} = e^{2\lambda}g_{ij}$, $l \rightarrow e^{\lambda}l$. Substituting this into eqn. 11.1 gives:

$$\phi_i \rightarrow \phi'_i = \phi_i + \lambda_{,i},$$

the familiar transformation law for the electromagnetic four-potential. Weyl thus identified the gauge-invariant, four-dimensional curl of the geometric quantity $\phi_i$ with the familiar electromagnetic field tensor.

For a given choice of gauge a comparison of the length of vectors at distant points can be effected by integrating eqn. 11.1 along a path connecting the points. This procedure will in general be path-independent just if the electromagnetic field tensor vanishes everywhere.

As is well known, despite his admiration for Weyl's theory, Einstein was soon to spot a serious difficulty with the non-integrability of length (Einstein 1918a). In the case of a static gravitational field, a clock undergoing a round-trip in space during which it encountered a spatially varying electromagnetic potential would return to its starting point ticking at a rate different from that of a second clock which had remained at the starting point and which was originally ticking at the same rate. An effect analogous to this 'second clock effect' would occur for the length of an infinitesimal rod under the same circumstances. But it is a fact of the world – and a highly fortunate one! – that the relative periods of clocks (and the relative lengths of rods) do not depend on their histories in this way.

Before looking at Weyl's reply to this conundrum, it is worth remarking that it was apparently only in 1983 that the question was asked: what became of Einstein's objection once the gauge principle found its natural home in quantum mechanics? C.N. Yang pointed out that because the non-integrable scale factor in quantum mechanics relates to phase, the second clock effect could be detected using wave functions rather than clocks, essentially what Aharonov and Bohm had discovered (Aharonov and Bohm 1959; see also Ehrenberg and Siday 1949).[8] We note that Yang's question can be inverted: is there a full analogue of the Aharonov–Bohm effect in Weyl's gauge theory? The answer is yes, and it indicates that there was a further sting in Einstein's objection to Weyl that he and his contemporaries failed to spot. The point is that the second clock effect obtains in Weyl's theory even when the electromagnetic field vanishes everywhere on the trajectory of the clocks, so long as the closed path of the clocks enclose some region in which there is a non-vanishing field. This circumstance highlights the difficulty one would face in providing a dynamical or 'constructive' account of the second clock effect in the spirit of the full Lorentzian pedagogy.[9] Weyl's theory seems to be bedevilled by non-locality of a very striking kind.

The precise nature of Weyl's response to Einstein's objection would vary in the years following 1918 as he went on to develop new formulations of his unified field theory based on the gauge principle (see Vizgin 1994). But the common element was Weyl's rejection of the view that the metric field could be assigned operational significance in terms of the behaviour of rods and clocks. His initial argument was an extension of the point he had made about the behaviour of clocks in SR: one cannot know how a clock will behave under accelerations and in the presence of electromagnetic fields until a full dynamical modelling of the clock under these circumstances is available. The price Weyl ultimately paid for the beauty of his gauge principle – quite apart from the complicated nature of his field equations – was the introduction of rather tentative speculations concerning a complicated dynamical adjustment of rods and clocks to the 'world curvature' so as to avoid the second clock effect and its analogue for rods.

We finish this section with a final observation on the nature of Weyl's theory, with an eye to issues in standard general relativity to be discussed shortly. We noted above that Weyl's connection is not a metric connection. It is a function not only of the metric and its first derivatives, but also of the electromagnetic gauge field; in particular, for a fixed choice of gauge, the covariant derivative of the metric does not vanish everywhere. What does this imply?

The vanishing of the covariant derivative of the metric – the condition of metric compatibility – is sometimes introduced perfunctorily in texts on general relativity, but Schrödinger was right to call it 'momentous' (Schrödinger 1985, p. 106). It means that the local Lorentz frames associated with a spacetime point $p$ (those for which, at $p$, the metric tensor takes the form $\mathrm{diag}(1, -1, -1, -1)$ and the first derivatives of all its components vanish) are also local inertial frames (relative to which the components of the connection vanish at $p$).[10] If the laws of physics of the non-gravitational interactions are assumed to take their standard special relativistic form at $p$ relative to such local Lorentz charts (the local validity of special relativity), then metric compatibility implies that gravity is not a force in the traditional sense – an agency causing deviation from natural motion – in so far as the worldlines of freely falling bodies are geodesics of the connection.

The full physical implications of the non-metric compatible connection in Weyl's theory remain obscure in the absence of a full-blown theory of matter. Weyl's hints at a solution to the Einstein objection seem to involve a violation of minimal coupling, i.e. a violation of the prohibition of curvature coupling in the non-gravitational equations, and hence of the local validity of special relativity. But it seems that the familiar insight into the special nature of the gravitational interaction provided by the strong equivalence principle – the encapsulation of the considerations given in the previous paragraph – is lost in the Weyl theory.

## 11.6 The case of general relativity

There is a recurrent, Helmholtzian theme in Einstein's writings concerning Euclidean geometry: he claims that, as it is understood in physics, it is the science 'des possibilités de deplacement des corps solides' (Einstein 1928), or of 'the possibilities of laying out or juxtaposing solid bodies' (Einstein 1982, p. 163; see also Einstein 1982, pp. 234–5.).

But consider a universe consisting of some large number of mass points interacting by way of the Newtonian gravitational potential. Few would deny that a well-defined theory of such objects can be constructed within the framework of Newtonian mechanics (or its recent Machian counterparts such as Barbour and Bertotti (1982)). In such a world, there is nothing remotely resembling rigid bodies or rulers which allow for a direct operational significance to be assigned to the interparticle distances. Yet these distances are taken to obey the algebraic relationships of Euclidean geometry; either because this is a foundational assumption of the theory (as in the Machian approach), or because this is true of the particles' co-ordinate differences when referred to the privileged co-ordinate systems with respect to which the laws take on a canonical form. Moreover, the Euclidean constraint on the instantaneous configuration of the particles is formally the same as in a more ramified pregeneral relativistic (quantum) theory of matter which in principle allows for non-gravitational forces as well, and hence for the possibility of the existence of stable, rigid bodies.

Einstein was not oblivious to this point. He stressed that the accepted theory of matter itself (even pre-special relativistic theory) rules out the possibility of completely rigid bodies, and that atomistic conceptions exclude 'the idea of sharply and statically defined bounding surfaces of solid bodies'. Einstein realized that such physics 'must make use of geometry in the establishment of its concepts', concluding that 'the empirical content of geometry can be stated and tested only in the framework of the whole of physics' (Einstein 1982, p. 163).

Rigid bodies furnish what is an already meaningful spatial geometry with an (approximate) direct operational significance, and the fact that they have this role is a consequence of the theory of matter. It should be noted here that a necessary condition for this state of affairs is that the dynamical equations governing the non-gravitational interactions satisfy the so-called Euclidean symmetries. But this more complicated and no doubt more correct way of looking at things surely weakens the literal reading of Einstein's original account of Euclidean geometry above as the science of the possible arrangement of rigid bodies in space. The behaviour of rigid bodies under displacements does not *define* so much as *instantiate* the spatial geometry which might even have primordial status in the foundations of the theory of such bodies. And this point leads to another observation which is of considerable relevance to this chapter.

The fact that rods or rulers function as surveying devices for the primordial Euclidean geometry is not because they 'interact' with it; the latter is not a dynamical player which couples to matter. In arranging themselves in space, rigid bodies do not 'feel' the background geometry. To put it another way, a rod is not a 'thermometer of space'. Nor is it in the intrinsic nature of such bodies to survey space. It is the theory of matter which in principle tells us what kind of entities, if any, can serve as accurate geometrical surveying devices, and our best theories tell us rigid bodies will do. One only has to consider the consequences of a violation of the Euclidean symmetries in the laws of physics to dispel any doubts in this connection. (All of this is to say that 'what is a ruler?' is as important a question in physics as 'what is a clock?' An answer to this last question ultimately depends on specifying very special devices which 'tick' in synchrony with an independently meaningful temporal metric – a metric that is specified by the *dynamics* of the total isolated system of which the devices are a part.)[11]

Turning now to the geometry of special relativity, what is of interest is the behaviour of rods and clocks in relative motion. While the Minkowski geometry does not play a primordial role in the dynamics of such entities, analogous to the role which might be attributed to the three-dimensional Euclidean geometry constraining relative distances, it is definable in terms of the Lorentz covariance of the fundamental dynamical laws.[12] Hence, spacetime geometry is equally not simply 'the science of the possible behaviour of physical rods and clocks in motion'. All of the qualifications analogous to those we were forced to consider in the case of Euclidean geometry apply. In the context of SR, rods and clocks *are* surveying devices for a four-dimensional geometric structure. But this is a structure defined in terms of the symmetries of dynamical laws. If matter and its interactions are removed from the picture, Minkowski spacetime is not left behind.[13] Rods and clocks do not interact with a background metric field: they are not thermometers of spacetime structure.

There is a temptation to extend this lesson to general relativity (GR) in the following way. One might want to say that it is the local validity of special relativity in GR – as defined at the end of the previous section – that accounts for the existence of a metric field of Lorentzian signature whose metric connection coincides with the connection defining the inertial (free-fall) trajectories. The real new dynamics in GR has to do not with the metrical properties of spacetime but with the (generally) curved connection field and its coupling to matter. In particular, a defender of the Lorentzian pedagogy might be forgiven for accepting the maxim: no matter, no metric (without, however, excluding the connection). As we have argued, this is surely the right maxim for special relativity. In a universe entirely bereft of matter fields, even if one were to accept the primordial existence of inertial frames, it is hard to attribute any meaning in special relativity to the claim that empty spacetime retains a Minkowski metric as an element of reality, or equivalently to the claim that the inertial frames are related by Lorentz transformations. (On this point Einstein seems to have been somewhat inconsistent. It is difficult to reconcile the remarks on special relativity in his *Autobiographical Notes*, where he warns about the reduction of physics to geometry, with the claim in Appendix 5 of his *Relativity* – which he added to the fifteenth edition of the book in 1952 – that with the removal of all 'matter and field' from Minkowski space, this space would be left behind.)[14]

Adopting this position in GR has fairly drastic consequences in the case of the vacuum (matter-free) solutions to Einstein's field equations, of which 'empty' Minkowski spacetime itself is one solution. It entails that while the *flatness* of spacetime in this case (and the curvature in other less trivial solutions) – essentially an affine notion – may be said to have physical meaning, the *metric* structure of the spacetime does not (Brown 1997). (The metric might retain a meaning if one could adopt Feynman's 1963 suggestion that such vacuum solutions are correctly viewed as obtained by taking solutions involving sources and matter-free regions and allowing these regions to become infinitely large (Feynman et al. 1995, pp. 133–4). Feynman's view was that gravitational fields without sources are probably unphysical, akin to the popular view that all electrodynamical radiation comes from charged sources. Now the analogy with electromagnetism is arguably not entirely happy given that gravity is itself a source of gravity. Moreover, such an interpretation seems clearly inapplicable to finite vacuum spacetimes. Be that as it may, Feynman seemed to be happy with the flat Minkowski vacuum solution, perhaps because he could not entirely rid himself of his alternative 'Venutian' view of gravity as a massless spin-2 field on an absolute Minkowski background.)

But the temptation to take the Lorentzian pedagogy this far should perhaps be resisted. It overlooks the simple fact that the metric field in GR (defined up to the diffeomorphic 'gauge' freedom) appears to be a *bona fide* dynamical player, on a par with, say, the electromagnetic field. Even if one accepts – possibly either for Machian reasons (see Barbour 1994) or with a view to quantum gravity – the 4-metric as less fundamental than the evolving curved 3-metrics of the Hamiltonian approach to GR, it is nonetheless surely coherent to attribute a metric field to spacetime whether the latter boasts matter fields or not. If absolute Euclidean distances can exist in Newtonian universes bereft of rigid bodies, so much more can the dynamical metric field in GR have claim to existence, even in the non-generic case of the universal vacuum. Any alternative interpretation of the metric field in GR would seem to require an account of the coupling of a connection field to matter which was not mediated by the metric field as it is in Einstein's field equations. We know of no reason to be optimistic that this can be achieved.[15]

Where does this leave the Lorentzian pedagogy in GR? In our opinion, it still plays a fundamental role in understanding in dynamical detail how rods and clocks survey the metric field. To see this, let us consider the following claim made by Torretti, in his magnificent 1983 foundational text on *Relativity and Geometry*. Torretti formulates the basic assumption of GR as:

> 'The phenomena of gravitation and inertia ... are to be accounted for by a symmetric (0, 2) tensor field **g** on the spacetime manifold, which is linked to the distribution of matter by the Einstein field equations and defines a Minkowski inner product on each tangent space'.

> (Torretti 1983, p. 240)

It follows immediately from this hypothesis that the Minkowski inner product on tangent spaces induces a local approximate Minkowski geometry on a small neighbourhood of each event. Torretti claims that one can thereby 'account for the Lorentz invariance of the laws of nature referred to local Lorentz charts'. The successes of special relativity follow, says Torretti, from this local Minkowski geometry (Torretti 1983, p. 240).

In our view, this claim is a *non sequitur*. It is mysterious to us how the existence of a local approximate Minkowski geometry entails the Lorentz covariance of the laws of the non-gravitational interactions. Theories postulating a Lorentzian metric but which violate minimal coupling would involve non-Lorentz covariant laws. Equally, the primordial Euclidean geometry in the Newtonian theory of mass points discussed at the start of this section does not entail that the corresponding 'laws of nature' (if there are any non-gravitational interactions in the theory!) satisfy the Euclidean symmetries. There is something missing in Torretti's account, and this problem reminds one to some degree of the plea in some of Grünbaum's writings for an account of why the **g** field has the operational significance that it does. (A critical analysis of Grünbaum's arguments, with detailed references, is given by Torretti (1983, pp. 242–7).)

It seems to us that the local validity of special relativity in GR cannot be derived from what Torretti takes to be the central hypothesis of GR above, but must be independently assumed. Indeed, it often appears in texts as part of the strong equivalence principle, taken as a postulate of GR (for example, Misner et al. 1973, p. 386). The assumption, which is intimately related to the postulate of minimal coupling in GR, is that relative to the local Lorentz frames, insofar as the effects of curvature (tidal forces) can be ignored, the laws for the non-gravitational interactions take their familiar special relativistic form; in particular, the laws are Lorentz covariant. It is here of course that the full Lorentzian pedagogy can in principle be used to infer the possibility of the existence of material devices which more or less accurately survey the local Minkowski geometry. In particular, it explains why ideal clocks, which are chosen initially on dynamical grounds, act as hodometers, or 'way-wisers' of local Minkowski spacetime, and hence measure the lengths of time-like curves over the extended regions in curved spacetime that they define. (A very clear account of how the 'hence' in this statement was probably first understood by Einstein is given by Torretti (1983, pp. 149–51). Elsewhere, Torretti (1983, p. 312, footnote 13) notes that as early as 1913 Einstein recognized that the operational significance of the **g** field, and in particular the significance of the null cones in the tangent spaces, required the 'separate postulate' of the local validity of special relativity. It seems from the above that Torretti did not wish to follow Einstein in this respect.)

To conclude, the fact that general relativistic spacetimes are locally Minkowskian only acquires its usual 'chronometric' operational significance because of the independent assumption concerning the local validity of special relativity. Our main claim in this section is that this point can only be understood correctly by an appeal to the Lorentzian pedagogy. Despite the fact that in GR one is led to attribute an independent real existence to the metric field, the general relativistic explanation of length contraction and time dilation is simply the dynamical one we have urged in the context of special relativity.

## Notes

We are grateful to the following people for helpful discussions related to issues arising in this paper: J. Armstrong, G. Bacciagaluppi, K. Brading, P. Holland, C. Isham, N. Maxwell, S. Saunders, D. Wallace and particularly J. Barbour. We are especially indebted to J. Butterfield for careful written comments on an earlier draft. Our thanks also go to the editors for the invitation to contribute to this volume. One of the authors (O.P.) acknowledges support from the UK Arts and Humanities Research Board.

1. For a recent account of the origins of length contraction, see Brown (1999).
2. We are grateful to P. Holland for emphasizing this point (private communication).
3. Details of this episode can be found in Miller (1981, Chapter 1).
4. Nor were the two principles together strictly sufficient. The isotropy of space was also a crucial, if less prominent, principle in the derivation. Detailed examinations of the logic of Einstein's derivation can be found in Brown and Maia (1993) and Brown (1993). For an investigation of the far-reaching implications of abandoning the principle of spatial isotropy in the derivation, see Budden (1997).
5. In further elucidating the principle theory versus constructive theory distinction, one might consider the Casimir effect (attraction between conducting plates in the vacuum). This effect is normally explained on the basis of vacuum fluctuations in QED – the plates merely serving as boundary conditions for the q-number photon field. But it can also be explained in terms of Schwinger's source theory, which uses a c-number electromagnetic field generated by the sources in the plates so that the effect is ultimately due to interactions between the

microscopic constituents of the plates. (For a recent analysis of the Casimir effect, see Rugh et al. (1999).) It might appear that the relationship between the first approach and the second is similar to that between Einstein's formulation of SR and the 'Maxwell–Lorentz theory': the QED approach is simpler than Schwinger's and makes no claims as to the microscopic constitution of the plates (other than the claim that they conduct). But this appearance is misleading. Both approaches are equally 'constructive' in Einstein's sense; it is just that one appeals to the quantum structure of the vacuum and the other to fluctuating dipole moments associated with atoms or molecules in the plates.

6. The entangling of kinematics and dynamics is not peculiar to the relativistic context: details of a similar dynamical derivation of the Galilean transformations due to the fictional Albert Keinstein in 1705 are given in Brown (1993).

7. It is worth noting at this point that Weyl could, and perhaps should have gone further! As the keen-eyed Einstein was to point out, it is in the spirit of Weyl's original geometric intuition to allow for the relation between tangent spaces to be a weaker affine mapping: why insist that it be a similarity mapping? Einstein made this point in a letter to Weyl in 1918. For details see Vizgin (1994, p. 102).

8. Yang (1984). Yang recounts this incident in Yang (1986, p. 18).

9. It is worth noting that in 1923, Lorentz himself wrote, in relation to the rod analogue of the second clock effect in the Weyl theory, that this 'would amount to an action of an electromagnetic field widely different from anything that could reasonably be expected' (Lorentz 1937, p. 375). But whether Lorentz was concerned with the dynamical problem of accounting for how Maxwell's electrodynamics could in principle have such an effect on physical bodies like rods – a consideration which one would not expect to be foreign to Lorentz's thinking! – or simply with the empirical fact that such an effect is non-existent, is not entirely clear from Lorentz's comments.

10. See Misner et al. (1973, p. 313), Ehlers (1973, pp. 16–20) and Stewart (1991, p. 57).

11. See, for example, Barbour (1994, Sections 3, 4, and 12).

12. A similar position is defended by DiSalle (1995, p. 326). Our analysis of Euclidean geometry, and of the role of spacetime geometry in general relativity (see below) differs, however, from that of DiSalle.

13. A more recent example in physics of an absolute geometrical structure with clear dynamical underpinnings is that of the projective Hilbert space (ray space) in quantum mechanics. A (curved) connection in this space can be defined for which the anholonomy associated with closed curves is the geometric phase of Aharonov and Anandan. This geometric phase encodes a universal (Hamiltonian-independent) feature of Schrödinger evolution around each closed path in ray space, in a manner analogous to that in which the Minkowski geometry in SR encodes the universal behaviour of ideal rods and clocks arising out of the Lorentz covariant nature of the laws of physics. For a review of geometric phase, see Anandan (1992); further comparisons with Minkowski geometry are spelt out in Anandan (1991).

14. Actually, Einstein says that what remains behind is 'inertial-space or, more accurately this space together with the associated time', but subsequent remarks seem to indicate that he meant the 'rigid four-dimensional space of special relativity' (1982, p. 171).

15. Although connection rather than metric variables are fundamental in what is now the main approach to canonical quantum gravity (Ashtekar 1986), the existence of a metric in this approach is still independent of the existence of matter.

# Part V

# Quantum Gravity and
the Interpretation of Quantum Mechanics

# Quantum spacetime without observers: Ontological clarity and the conceptual foundations of quantum gravity

## Sheldon Goldstein and Stefan Teufel

### 12.1 Introduction

'The term "3-geometry" makes sense as well in quantum geometrodynamics as in classical theory. So does superspace. But space-time does not. Give a 3-geometry, and give its time rate of change. That is enough, under typical circumstances to fix the whole time-evolution of the geometry; enough in other words, to determine the entire four-dimensional space-time geometry, provided one is considering the problem in the context of classical physics. In the real world of quantum physics, however, one cannot give both a dynamic variable and its time-rate of change. The principle of complementarity forbids. Given the precise 3-geometry at one instant, one cannot also know at that instant the time-rate of change of the 3-geometry. . . . The uncertainty principle thus deprives one of any way whatsoever to predict, or even to give meaning to, "the deterministic classical history of space evolving in time". *No prediction of spacetime, therefore no meaning for spacetime,* is the verdict of the quantum principle'.

(Misner, Thorne, and Wheeler 1973)

One of the few propositions about quantum gravity that most physicists in the field would agree upon, that our notion of spacetime must, at best, be altered considerably in any theory conjoining the basic principles of quantum mechanics with those of general relativity, will be questioned in this chapter. We will argue, in fact, that most, if not all, of the conceptual problems in quantum gravity arise from the sort of thinking on display in the preceding quotation.

It is also widely agreed, almost forty years after the first attempts to quantize general relativity, that there is still no single set of ideas on how to proceed, and certainly no physical theory successfully concluding this program. Rather, there are a great variety of approaches to quantum gravity; for a detailed overview, see, for example Rovelli (1998). While the different approaches to quantum gravity often have little in common, they all are intended ultimately to provide us with a consistent quantum theory agreeing in its predictions with general relativity in the appropriate physical domain. Although we will focus here on the conceptual problems faced by those approaches which amount to a canonical quantization of classical general relativity, the main lessons will apply to most of the other approaches as well.

This is because – as we shall argue – many of these difficulties arise from the subjectivity and the ontological vagueness inherent in the very framework of orthodox quantum theory – a framework taken for granted by almost all approaches to quantum gravity. We shall sketch how most, and perhaps all, of the conceptual problems of canonical quantum gravity vanish if we insist upon formulating our cosmological theories in such a manner that it is reasonably clear what they are about – if we insist, that is, upon ontological clarity – and, at the same time, avoid any reference to such vague notions as measurement, observers, and observables.

The earliest approach, canonical quantum gravity, amounts to quantizing general relativity according to the usual rules of canonical quantization. However, to apply canonical quantization to general relativity, the latter must first be cast into canonical form. Since the quantization of the standard canonical formulation of general relativity, the Arnowitt, Deser, Misner formulation (Arnowitt, Deser, and Misner 1962) has led to severe conceptual and technical difficulties, non-standard choices of canonical variables, such as in the Ashtekar formulation (Ashtekar 1987) and in loop quantum gravity (Rovelli and Smolin 1990), have been used as starting points for quantization. While some of the technical problems have been resolved by these new ideas, the basic conceptual problems have not been addressed.

After the great empirical success of the standard model in particle physics, the hope arose that the gravitational interaction could also be incorporated in a similar model. The search for such a unified theory led to string theory, which apparently reproduces not only the standard model but also general relativity in a suitable low energy limit. However, since string theory is, after all, a quantum theory, it retains all the conceptual difficulties of quantum theory, and our criticisms and conclusions pertaining to quantum theory in general, in Sections 12.3 and 12.4 of this chapter, will apply to it as well. Nonetheless, our focus, again, will be on the canonical approaches, restricted for simplicity to pure gravity, ignoring matter.

This chapter is organized as follows: In Section 12.2 we will sketch the fundamental conceptual problems faced by most approaches to quantum gravity. The seemingly unrelated problems in the foundations of orthodox quantum theory will be touched upon in Section 12.3. Approaches to the resolution of these problems based upon the demand for ontological clarity will be discussed in Section 12.4, where we will focus on the simplest such approach, the de Broglie–Bohm theory or Bohmian mechanics. Our central point will be made in Section 12.5, where we indicate how the conceptual problems of canonical quantum gravity disappear when the main insights of the Bohmian approach to quantum theory are applied.

Finally, in Section 12.6, we will discuss how the status and significance of the wave function, in Bohmian mechanics as well as in orthodox quantum theory, is radically altered when we adopt a universal perspective. This altered status of the wave function, together with the very stringent symmetry demands so central to general relativity, suggests the possibility – though by no means the inevitability – of finding an answer to the question, Why should the universe be governed by laws so apparently peculiar as those of quantum mechanics?

## 12.2 The conceptual problems of quantum gravity

In the canonical approach to quantum gravity one must first reformulate general relativity as a Hamiltonian dynamical system. This was done by Arnowitt et al. (1962), using the 3-metric $g_{ij}(x^a)$ on a space-like hypersurface $\Sigma$ as the configurational variable and the extrinsic curvature of the hypersurface as its conjugate momentum $\pi^{ij}(x^a)$.[1] The real time parameter of usual Hamiltonian systems is replaced by a 'multi-fingered time' corresponding to arbitrary deformations $d\Sigma$ of the hypersurface. These deformations are split into two groups: those changing only the three dimensional co-ordinate system $x^a$ on the hypersurface (with which, as part of what is meant by the hypersurface, it is assumed to be equipped); and deformations of the hypersurface along its normal vector field. While the changes of the canonical variables under both kind of deformations are generated by Hamiltonian functions on phase space, $H_i(g, \pi)$ for spatial diffeomorphisms and $H(g, \pi)$ for normal deformations, their changes under pure co-ordinate transformations on the hypersurfaces are dictated by their geometrical meaning. The dynamics of the theory is therefore determined by the Hamiltonian functions $H(g, \pi)$ generating changes under normal deformations of the hypersurface.

Denote by $N(x^a)$ the freely specifiable lapse function that determines how far, in terms of proper length, one moves the space-like hypersurface at the point $x = (x^a)$ along its normal vector: This distance is $N(x^a)\, d\tau$, where $\tau$ is a parameter labelling the successive hypersurfaces arrived at under the deformation (and defined by this equation). The infinitesimal changes of the canonical variables are then generated by the Hamiltonian $H_N$ associated with $N$ (an integral over $\Sigma$ of the product of $N$ with a Hamiltonian density $H(g, \pi; x^a)$):

$$dg_{ij}(x^a) = \frac{\delta H_N(g, \pi)}{\delta \pi^{ij}(x^a)}\, d\tau$$

$$d\pi^{ij}(x^a) = -\frac{\delta H_N(g, \pi)}{\delta g_{ij}(x^a)}\, d\tau. \tag{12.1}$$

In what follows we shall denote by $H(g, \pi)$ the collection $\{H_N(g, \pi)\}$ of all such Hamiltonians (or, what comes pretty much to the same thing, the collection $\{H(g, \pi; x)\}$ for all points $x \in \Sigma$) and similarly for $H_i$.

It is important to stress that the theory can be formulated completely in terms of geometrical objects on a three-dimensional manifold, with no *a priori* need to refer to its embedding into a spacetime. A solution of eqn. 12.1 is a family of 3-metrics $g(\tau)$ that can be glued together to build up a 4-metric using the lapse function $N$ (to determine the transverse geometry). In this way the spacetime metric emerges dynamically when one evolves the canonical variables with respect to multi-fingered time.

However, the initial canonical data cannot be chosen arbitrarily, but must obey certain constraints: Only for initial conditions that lie in the submanifold of phase space on which $H_i(g, \pi)$ and $H(g, \pi)$ vanish do the solutions (spacetime metrics $g_{\mu\nu}(x^\mu)$) also satisfy Einstein's equations. In fact, away from this so-called constraint manifold the theory is not even well defined, at least not as a theory involving a multi-fingered time, since the solutions would depend on the special way we choose to evolve the space-like hypersurface, i.e. on the choice of $N(x^a)$, to build up spacetime. Of course, a theory based on a single choice, for example $N(x^a) = 1$, would be well defined, at least formally.

By the same token, the invariance of the theory under spacetime diffeomorphisms is no longer so obvious as in the formulation in terms of Einstein's equations: In the ADM formulation 4-diffeomorphism invariance amounts to the requirement that one ends up with the same spacetime, up to co-ordinate transformations, regardless of which path in multi-fingered time is followed, i.e. which lapse function $N$, or $\tau$-dependent sequence of lapse functions $N(\tau)$, is used. This says that for the spacetime built up from any particular choice of multi-fingered time, the dynamical equations 12.1 will be satisfied for *any* foliation of the resulting spacetime into space-like hypersurfaces – using in eqn. 12.1 the lapse function $N(\tau)$ associated with that foliation – and not just for the foliation associated with that particular choice.

Formally, it is now straightforward to quantize this constrained Hamiltonian theory using Dirac's rules for the quantization of constrained systems (Dirac 1964). First, one must replace the canonical variables $g_{ij}$ and $\pi^{ij}$ by operators $\hat{g}_{ij}$ and $\hat{\pi}^{ij} = -i\,\delta/\delta g_{ij}$ satisfying the canonical commutation relations.[2] One then formally inserts these into the Hamiltonian functions $H(g, \pi)$ and $H_i(g, \pi)$ of the classical theory to obtain operators $\hat{H}(\hat{g}, \hat{\pi})$ and $\hat{H}_i(\hat{g}, \hat{\pi})$ acting on functionals $\Psi(g)$ on the configuration space of 3-metrics. Since the Hamiltonians were constrained in the classical theory one demands that the corresponding operators annihilate the physical states in the corresponding quantum theory:

$$\hat{H}\Psi = 0 \tag{12.2}$$

$$\hat{H}_i\Psi = 0. \tag{12.3}$$

Equation 12.3 has a simple meaning, namely that $\Psi(g)$ be invariant under 3-diffeomorphisms (co-ordinate changes on the 3-manifold), so that it depends on the 3-metric $g$ only through the 3-geometry. However, the interpretation of the *Wheeler–DeWitt equation* (eqn. 12.2) is not at all clear.

Before discussing the several problems which arise in attempts to give a physical meaning to the approach just described, a few remarks are in order: While we have omitted many technical details and problems from our schematic description of the 'Dirac constraint quantization' of gravity, these problems either do not concern, or are consequences of, the main conceptual problems of canonical quantum gravity. Other approaches, such as the canonical quantization of the Ashtekar formulation of classical general relativity and its further development into loop quantum gravity, resolve some of the technical problems faced by canonical quantization in the metric representation, but leave the main conceptual problems untouched.

Suppose now that we have found a solution $\Psi(g)$ to eqns. 12.2 and 12.3. What physical predictions would be implied? In orthodox quantum theory a solution $\Psi_t$ of the time-dependent Schrödinger equation provides us with a time-dependent probability distribution $|\Psi_t|^2$, as well as with the absolute square of other time-dependent probability amplitudes. The measurement problem and the like aside, the physical meaning of these is reasonably clear: they are probabilities for the results of the measurement of the configuration or of other observables. But any attempt to interpret canonical quantum gravity along orthodox lines immediately faces the following problems:

### 12.2.1 The problem of time

In canonical quantum gravity there is no time-dependent Schrödinger equation; it was replaced by the time-independent Wheeler–DeWitt equation. The Hamiltonians – the generators of multi-fingered-time evolution in the classical case – annihilate the state vector and therefore cease to generate any evolution at all. The theory provides us with only a timeless wave function on the configuration space of 3-metrics, i.e. on the possible configurations of space, not of spacetime. But how can a theory that provides us (at best) with a single fixed probability distribution for configurations of space ever be able to describe the always changing world in which we live? This, in a nutshell, is the problem of time in canonical quantum gravity.

### 12.2.2 The problem of 4-diffeomorphism invariance

The fundamental symmetry at the heart of general relativity is its invariance under general co-ordinate transformations of spacetime. It is important to stress that almost any theory can be formulated in such a 4-diffeomorphism invariant manner by adding further structure to the theory (e.g. a preferred foliation of spacetime as a dynamical object). General relativity has what is sometimes called serious diffeomorphism-invariance, meaning that it involves no spacetime structure beyond the 4-metric and, in particular, singles out no special foliation of spacetime. In canonical quantum gravity, while the invariance under co-ordinate transformations of space is retained, it is not at all clear what 4-diffeomorphism invariance could possibly mean. Therefore the basic symmetry, and arguably the essence, of general relativity seems to be lost in quantization.

### 12.2.3 The problem of 'no outside observer'

One of the most fascinating applications of quantum gravity is to quantum cosmology. Orthodox quantum theory attains physical meaning only via its predictions about the statistics of outcomes of measurements of observables, performed by observers that are not part of the system under consideration, and seems to make no clear physical statements about the behaviour of a closed system, not under observation. The quantum formalism concerns the interplay between – and requires for its very meaning – two kinds of objects: a quantum system, and a more or less classical apparatus. It is hardly imaginable how one could make any sense out of this formalism for quantum cosmology, for which the system of interest is the whole universe, a closed system if there ever was one.

### 12.2.4 The problem of diffeomorphism-invariant observables

Even if we pretend for the moment that we are able to give meaning to the quantum formalism without referring to an observer located outside of the universe, we encounter a more subtle difficulty. Classical general relativity is fundamentally diffeomorphism invariant. It is only the spacetime geometry, not the 4-metric nor the identity of the individual points in the spacetime manifold, that has physical significance. Therefore the physical observables in general relativity should be independent of special co-ordinate systems; they must be invariant under 4-diffeomorphisms, which are in effect generated by the Hamiltonians $H$ and $H_i$. Since the quantum observables are constructed, via quantization, from the classical ones, it would seem that they must commute with the Hamiltonians $\hat{H}$ and $\hat{H}_i$. But such diffeomorphism-invariant quantum observables are extremely hard to come by, and there are certainly far too few of them to even begin to account for the bewildering variety of our experience which it is the purpose of quantum theory to explain. (For a discussion of the question of existence of diffeomorphism-invariant observables, see Kuchař 1992.)

These conceptual problems, and the attempts to solve them, have led to a variety of technical problems that are discussed in much detail in, for example Kuchař (1992, 1993), and Isham (1994). However, since we are not aware of any orthodox proposals successfully resolving the conceptual problems, we shall not discuss such details here. Rather, we shall proceed in the opposite direction, toward their common cause, and argue that they originate in a deficiency shared with, and inherited from, orthodox quantum mechanics: the lack of a coherent ontology.

Regarding the first two problems of canonical quantum gravity, it is not hard to discern their origin: the theory is concerned only with configurations of and on space, the notion of a spacetime having entirely disappeared. It is true that even with classical general relativity, Newton's external absolute time is abandoned. But a notion of time, for an observer located somewhere in spacetime and employing a co-ordinate system of his or her convenience, is retained, emerging from spacetime. The problem of time in canonical quantum gravity is a direct consequence of the fact that in an orthodox quantum theory for spacetime itself we must insist on its non-existence (compare the quote at the beginning of this article). Similarly, the problem of diffeomorphism invariance or, better, the problem of not even being able to address this question properly, is an immediate consequence of having no notion of spacetime in orthodox quantum gravity.

### 12.3 The basic problem of orthodox quantum theory: the lack of a coherent ontology

Despite its extraordinary predictive successes, quantum theory has, since its inception some seventy-five years ago, been plagued by severe conceptual difficulties. The most widely cited of these is the measurement problem, best known as the paradox of Schrödinger's cat. For many physicists the measurement problem is, in fact, not *a* but *the* conceptual difficulty of quantum theory.

In orthodox quantum theory the wave function of a physical system is regarded as providing its complete description. But when we analyse the process of measurement itself in quantum mechanical terms, we find that the after-measurement wave function for system and apparatus arising from Schrödinger's equation for the composite system typically involves a superposition over terms corresponding to what we would like to regard as the various possible results of the measurement – e.g. different pointer orientations. Since it seems rather important that the actual result of the measurement be a part of the description of the after-measurement situation, it is difficult to believe that the wave function alone provides the complete description of that situation.

The usual collapse postulate for quantum measurement solves this problem for all practical purposes, but only at the very steep price of the introduction of an *observer* or *classical measurement apparatus* as an irreducible, unanalysable element of the theory. This leads to a variety of further problems. The unobserved physical reality becomes drastically different from the observed, even on the macroscopic level of daily life. Even worse, with the introduction at a fundamental level of such vague notions as classical measurement apparatus, the physical theory itself becomes unprofessionally vague and ill-defined. The notions of observation and measurement can hardly be captured in a manner appropriate to the standards of rigour and clarity that should be demanded of a fundamental physical theory. And in quantum cosmology the notion of an external observer is of course entirely obscure.

The collapse postulate is, in effect, an unsuccessful attempt to evade the measurement problem without taking seriously its obvious implication: that the wave function does not provide a complete description of physical reality. If we do accept this conclusion, we must naturally enquire about the nature of the more complete description with which a less problematical formulation of quantum theory should be concerned. We must ask, which theoretical entities, in addition to the wave function, might the theory describe? What mathematical objects and structures represent entities that, according to the theory, simply *are*, regardless of whether or not they are observed? We must ask, in other words, about the *primitive ontology* of the theory, what the theory is fundamentally about (see Goldstein 1998, 1998a). And when we know what the theory is really about, measurement and observation become secondary phenomenological concepts that, like anything else in a world governed by the theory, can be analysed in terms of the behaviour of its primitive ontology.

By far the simplest possibility for the primitive ontology is that of particles described by their positions. The corresponding theory, for non-relativistic particles, is Bohmian mechanics.

## 12.4 **Bohmian mechanics**

According to Bohmian mechanics the complete description of an $n$-particle system is provided by its wave function $\Psi$ together with its configuration $Q = (\mathbf{Q}_1, \ldots, \mathbf{Q}_n)$, where the $\mathbf{Q}_k$ are the positions of its particles. The wave function, which evolves according to Schrödinger's equation, choreographs the motion of the particles: these evolve – in the simplest manner possible – according to a first-order ordinary differential equation

$$\frac{dQ}{dt} = v^{\Psi}(Q)$$

whose right-hand side, a velocity vector field on configuration space, is generated by the wave function. Considerations of simplicity and spacetime symmetry – Galilean and time-reversal invariance – then determine the form of $v^{\Psi}$, yielding the defining (evolution) equations of Bohmian mechanics (for spinless particles):

$$\frac{d\mathbf{Q}_k}{dt} = \mathbf{v}_k^{\Psi}(\mathbf{Q}_1, \ldots, \mathbf{Q}_n) = \frac{\hbar}{m_k} \operatorname{Im} \frac{\nabla_{q_k}\Psi}{\Psi}(\mathbf{Q}_1, \ldots, \mathbf{Q}_n) \tag{12.4}$$

and

$$i\hbar \frac{\partial \Psi}{\partial t} = \hat{H}\Psi, \tag{12.5}$$

where $\hat{H}$ is the usual Schrödinger Hamiltonian, containing as parameters the masses $m_1, \ldots, m_n$ of the particles as well as the potential energy function $V$ of the system. For an $n$-particle universe, these two equations form a complete specification of the theory. There is no need, and indeed no room, for any further axioms, describing either the behaviour of other 'observables' or the effects of 'measurement'.

Bohmian mechanics is the most naively obvious embedding imaginable of Schrödinger's equation into a completely coherent physical theory! If one did not already know better, one would naturally conclude that it cannot 'work,' i.e. that it cannot account for quantum phenomena. After all, if something so obvious and, indeed, so trivial works, great physicists – so it would seem – would never have insisted, as they have and as they continue to do, that quantum theory demands radical epistemological and metaphysical innovations.

Moreover, it is hard to avoid wondering how Bohmian mechanics *could* have much to do with quantum theory? Where is quantum randomness in this deterministic theory? Where is quantum uncertainty? Where are operators as observables and all the rest?

Be that as it may, Bohmian mechanics is certainly *a* theory. It describes a world in which particles participate in a highly non-Newtonian motion, and it would do so even if this motion had absolutely nothing to do with quantum mechanics.

It turns out, however, as a surprising consequence of the eqns. 12.4 and 12.5, that when a system has wave function $\Psi$, its configuration is typically random, with probability density $\rho$ given by $\rho = |\Psi|^2$, the *quantum equilibrium* distribution. In other words, it turns out that systems are somehow typically in quantum equilibrium. Moreover, this conclusion comes together with the clarification of what precisely this means, and also implies that a Bohmian universe embodies an absolute uncertainty which can itself be regarded as the origin of the uncertainty principle. We shall not go into these matters here, since we have discussed them at length elsewhere (Dürr, Goldstein, and Zanghì 1992). We note, however, that nowadays, with chaos theory and non-linear dynamics so fashionable, it is not generally regarded as terribly astonishing for an appearance of randomness to emerge from a deterministic dynamical system.

It also turns out that the entire quantum formalism, operators as observables and all the rest, is a consequence of Bohmian mechanics, emerging from an analysis of idealized measurement-like situations (for details, see Daumer et al. 1997, 1999; see also Bohm 1952). There is no measurement problem in Bohmian mechanics because the complete description of the after-measurement situation includes, in addition to the wave function, the definite configuration of the system and apparatus. While the wave function may still be a superposition of states corresponding to macroscopically different possible outcomes, the actual configuration singles out the outcome that has occurred.

Why have we elaborated in such detail on non-relativistic quantum mechanics and Bohmian mechanics if our main concern here is with quantum gravity? Because there are two important lessons to be learned from a Bohmian perspective on quantum theory. First of all, the existence of Bohmian mechanics demonstrates that the characteristic features of quantum theory, usually viewed as fundamental – intrinsic randomness, operators as observables, non-commutativity, and uncertainty – need play no role whatsoever in the formulation of a quantum theory, naturally emerging instead, as a consequence of the theory, in special measurement-like situations. Therefore we should perhaps not be too surprised when approaches to quantum gravity that regard these features as fundamental encounter fundamental conceptual difficulties. Second, the main point of our chapter is made transparent in the simple example of Bohmian mechanics. If we base our theory on a coherent ontology, the conceptual problems may disappear, and, what may be even more important, a genuine understanding of the features that have seemed most puzzling might be achieved.

We shall now turn to what one might call a Bohmian approach to quantum gravity.

## 12.5 Bohmian quantum gravity

The transition from quantum mechanics to Bohmian mechanics is very simple, if not trivial: one simply incorporates the actual configuration into the theory as the basic variable, and stipulates that this evolve in a natural way, suggested by symmetry and by Schrödinger's equation. The velocity field $v^{\Psi_t}$ is, in fact, related to the quantum probability current $j^{\Psi_t}$ by

$$v^{\Psi_t} = \frac{j^{\Psi_t}}{|\Psi_t|^2},$$

suggesting, since $\rho^{\Psi} = |\Psi|^2$ satisfies the continuity equation with $j^{\Psi_t} = \rho^{\Psi_t} v^{\Psi_t}$, that the empirical predictions of Bohmian mechanics, for positions and ultimately, in fact, for other 'observables' as well, agree with those of quantum mechanics (as in fact they do; see Dürr et al. 1992).

Formally, one can follow exactly the same procedure in canonical quantum gravity, where the configuration space is the space of (positive-definite) 3-metrics (on an appropriate fixed manifold). The basic variable in Bohmian quantum gravity is therefore the 3-metric $g$ (representing the geometry on a space-like hypersurface of the spacetime to be generated by the dynamics) and its change under (what will become) normal deformations is given by a vector field on configuration space generated by the wave function $\Psi(g)$. Considerations analogous to those for non-relativistic particles lead to the following form for the Bohmian equation of motion:

$$dg_{ij}(x^a) = G_{ijab}(x^a) \, \mathrm{Im}\left( \Psi(g)^{-1} \frac{\delta \Psi(g)}{\delta g_{ab}(x^a)} \right) N(x^a) \, d\tau. \tag{12.6}$$

The wave function $\Psi(g)$ is a solution of the timeless Wheeler–DeWitt equation (eqn. 12.2) and therefore does not evolve. But the vector field on the right-hand side of eqn. 12.6 that it generates is typically non-vanishing if $\Psi(g)$ is complex, leading to a non-trivial evolution $g(\tau)$ of the 3-metric. Suitably gluing together the 3-metrics $g(\tau)$, we obtain a spacetime (see the paragraph following eqn. 12.1). Interpretations of canonical quantum gravity along these lines have been proposed by, for example Holland (1993) and discussed, for example by Shtanov (1996). Minisuperspace Bohmian cosmologies have been considered by Kowalski-Glikman and Vink (1990), Squires (1992), and Callender and Weingard (1994).

However, there is a crucial point which is often overlooked or, at least, not made sufficiently clear in the literature. A spacetime generated by a solution of eqn. 12.2 via eqn. 12.6 will in general depend on the choice of lapse function $N$ (or $N(\tau)$). Thus, the theory is not well defined as so far formulated. There are essentially two ways to complete the theory. Either one chooses a special lapse function $N$, e.g. $N = 1$, or one employs only special solutions $\Psi$ of eqn. 12.2, those yielding a vector field that generates an $N$-independent spacetime. In the first case, with special $N$ but general solution $\Psi$ of eqn. 12.2, the general covariance of the theory will typically be broken, the theory incorporating a special foliation (see the paragraph before that containing eqn. 12.2). The possible existence of special solutions giving rise to a covariant dynamics will be discussed in more detail elsewhere (Goldstein and Teufel 1999), and will be touched upon towards the end of Section 12.6. However, most of the following discussion, especially in the first part of Section 12.6, does not depend upon whether or not the theory incorporates a special foliation.

Let us now examine the impact of the Bohmian formulation of canonical quantum gravity on the basic conceptual problems of orthodox canonical quantum gravity. Since a solution to the equations of Bohmian quantum gravity defines a spacetime, the problem of time is resolved in the most satisfactory way. Time plays exactly the same role as in classical general relativity; there is no need whatsoever for an external absolute time, which has seemed so essential to orthodox quantum theory. The problem of diffeomorphism invariance is ameliorated, in that in this formulation it is at least clear what diffeomorphism invariance means. But, as explained above, general covariance can be expected at most for special solutions of eqn. 12.2. If it should turn out, however, that we must abandon general covariance on the fundamental level by introducing a special foliation of spacetime, it may still be possible to retain it on the observational level (see, e.g. Münch-Berndl et al. 1999, where it is also argued, however, that a special, dynamical, foliation of spacetime need not be regarded as incompatible with serious covariance).

A short answer to the problems connected with the role of observers and observables is this: There can be no such problems in the Bohmian formulation of canonical quantum gravity since observers and observables play no role in this formulation. But this is perhaps too short. What, after all, is wrong with the observation that, since individual spacetime points have no physical meaning, physically significant quantities must correspond to diffeomorphism-invariant observables, of which there are far too few to describe very much of what we most care about?

The basic answer, we believe, is this: We ourselves are not – or, at least, need not be – diffeomorphism invariant.[3] Most physical questions of relevance to us are not formulated in a diffeomorphism-invariant manner because, naturally enough, they refer to our own special location in spacetime. Nonetheless, we know very well what they mean – we know for example what it means to ask where and when something happens with respect to our own point of view. Such questions can be addressed, in fact because of diffeomorphism invariance, by taking into account the details of our environment and asking about the local predictions of the theory conditioned on such an environment, past and present.

The observer who sets the frame of reference for his or her physical predictions is part of and located inside the system – the universe. In classical general relativity this is not at all problematical, since that theory provides us with a coherent ontology, a potentially complete description of spacetime and, if we wish, a description taking into account our special point of view in the universe. But once the step to quantum theory is taken, the coherent spacetime ontology is replaced by an incoherent 'ontology' of quantum observables. In orthodox quantum theory this problem can be talked away by introducing an outside observer actually serving two purposes: the observer sets the frame of reference with respect to which the predictions are to be understood, a totally legitimate and sensible purpose. But of course the main reason for the focus on observers in quantum theory is that it is only with respect to them that the intrinsically incoherent quantum description of the system under observation can be given any meaning. In quantum cosmology, however, no outside observer is at hand, neither for setting a frame of reference nor for transforming the incoherent quantum picture into a coherent one.

In Bohmian quantum gravity, again, both problems disappear. Since we have a coherent description of the system itself, in this case the universe, there is no need for an outside observer in order to give meaning to the theory. Nor do we have to worry about the diffeomorphism invariance of observables, since we are free to refer to observers who are themselves part of the system.

There is, however, an important aspect of the problem of time that we have not yet addressed. From a Bohmian perspective, as we have seen, a time-dependent wave function, satisfying Schrödinger's equation, is by no means necessary to understand the possibility of what we call change. Nonetheless, a great deal of physics is, in fact, described by such time-dependent wave functions. We shall see in the next section how these also naturally emerge from the structure of Bohmian quantum gravity, which fundamentally has only a timeless universal wave function.

## 12.6 A universal Bohmian theory

When Bohmian mechanics is viewed from a universal perspective, the status of the wave function is altered dramatically. To appreciate what we have in mind here, it might help to consider two very common objections to Bohmian mechanics.

Bohmian mechanics violates the action–reaction principle that is central to all of modern physics, both classical and (non-Bohmian) quantum. In Bohmian mechanics there is no back-action of the configuration upon the wave function, which evolves, autonomously, according to Schrödinger's equation. And the wave function, which is part of the state description of – and hence presumably part of the reality comprising – a Bohmian universe, is not the usual sort of physical field on physical space (like the electromagnetic field) to which we are accustomed, but rather a field on the abstract space of all possible configurations, a space of enormous dimension, a space constructed, it would seem, by physicists as a matter of convenience.

It should be clear by now what, from a universal viewpoint, the answer to these objections must be. As first suggested by Dürr, Goldstein, and Zanghì (1997), the wave function $\Psi$ of the universe should be regarded as a representation, not of substantial physical reality, but of physical law. In a universal Bohmian theory $\Psi$ should be a functional of the configurations of all elements of physical reality: geometry, particle positions, field or string configurations, or whatever primitive ontology turns out to describe nature best. As in the case of pure quantum gravity, $\Psi$ should be a (special) solution of some fundamental equation (such as the Wheeler–DeWitt equation (eqn. 12.2) with additional terms for particles, fields, etc.). Such a universal wave function would be static – a wave function whose timelessness constitutes the problem of time in canonical quantum gravity – and, insofar as our universe is concerned, unique. But this does not mean, as we have already seen, that the world it describes would be static and timeless. No longer part of the state description, the universal wave function $\Psi$ provides a representation of dynamical law, via the vector field on configuration space that it defines. As such, the wave function plays a role analogous to that of the Hamiltonian function $H = H(Q, P) \equiv H(\xi)$ in classical mechanics – a function on phase space, a space even more abstract than configuration space. In fact, the wave function and the Hamiltonian function generate motions in pretty much the same way

$$\frac{d\xi}{dt} = \text{Der } H \longleftrightarrow \frac{dQ}{dt} = \text{Der}(\log \Psi),$$

with Der a derivation. And few would be tempted to regard the Hamiltonian function $H$ as a real physical field, or expect any back-action of particle configurations on this Hamiltonian function.

Once we recognize that the role of the wave function is thus nomological, two important questions naturally arise: Why and how should a formalism involving time-dependent wave functions obeying Schrödinger's equation emerge from a theory involving a fixed timeless universal wave function? And which principle singles out the special unique wave function $\Psi$ that governs the motion in our universe? Our answers to these questions are somewhat speculative, but they do provide further insight into the role of the wave function in quantum mechanics and might even explain why, in fact, our world is quantum mechanical.

In order to understand the emergence of a time-dependent wave function, we must ask the right question, which is this: Is it ever possible to find a simple effective theory governing the behaviour of suitable subsystems of a Bohmian universe? Suppose, then, that the configuration of the universe has a decomposition of the form $q = (x, y)$, where $x$ describes the degrees of freedom with which we are somehow most directly concerned (defining the *subsystem*, the '*x*-system') and $y$ describes the remaining degrees of freedom (the subsystem's *environment*, the '*y*-system'). For example, $x$ might be the configuration of all the degrees of freedom governed by standard quantum field theory, describing the fermionic matter fields as well as the bosonic force fields, while $y$ refers to the gravitational degrees of freedom.

Suppose further that we have, corresponding to this decomposition, a solution $Q(\tau) = (X(\tau), Y(\tau))$ of the appropriate (yet to be defined) extension of eqn. 12.6, where the real continuous parameter $\tau$ labels the slices in a suitable foliation of spacetime.

Focus now on the *conditional wave function*

$$\psi_\tau(x) = \Psi(x, Y(\tau)) \tag{12.7}$$

of the subsystem, governing its motion, and ask whether $\psi_\tau(x)$ could be – and might under suitable conditions be expected to be – governed by a simple law that does not refer directly to its environment. (The conditional wave function of the $x$-system should be regarded as defined only up to a factor that does not depend upon $x$.)

Suppose that $\Psi$ satisfies an equation of the form of eqn. 12.2, with $\hat{H} = \{\hat{H}_N\}$. Suppose further that for $y$ in some '$y$-region' of configuration space and for some choice of lapse function $N$ we have that $\hat{H}_N \simeq \hat{H}_N^{(x)} + \hat{H}_N^{(y)}$ and can write

$$\Psi(x, y) = e^{-i\tau \hat{H}_N} \Psi(x, y) \simeq e^{-i\tau \hat{H}_N} \sum_\alpha \psi_0^\alpha(x) \phi_0^\alpha(y)$$

$$\simeq \sum_\alpha \left( e^{-i\tau \hat{H}_N^{(x)}} \psi_0^\alpha(x) \right) \left( e^{-i\tau \hat{H}_N^{(y)}} \phi_0^\alpha(y) \right)$$

$$=: \sum_\alpha \psi_\tau^\alpha(x) \phi_\tau^\alpha(y), \tag{12.8}$$

where the $\phi_0^\alpha$ are 'narrow disjoint wave packets' and remain approximately so as long as $\tau$ is not too large. Suppose (as would be the case for Bohmian mechanics) that the motion is such that if the configuration $Y(0)$ lies in the support of one $\phi_0^{\alpha'}$, then $Y(\tau)$ will keep up with $\phi_\tau^{\alpha'}$ as long as the above conditions are satisfied. It then follows from eqn. 12.8 that for the conditional wave function of the subsystem we have

$$\psi_\tau(x) \approx \psi_\tau^{\alpha'},$$

and it thus approximately satisfies the time-dependent Schrödinger equation

$$i \frac{\partial \psi}{\partial \tau} = \hat{H}_N^{(x)} \psi. \tag{12.9}$$

(In the case of (an extension of) Bohmian quantum gravity with preferred foliation, this foliation must correspond to the lapse function $N$ in eqn. 12.8.)

We may allow here for an interaction $\hat{W}_N(x, y)$ between the subsystem and its environment in the Hamiltonian in eqn. 12.8, provided that the influence of the $x$-system on the $y$-system is negligible. In this case we can replace $\hat{H}_N^{(x)}$ in eqn. 12.8 and eqn. 12.9 by $\hat{H}_N^{(x)}(Y(\tau)) \equiv \hat{H}_N^{(x)} + \hat{W}_N(x, Y(\tau))$, since the wave packets $\phi^\alpha(y)$ are assumed to be narrow. Think, for the simplest example, of the case in which the $y$-system is the gravitational field and the $x$-system consists of very light particles.

Now one physical situation (which can be regarded as corresponding to a region of configuration space) in which eqn. 12.8, and hence the Schrödinger evolution (eqn. 12.9), should obtain is when the $y$-system behaves semiclassically. In the semiclassical regime, one expects an initial collection of narrow and approximately disjoint wave packets $\phi_0^\alpha(y)$ to remain so under their (approximately classical) evolution.

As a matter of fact, the emergence of Schrödinger's equation in the semiclassical regime for gravity can be justified in a more systematic way, using perturbation theory, by expanding $\Psi$ in powers of the gravitational constant $\kappa$. Then for a 'semiclassical wave function' $\Psi$, the phase $S$ of $\Psi$, to leading order, $\kappa^{-1}$, depends only on the 3-metric and obeys the classical Einstein–Hamilton–Jacobi equation, so that the metric evolves approximately classically, with the conditional wave function for the matter degrees of freedom satisfying, to leading (zeroth) order, Schrödinger's equation for, say, quantum field theory on a given evolving background. The relevant analysis was done by Banks (1985) for canonical quantum gravity, but the significance of that analysis is rather obscure from an orthodox perspective.

The semiclassical limit has been proposed as a solution to the problem of time in quantum gravity, and as such has been severely criticized by Kuchař (1992), who concludes his critique by observing that 'the semiclassical interpretation does not solve the standard problems of time. It merely obscures them by the approximation procedure and, along the way, creates more problems'. Perhaps the main difficulty is that, within the orthodox framework, the classical evolution of the metric is not really an approximation at all. Rather, it is put in by hand, and can in no way be justified on the basis of an entirely quantum mechanical treatment, even as an approximation. This is in stark contrast with the status of the semiclassical approximation within a Bohmian framework, for which there is no problem of time. In this approach, the classical evolution of the metric is indeed merely an approximation to its exact evolution, corresponding to the exact phase of the wave function (i.e. to eqn. 12.6). To the extent that this approximation is valid, the appropriate conclusions can be drawn, but the theory makes sense, and suffers from no conceptual problems, even when the approximation is not valid.

Now to our second question. Suppose that we demand of a universal dynamics that it be first-order for the variables describing the primitive ontology (the simplest possibility for a dynamics) and covariant – involving no preferred foliation, no special choice of lapse function $N$, in its formulation. This places a very strong constraint on the vector field defining the law of motion – and on the universal wave function, should this motion be generated by a wave function. The set of wave functions satisfying this constraint should be very small, far smaller than the set of wave functions we normally consider as possible initial states for a quantum system. However, according to our conception of the wave function as nomological, this very fact might well be a distinct virtue.

We have begun to investigate the possibility of a first-order covariant geometrodynamics in connection with Bohmian quantum gravity, and have found that the constraint for general covariance is captured by the Dirac algebra (see also Hojman, Kuchař, and Teitelboim 1976), which expresses the relation between successive infinitesimal deformations of hypersurfaces, taken in different orders: We have shown (see Goldstein and Teufel 1999) that defining a representation of the Dirac algebra is more or less the necessary and sufficient condition for a vector field on the space of 3-metrics to yield a generally covariant dynamics, generating a 4-geometry involving no dynamically distinguished hypersurfaces.

This work is very much in its infancy. In addition to the problem of finding a mathematically rigorous proof of the result just mentioned, there remains the difficult question of the possible representations of the Dirac algebra, both for pure gravity and for gravity plus matter. For pure gravity it seems that a first-order generally covariant geometrodynamics is achievable, but only with vector fields that generate classical 4-geometries – solutions of the Einstein equations with a possible cosmological constant. How this situation might be affected by the inclusion of matter is not easy to say.

Even a negative result – to the effect that a generally covariant Bohmian theory must involve additional spacetime structure – would be illuminating. A positive result – to the effect that a first-order dynamics, for geometry plus matter, that does not invoke additional spacetime structure can be generally covariant more or less only when the vector field defining this dynamics arises from an essentially unique wave function of the universe that happens to satisfy an equation like the Wheeler–DeWitt equation (and from which a time-dependent Schrödinger equation emerges, in the manner we have described, as part of a phenomenological description governing the behavior of appropriate subsystems) – would be profound. For then we would know, not just what quantum mechanics is, but why it is.

### Notes

1. Actually, the extrinsic curvature is given by $K_{ij} = G_{ijab}\pi^{ab}$ where $G_{ijab}$ is the so-called supermetric, which is itself a function of $g_{ij}$. This distinction is, however, not relevant to our discussion.
2. We choose units in which $\hbar$ and $c$ are 1.
3. In some models of quantum cosmology, e.g. those permitting the definition of a global time function, it may well be possible to pick ourselves out in a diffeomorphism-invariant manner.

# 13    On gravity's role in quantum state reduction

Roger Penrose

## 13.1 The problem of quantum state reduction

The fundamental problem of quantum mechanics, as that theory is presently understood, is to make sense of the *reduction of the state vector* (i.e. *collapse of the wave function*), denoted here by R. This issue is usually addressed in terms of the 'quantum measurement problem', which is to comprehend how, upon *measurement* of a quantum system, this (seemingly) discontinuous R-process can come about. A measurement, after all, merely consists of the quantum state under consideration becoming entangled with a more extended part of the physical universe, e.g. with a measuring apparatus. This measuring apparatus – together with the observing physicist and their common environment – should, according to conventional understanding, all also have some quantum description. Accordingly, there should be a quantum description of this *entire* quantum state, involving not only the original system under consideration but also the apparatus, physicist, and remaining environment – and this entire state would be expected to evolve continuously, solely according to the Schrödinger equation (unitary evolution), here denoted by the symbol U.

Numerous different attitudes to R have been expressed over many years, ever since quantum mechanics was first clearly formulated. The most influential viewpoint has been the 'Copenhagen interpretation' of Niels Bohr, according to which the state vector $|\psi\rangle$ is not to be taken seriously as describing a quantum-level physical reality, but is to be regarded as merely referring to our (maximal) 'knowledge' of a physical system, and whose ultimate role is simply to provide us with a means to calculate probabilities when a measurement is performed on the system. That 'measurement' would be taken to come about when that system interacts with a *classical* measuring apparatus. Since 'our knowledge' of a physical system can undergo discontinuous jumps, there is no reason to be surprised – so it would be argued – when $|\psi\rangle$ undergoes discontinuous jumps also!

Closely related is the environment–entanglement 'FAPP' ('for all practical purposes') point of view (e.g. Zurek 1991; cf. Bell 1990), according to which the $R$-process is taken to be some kind of approximation to the $U$-evolution of the system together with its environment, and $R$ is viewed as having taken place in the system itself 'for all practical purposes' (Bell 1990). The essential idea is that the environment involves enormously numerous random degrees of freedom, and these become entangled with the limited number of degrees of freedom in the system itself. Accordingly, the delicate phase relations between the system's degrees of freedom become irretrievably lost in these entanglements with the environment. It is argued that this effective loss of phase coherence in the system itself gives rise to $R$, FAPP, although certain additional assumptions are needed in order for this conclusion to be entertained.

There are also viewpoints – referred to as the 'many worlds', or 'many minds', or the Everett interpretation (Everett 1957; cf. DeWitt and Graham 1973) – whereby it is accepted that all the different macroscopic alternatives which (according to $U$) must remain superposed, actually coexist *in reality*. It can be convincingly argued that if $|\psi\rangle$ is indeed taken to represent an *actual* reality at all levels, and if $|\psi\rangle$ evolves *precisely* according to $U$, then something of this nature must hold. However, by itself, and without further assumptions, this would offer no explanation as to why the 'illusion' of merely one world presents itself to our consciousnesses, nor why the correct quantum-mechanical probabilities come about (Squires 1990, 1992a; cf. also Penrose 1994a, Sections 6.2, 6.7).

Other alternative interpretations have been put forward, such as those of de Broglie (1956), Bohm (1952), Bohm and Hiley (1994), Haag (1992), Griffiths (1984), Omnès (1992), and Gell-Mann and Hartle (1993), in which the standard quantum procedures are reformulated in a different mathematical framework. The authors of these proposals do not normally take the view that any experimentally testable deviations from standard quantum mechanics can arise within these schemes.

Set against all these are proposals of a different nature, according to which it is argued that present-day quantum mechanics is a limiting case of some more unified scheme, whereby the $U$ and $R$ procedures are *both* to be approximations to some new theory of physical reality. Such a theory would have to provide, as an appropriate limit, something equivalent to a unitarily evolving state vector $|\psi\rangle$. Indeed, many of these schemes, such as that of Pearle (1986, 1989), Bialynicki-Biruta and Mycielski (1976), Ghirardi, Rimini, and Weber (1986), and Weinberg (1989), use a quantum-state description $|\psi\rangle$, just as in standard quantum mechanics, but where the evolution of the state deviates by a tiny amount from the precise Schrödinger (or Heisenberg) evolution $U$. Related to these are proposals which posit that it is in the behaviour of conscious beings that deviations from precise $U$-evolution are to be found (cf. Wigner 1961). But in the more 'physical' such schemes, the suggested deviations from standard $U$-evolution become noticeable merely when the system becomes 'large', in some appropriate sense. This 'largeness' need not refer to physical dimension, but it might, for example, be the number of particles in the system that is relevant (such as in the specific scheme put forward in Ghirardi et al. 1986). In other proposals, it is considered that it is the mass, or mass distribution, that is all important. In schemes of the latter nature, it is normally taken that it is *gravity* that provides the influence that introduces deviations from the standard quantum rules. (See Károlyházy 1966, 1974, Károlyházy, Frenkel, and Lukács 1986, Kibble 1981, Komar 1969, Diósi 1989, Ghirardi, Grassi, and Rimini 1990, Pearle and Squires 1996, Percival 1995, Penrose 1981, 1986, 1987, 1989, 1993, 1994, 1994a, for proposals of this nature.)

In this chapter, a new argument is given that explicitly supports a gravitational role in state-vector reduction. Most particularly, it is consistent with a particular criterion, explicitly put forward in (Penrose 1994, 1994a), according to which a macroscopic quantum superposition of two differing mass distributions is *unstable* (analogous to an unstable particle). Accordingly, such a state would decay, after a characteristic lifetime $T$, into one or the other of the two states. To compute $T$, we take the difference between the two mass distributions under consideration (so that one counts positively and the other negatively) and compute its gravitational self-energy $E_\Delta$. The idea is that we then have

$$T \simeq \frac{\hbar}{E_\Delta}.$$

This criterion is also very close to one put forward earlier by Diósi (1989), and as modified by Ghirardi et al. (1990), but their point of view is different from the one used here, proposed by Penrose (1993, 1994, 1994a). In particular, in their type of scheme, there are violations of energy conservation which could – in principle or in practice – be experimentally detected. With the type of proposal in mind here, the idea would be that such violations of energy conservation ought to be absent, owing to the fundamental involvement of some of the basic principles of general relativity. (There is a brief discussion of this point in Penrose 1994a, p. 345.) In this chapter it is indicated that there is a basic conflict between Einstein's general covariance principle and the basic principles of quantum theory, as they relate to stationary states of superposed gravitational fields. It is argued that this conflict can be resolved within the framework of the specific state-reduction proposal of Penrose (1994, 1994a), according to which such superposed gravitational fields are essentially *unstable*.

It should be made clear, however, that this proposal does not provide a *theory* of quantum state reduction. It merely indicates the level at which deviations from standard linear Schrödinger (unitary) evolution are to be expected owing to gravitational effects. Indeed, it is this author's personal opinion that the correct theory uniting general relativity with quantum mechanics will involve a major change in our physical world-view – of a magnitude at least comparable with that involved in the shift from Newtonian to Einsteinian gravitational physics. The present chapter makes no pretensions about even aiming us in the right direction in this regard. Its purpose is a different one, namely to show that even within the framework of completely conventional quantum theory, there is a fundamental issue to be faced, when gravitational effects begin to become important. Standard theory does not provide a clear answer; moreover, it allows room for the type of instability in superposed states that would be consistent with the proposals of Diósi (1989), Ghirardi et al. (1990) and Penrose (1993, 1994, 1994a).

### 13.2 Stationary states

Let us consider the following situation. We suppose that a quantum superposition of two states has been set up, where each individual state has a well-defined static mass distribution, but where the mass distributions differ from one state to the other. For example, we could have a rigid lump of material which we contrive to place in a quantum superposition of two different locations. Such a superposed state could be achieved by having a photon simultaneously transmitted through and reflected off a half-silvered mirror placed at a distance from the lump, where the transmitted part of the photon's state then triggers a device which slowly moves the lump from its initial location to somewhere nearby, but the reflected part leaves the lump alone. (This is an inanimate version of 'Schrödinger's cat'.)

If we ignore any gravitational effects, the two alternative locations of the lump will each be stationary states, and will therefore have wave functions, say $|\psi\rangle$ and $|\chi\rangle$, that are eigenfunctions of the energy operator $H$:

$$H|\psi\rangle = v|\psi\rangle, \qquad H|\chi\rangle = v|\chi\rangle.$$

Here, the energy eigenvalue $v$ is the same in each case, because the mass is simply displaced from one location to the other. (We ignore the energy of the photon in this, the lump itself being supposed to have enormously greater mass-energy, and also any energy involvement in the device that moves the lump.) It is clear from this that there is a complete degeneracy for linear superpositions of $|\psi\rangle$ and $|\chi\rangle$, and any linear combination

$$\lambda|\psi\rangle + \mu|\chi\rangle$$

(with $\lambda$ and $\mu$ complex constants) will also be an eigenstate of $H$ with the same eigenvalue $v$. Each such linear combination will also be a stationary state, with the same energy eigenvalue $v$. Thus, any one of these combinations would be just as stable as the original two, and must therefore persist unchanged for all time. We shall shortly consider the delicate theoretical issues that arise when one considers the (superposed) *gravitational fields* of each of the two instances of the lump also to be involved in the superposition. In accordance with this, it will be appropriate to be rather careful about the kinds of issues that become relevant when quantum theory and general relativity are considered together.

### 13.3 Preliminary considerations

In particular, there is a point of subtlety which needs to be addressed, even before we consider the details of any gravitational effects. We take note of the fact that, in the absence of any spatial inhomogeneity in the background potentials (gravitational or otherwise), there is nothing in the intrinsic nature of one lump location that allows us to distinguish it from any other lump location. Thus, we might choose to adopt the standpoint that there is really no physical difference at all between the various states of location of the lump and, accordingly, take the view that all these seemingly different states of the lump are actually all the same state! Indeed, this would be the normal standpoint of (quantum) general relativity. Each spacetime geometry arising from each separate lump location would be identical. The principle of general covariance forbids us to assign a meaningful (co-ordinate) label to each individual point, and there is no co-ordinate-independent way of saying that the lump occupies a different location in each of the two configurations under consideration.

However, this is not the standard attitude in ordinary quantum mechanics. If the various position states of a single quantum particle were considered to be all the same, then we would not be able to construct all the many different wave functions for that particle which are, after all, simply superpositions of states in which the particle occupies different locations. In standard quantum mechanics, all the different particle locations correspond to different quantum states whose superpositions can be independently involved.

It is worth while, in this context, to consider how standard quantum mechanics treats *composite* particles. Here one may choose to factor out by this translational freedom. For example, the wave function of a hydrogen atom, in an eigenstate of the energy and angular momentum operators, is still completely degenerate with respect to the spatial location of its centre of mass. In considering the quantum mechanics of a hydrogen atom, the normal procedure is (implicitly or otherwise) to factor out by the degrees of freedom that specify the location of the mass centre (or, as a simplifying approximation, merely the location of its proton), and consider the quantum mechanics only of what remains: essentially the location of the orbiting electron, this being what is fixed by the energy and angular momentum.

In the present context, with the two superposed lump locations, we are trying to examine precisely the translational degrees of freedom that had been 'factored out' in the case of the hydrogen atom. In fact, there is something of a physical inconsistency in simply considering our lump to be displaced from one location to another in the two states under superposition. There is a conservation law which requires that the mass centre remain fixed in space (or move uniformly in a straight line). This would have been violated unless the lump displacement is compensated by the displacement of some other massive object in the opposite direction. We shall consider that there is such another object, and that it is enormously more massive than is the lump itself. Let us call this other object 'the Earth'. To move the lump from one location to the other, we simply allow the Earth to move by a very tiny amount, so as to allow the mass centre to remain fixed; and since the Earth is so very much more massive than the lump, we can consider that in practice the Earth does not move at all.

The presence of the Earth in these considerations allows us to circumvent the problem that we had previously considered. For the Earth (assumed to have some large but static finite irregular shape) serves to establish a 'frame of reference' against which the motion of the lump can be considered to be taking place. The spacetime geometries corresponding to all the various lump locations are now all different from one another, and 'general covariance' does not prevent us from considering these various states to be all distinct.

At this point, another complication appears to arise because the Earth's gravitational field has to be taken into account. In fact, we shall not be much concerned with the Earth's field. It will be the differences in the gravitational fields in the various *lump* locations that will have essential relevance. In fact, there is no problem, in the considerations which follow, in allowing the Earth's gravitational field to be involved also. But we may prefer to avoid the complication of the Earth's field, if we choose, by supposing that our 'experiment' is to be set up within a spherical cavity situated at the centre of the Earth. Then, the Earth's actual field could be eliminated completely. Yet the Earth could still serve to establish a 'frame of reference' against which the motion of the lump could be considered to be taking place.

On the other hand, we shall prefer to consider that the lumps are actually sitting on the surface of the Earth. The Earth's (effectively) constant gravitational field does not have any significant influence on the considerations of relevance here. There is no problem about having the lump in a stationary state on the surface of the Earth, provided that some appropriate upward forces are introduced on the lump, to allow it to remain at rest with respect to the Earth.

If we were to consider a situation in which the two superposed lump positions are at different heights from the ground, then we would have to consider the effects of the energy expended in raising the lump in the Earth's field. This would introduce an additional factor, but would not seriously affect the situation that we are concerned with here. If desired, all these considerations can be evaded if we regard our lump movements to be taking place entirely within our spherical cavity at the centre of the Earth. However, it will not be necessary to pass to this extreme idealization for the considerations of this chapter.

### 13.4 Superposed gravitational fields

Let us now try to consider how the gravitational field of the lump itself affects our superposed state. Each lump location is accompanied by the static gravitational field produced by the lump in that location. We must envisage that the superposed state is now an entangled one,

$$\lambda|\psi\rangle|G_\psi\rangle + \mu|\chi\rangle|G_\chi\rangle,$$

where $|G_\psi\rangle$ and $|G_\chi\rangle$ are the quantum states of the gravitational fields of the lump locations corresponding to $|\psi\rangle$ and $|\chi\rangle$, respectively. We might choose to think of $|G_\psi\rangle$ and $|G_\chi\rangle$ as *coherent states*, if that is the appropriate description. In any case, whatever is to be meant by the quantum state of a stationary gravitational field – including all the internal degrees of freedom of the field – would be supposed to be incorporated into $|G_\psi\rangle$ or $|G_\chi\rangle$. But it does not greatly matter, for our present purposes, what the 'correct' quantum mechanical descriptions of macroscopic gravitational fields actually are. We must suppose, however, that whatever this description is, it closely accords in its physical interpretation, for a *single* unsuperposed lump, with the classical gravitational field of that lump according to the description of Einstein's general relativity.

In fact, the 'entangled' nature of the superposed state will not be of importance for us here. We simply take the two states under superposition to be

$$|\Psi\rangle = |\psi\rangle|G_\psi\rangle \quad \text{and} \quad |X\rangle = |\chi\rangle|G_\chi\rangle.$$

The essential point is that each of the two states concerned must involve a reasonably well-defined (stationary) spacetime geometry, where these two spacetime geometries differ significantly from each other. We must now raise the question: is this *superposed* state still a stationary state?

We have to consider carefully what a 'stationary state' means in a context such as this. In a stationary spacetime, we have a well-defined concept of 'stationary' for a quantum state in that background, because there is a Killing vector $T$ in the spacetime that generates the time-translations. Regarding $T$ as a differential operator (the '$\partial/\partial t$' for the spacetime), we simply ask for the quantum states that are eigenstates of $T$, and these will be the stationary states, i.e. states with well-defined energy values. In each case the energy value of the state in question, $|\Psi\rangle$, would be (essentially) the eigenvalue $E_\Psi$ of $T$ corresponding to that state:

$$T|\Psi\rangle = -i\hbar E_\Psi |\Psi\rangle.$$

However, for the superposed state we are considering here we have a serious problem. For we do not now have a specific spacetime, but a superposition of two slightly differing spacetimes. How are we to regard such a 'superposition of spacetimes'? Is there an operator that we can use to describe 'time-translation' in such a superposed spacetime? Such an operator would be needed so that we can identify the 'stationary states' as its eigenvectors, these being the states with definite energy. It will be shown that there is a fundamental difficulty with these concepts, and that the notion of time-translation operator is *essentially* ill-defined. Moreover, it will be possible to define a clear-cut measure of the *degree* of this ill-definedness for such a superposed state. Accordingly, there is, in particular, an essential uncertainty or 'fuzziness' in the very concept of energy for such a state, and the degree of this uncertainty can be estimated in a clear-cut way. This is consistent with the view that such a superposed state is 'unstable', and the lifetime of the state will be given by $\hbar$ divided by this measure of energy uncertainty, in accordance with the way in which Heisenberg's uncertainty principle is employed in the theory of unstable particles, where the particle's lifetime is related to its mass-energy uncertainty.

How are we to regard a quantum superposition of two spacetimes? It is not sufficient to take a completely formal attitude to such matters, as is common in discussions of quantum gravity. According to the sorts of procedure that are often adopted in quantum gravity, the superposition of different spacetimes is indeed treated in a very formal way, in terms of complex functions on the space of 3-geometries (or 4-geometries), for example, where there is no pretence at a point-wise identification of the different geometries under superposition. A difficulty with such formal procedures arises, however, if we attempt to discuss the physics that takes place *within* such a formal superposition of spaces, as is the case with the type of situation under consideration here (cf. also Anandan 1994).

Indeed, in the case of the two minutely differing spacetimes that occur in our situation with the two superposed lump locations, there would be no obvious way to register the fact that the lump is actually in a different place, in each of the two configurations under superposition, unless there is some sort of (approximate) identification. (Of course, in this identification, since the lump itself is supposed to be 'moved', corresponding points of the lump are not identified, but the corresponding points of the Earth are, since the Earth is not considered to have significantly moved.)

Nevertheless, it is clear from the principles of general relativity that it is not appropriate, in general, to make a precise identification between points of one spacetime and corresponding points of the other. The *gauge freedom* of general relativity – as reflected in 'the principle of general covariance' (or, equivalently, 'diffeomorphism invariance') – is precisely the freedom that forbids a meaningful precise labelling of individual points in a spacetime. Accordingly, there is generally no precise meaningful pointwise identification between different spacetimes. (In special cases, it may be possible to circumvent this problem; see Anandan 1994, 1996, where issues of this nature are examined, and interesting effects are anticipated, in relation, in particular, to the quantization of a cosmic string.) In the general case, all that we can expect will be some kind of *approximate* pointwise identification.

### 13.5 The semiclassical approach

One possible way to address this difficulty might be to adopt a 'semiclassical approximation' to quantum gravity, according to which it would be the *expectation value* of the (quantum) energy–momentum tensor that serves to specify the right-hand side of Einstein's equations

$$R_{ab} - \frac{R}{2}g_{ab} = -8\pi G T_{ab}.$$

In such an approximation the superposition of our pair of states in which one lump is in two spatially displaced locations would have a gravitational field which is merely the average field of the fields of each lump individually.

In fact, the semiclassical approximation is not really physically consistent. In particular, it allows superluminary communication (see Pearle and Squires 1996); and, in a certain interpretation, it is grossly inconsistent with observation (Page and Geilker 1981). These difficulties might be avoided if the semiclassical interpretation can be combined with some scheme of gravitationally induced state-vector reduction, as is argued by Pearle and Squires (1996; see also Kibble 1981).

From the point of view of the present chapter, there is an additional difficulty with the semiclassical approximation. The semiclassical description would provide a spacetime containing the mass distributions of two spatially displaced lumps with vacuum ($R_{ab} = 0$) between them. This gravitational field – of a pair of spatially displaced lumps – is not really the same as a linear superposition of the two fields, each describing one of the two lump locations individually. The gravitational interaction effects between the pair of lumps would have to be taken into account, according to the non-linear effects of general relativity. Assume that the actual lump, in its two displaced locations, sits on a smooth horizontal table ('the Earth'). The 'semiclassical' state would represent a pair of spatially separated massive lumps, and would not actually possess an exact Killing vector which could play the role of $T$ ($= \partial/\partial t$) because the 'semiclassical' lumps would fall towards each other along the table in accordance with their gravitational attractions, leading to a non-static spacetime geometry.

For reasons such as these, it is proposed not to follow a semiclassical description here. Instead, we shall follow the route of supposing that there is some kind of *approximate sense* in which the two superposed spacetimes can be pointwise identified. Without some kind of (approximate) identification between the spacetimes, we do not seem even to be able to express the fact that the quantum state of the lump is a superposition of two distinct locations, in which the degree of displacement between the two lump locations is reasonably well defined.

### 13.6 Approximate spacetime point identification

The basic principles of general relativity – as encompassed in the term 'the principle of general covariance' (and also 'principle of equivalence') – tell us that there is no natural way to identify the points of one spacetime with corresponding points of another. Consider our quantum superposition between two different spacetimes (here, the fields of two alternative lump locations). If we are to attempt to make a pointwise identification between these two spacetimes, we can do this in a way that would be only *approximately* meaningful. Let us try to obtain some measure of this degree of approximation.

In order to proceed to a reasonably explicit expression of this measure, it will be helpful to make the assumption that a Newtonian approximation to the gravitational fields of each lump location is adequate. Indeed, in any plausible practical situation in which a quantum superposition of lump locations might have to be considered, this would certainly be the case. Thus, there will be spatial sections of the two spacetimes in question which are Euclidean 3-spaces, and each spacetime will possess a well-defined time co-ordinate whose constant values define these Euclidean spatial sections. We shall suppose that the time co-ordinates for each spacetime can be naturally identified with each other, so we have a single time parameter $t$ common to the two spacetimes. This would seem reasonable for the Newtonian situation under consideration, although there would be essential subtleties arising when the two gravitational fields are treated according to full general relativity.

It should be made clear, however, that passing to the Newtonian limit does not remove the difficulties that the principle of general covariance – or the principle of equivalence – presents in relation to the quantum superposition of gravitational fields. As may be recalled, there is a spacetime formulation of Newtonian gravitational theory originally provided by Cartan (1923, 1924), and further studied independently by Friedrichs (1927) and Trautman (1965), in which the Newtonian version of the equivalence principle is directly incorporated into the geometrical description. (See Ehlers 1991 for an up-to-date account.) The Cartan geometric formulation of Newtonian gravitational theory is indeed the appropriate one for taking into account the subtleties of (what remains of) the principle of general covariance and the principle of equivalence. (For example, the Newton–Cartan spacetime of a *constant* non-zero Newtonian gravitational field has an identical geometry to that of the zero gravitational field, but it differs from a non-uniform Newtonian field which produces tidal effects.) It has been pointed out by Christian (1995) that the Newton–Cartan framework provides a valuable setting for exploring some of the fundamental problems of unifying quantum theory with gravitational theory without, at this stage, the more severe difficulties of general relativity having to be faced. Christian argues that this framework indeed sheds important light on the role of gravity in the measurement problem.

As it turns out, the criterion for quantum state reduction that we shall be led to here is independent of the value of the speed of light $c$. It will thus have a well-defined non-trivial Newtonian limit, and it can be expressed within the Newton–Cartan framework. However, although it will be valuable to bear this framework in mind here, we shall not actually use its specific mathematical details.

The essential point about superposing a pair of Newton–Cartan spacetimes is that whereas we are allowing that the time co-ordinate $t$ can be identified in the two spacetimes – and so there is a canonical correspondence between the various space sections of one spacetime with those of the other – there is no canonical way of identifying the individual *points* of a section of one spacetime with corresponding points of the other. It is this lack of a definite pointwise identification between the spatial sections of the two spacetimes which will lead us to an essential ill-definedness of the notion of time-translation – and therefore of the notion of stationarity – for the quantum-superposed state.

The reader might wonder why it is *time-translation* that should encounter problems, when, in our Newtonian context, there is no problem with the time co-ordinate $t$. However, this is how it should be; for 'time-translation' is something which is represented by the operator '$\partial/\partial t$', and the meaning of this operator is really concerned more with the choice of the *remaining* variables, $x$, $y$, $z$ (those parameters to be held fixed in the definition of $\partial/\partial t$), than it is with $t$ itself. An uncertainty in the *pointwise* identification of the spatial sections of one spacetime with the spatial sections of the other will indeed show up in an uncertainty in the definition of time-translation. Suppose that we use the co-ordinates $x$, $y$, $z$, and $t$ for the points of one spacetime, and $x'$, $y'$, $z'$, and $t'$ for the points of the other (where, of course, $t = t'$); then we have

$$\frac{\partial}{\partial t'} = \frac{\partial}{\partial t} + \frac{\partial x}{\partial t'}\frac{\partial}{\partial x} + \frac{\partial y}{\partial t'}\frac{\partial}{\partial y} + \frac{\partial z}{\partial t'}\frac{\partial}{\partial z}$$

$$= \frac{\partial}{\partial t} + \mathbf{v} \cdot \nabla,$$

where $\mathbf{v}$ is the spatial velocity, as described with respect to the unprimed co-ordinate system, of a point fixed in the primed co-ordinate system. (In the $(x,y,z,t)$-system, the 3-vector $\mathbf{v}$ has components $(\partial x/\partial t', \partial y/\partial t', \partial z/\partial t')$, i.e. $(-\partial x'/\partial t, -\partial y'/\partial t, -\partial z'/\partial t)$, and the operator $\nabla$ has components $(\partial/\partial x, \partial/\partial y, \partial/\partial z)$.) If we were to attempt to identify the point with coordinates $(x, y, z, t)$ in one of the spacetimes under consideration, with the point with co-ordinates $(x', y', z', t')$ in the other, then we should encounter an incompatibility between their notions of time-evolution unless $v$ vanishes everywhere.

Of course, in this Newtonian limit, it is possible to arrange that $\mathbf{v} = 0$ everywhere, simply by taking ordinary static, non-rotating Newtonian/Cartesian co-ordinates for the two spacetimes, related by a constant spatial displacement between them. But this would be to go against the spirit of what is entailed, in the present context, by Einstein's principle of general covariance. It is the special nature of the Newtonian limit that provides us with flat Euclidean spatial sections whose local motions are determined directly by what happens at infinity. In this Newtonian limit, the 'co-ordinate freedom' can indeed be eliminated if we nail things down at infinity, because in Newton–Cartan geometry, there is exact spatial rigidity. However, within the more general context of curved (and not necessarily stationary) spacetimes, this is not at all appropriate. Our physical expectations would be for the criteria characterizing the nature of a localized quantum superposition to depend on reasonably local criteria. The principles of general relativity are antagonistic towards the idea of identifying individual local points in a precise way, in terms of the situation at infinity.

Similar remarks also apply to certain proposals within the general programme of quantizing general relativity proper, according to which the time translation operator would not be taken to have local relevance, but to refer merely to the symmetries at spatial infinity – it being supposed that only asymptotically flat spacetimes are to be considered in quantum superposition. It seems, however, for situations such as those under consideration here where we are concerned with a reasonably local problem, that it is quite inappropriate to define the notion of time translation merely in terms of what is going on at spatial infinity. We do not know, after all, what the geometry of the actual universe is on a very large scale (spatially asymptotically spherical or hyperbolic, for example, either of which would lead to different notions of 'time-translation' from that which is normally considered). It is most unlikely that, for a reasonably local quantum problem such as this, Nature should really 'care' what is going on at infinity.

We shall suppose instead that, in the particular Newtonian quantum superposition under consideration, it is appropriate to demand an approximate spatial identification between the two spacetimes, where the degree of approximation is governed by local (or quasi-local) considerations in the vicinity of the region where the identification occurs. In accordance with the principle of equivalence, it is the notion of free fall which is locally defined, so the most natural local identification between a local region of one spacetime and a corresponding local region of the other would be that in which the free falls (i.e. spacetime geodesics) agree. However, in the superposition under consideration, there is no way to make the spatial identifications so that the free falls agree everywhere throughout the spacetimes. The best that one can do is to try to minimize the amount of the *difference* between free fall motions.

How are we to express this difference mathematically? (Let us assume that the two quantum amplitudes assigned to the two superposed states are about the same size. Then we may take it that neither spacetime's geometry dominates the other.) Let $\mathbf{f}$ and $\mathbf{f}'$ be the acceleration 3-vectors of the free-fall motions in the respective spacetimes, at some identified point (where the accelerations can be taken with respect to the appropriate local identified co-ordinates). In fact, $\mathbf{f}$ and $\mathbf{f}'$ will be the Newtonian gravitational force-per-unit-test-mass, at that point, in each spacetime. Let us take the scalar quantity

$$(\mathbf{f} - \mathbf{f}')^2 = (\mathbf{f} - \mathbf{f}') \cdot (\mathbf{f} - \mathbf{f}')$$

as the measure of incompatibility of the identification – and, accordingly, of the 'uncertainty' involved in this identification. It may be noted that if we apply an *acceleration* to the common co-ordinates to the two spacetimes, while keeping the actual point identification unaltered, then the difference $\mathbf{f} - \mathbf{f}'$ does not change. In fact, the expression $(\mathbf{f} - \mathbf{f}')^2$ is co-ordinate-independent (assuming that the 'dot' · refers to the actual three-dimensional spatial metric).

For the total measure of incompatibility (or 'uncertainty') $\Delta$ at a particular instant, given by a particular $t$-value, it is proposed to integrate this quantity (with respect to the spatial 3-volume element $d^3x$) over the $t = $ constant 3-space:

$$\Delta = \int (\mathbf{f} - \mathbf{f}')^2 \, d^3x$$

$$= \int (\nabla\Phi - \nabla\Phi')^2 \, d^3x$$

$$= \int (\nabla\Phi - \nabla\Phi') \cdot (\nabla\Phi - \nabla\Phi') \, d^3x$$

$$= -\int (\Phi - \Phi')(\nabla^2\Phi - \nabla^2\Phi') \, d^3x,$$

where $\mathbf{f} = -\nabla\Phi$ and $\mathbf{f}' = -\nabla\Phi'$, $\Phi$ and $\Phi'$ being the respective gravitational potentials for the two spacetimes. By Poisson's formula

$$\nabla^2\Phi = -4\pi G\rho,$$

we obtain

$$\Delta = 4\pi G \int (\Phi - \Phi')(\rho - \rho') \, d^3x,$$

where $\rho$ and $\rho'$ are the respective mass densities. Using the integral formula

$$\Phi(x) = -\int \frac{\rho(y)}{|x - y|} \, d^3y,$$

we get

$$\Delta = -4\pi G \iint \frac{(\rho(x) - \rho'(x))(\rho(y) - \rho'(y))}{|x - y|} \, d^3x \, d^3y,$$

which is basically just the gravitational self-energy of the *difference* between the mass distributions of each of the two lump locations.

How does this relate to the uncertainty in the time-translation Killing vector referred to at the beginning of this section? That there should be some relationship follows from general considerations, but the exact form that this relationship should take seems to depend upon the specific model that is used to describe the uncertainty of spacetime identification. (As long ago as 1966, Károlyházy provided some fairly closely related considerations, cf. Károlyházy, 1966, and also Károlyházy 1974,

Károlyházy et al. 1986.) To understand what is involved in the present context, we consider three successive time slices, given by $t = \tau$, $t = \tau + \Delta t$, and $t = \tau + 2\Delta t$. We imagine that there is some uncertainty in the identification between the two spacetimes under superposition at each of these three times. The measure of relative 'acceleration uncertainty' that we have just been considering, and which seems to be forced upon us from consideration of the principles of equivalence and general covariance, has to do with the difference between the error in spatial identifications at $t = \tau + \Delta t$ and the average of those at $t = \tau$ and at $t = \tau + 2\Delta t$. The overall uncertainty in the quantity $\mathbf{v} \cdot \nabla$, which appears as the uncertainty in the time-translation Killing vector for the superposed spacetimes is a 'velocity uncertainty', and it has to do with the difference between the error in spatial identification at $t = \tau + \Delta t$ and that at $t = \tau$ (or else between the error at $t = \tau + 2\Delta t$ and that at $t = \tau + \Delta t$). The precise relationship between the 'acceleration uncertainty' and 'velocity uncertainty' seems to depend upon the way that this uncertainty is actually modelled. The direct 'position uncertainty' does not feature in these considerations; it is the time-evolution of the error in spatial identification that has importance for us here.

A further ingredient of possible importance is an error in the identification in the actual time co-ordinate $t$ for the two spacetimes. We have ignored this complication here, but it clearly has relevance in full general relativity. Perhaps it should also be taken into account in the Newtonian limit.

In view of these various complicating issues, no attempt will be made to formulate a definitive statement of the precise measure of uncertainty that is to be assigned to the 'superposed Killing vector' and to the corresponding notion of 'stationarity' for the superposed spacetime. However, it is strongly indicated by the above considerations that the quantity $\Delta$, as defined above, gives a very plausible (though provisional) estimate of this uncertainty. Hence, we can use $\Delta$, or some simple multiple of this quantity, as a measure of the fundamental energy uncertainty '$E_\Delta$' of the superposed state. Accordingly, the superposed state would not be exactly stationary, but it is consistent with the above considerations that it should have a lifetime of the general order of $\hbar / E_\Delta$.

It is reassuring that basically the same expression is seen to arise from certain other considerations. For example, it was pointed out in Penrose (1994, 1994a) that the gravitational self-energy involved in a quantum superposition of a pair of differing spacetime geometries should involve an essential uncertainty, owing to the fact that even in classical general relativity there is difficulty with the energy concept for gravity. (There is no local expression for gravitational energy.) Moreover, considerations of the symplectic structure of linearized gravity (Penrose 1993) lead to something very similar, and so also does the earlier model of Diósi (1989), although in these papers the suggestion is that the expression

$$G \iint \frac{\rho(x)\rho'(y)}{|x - y|} \, d^3x \, d^3y,$$

which is the gravitational interaction energy, should be the relevant measure of the required 'energy uncertainty' $E_\Delta$. In fact (apart from the factor $4\pi$), it does not make any difference which expression is used, provided that – as is the case here – the two individual states under superposition each have the same gravitational self-energy. (A situation for which the difference in the expressions might be important would be in a cloud chamber, where the two states in superposition might be 'droplet forming' and 'droplet not forming'. In this case the gravitational self-energy of the droplet would contribute a difference between the two expressions.)

### 13.7 Further considerations

It is clear that the above arguments provide only a very preliminary analysis of the difficulties involved in the notion of 'stationarity' for a quantum superposition of differing stationary spacetimes. However, even just these preliminary considerations seem to suggest that such a superposition would be unstable, and that a very plausible expression for the order of magnitude of the lifetime of such a superposed state (where the two relevant amplitudes are roughly equal) is $\hbar/E_\Delta$, where $E_\Delta$ is the gravitational self-energy of the *difference* between the two mass distributions involved.

In any case, it should be clear from what has been said above that *conventional quantum theory* provides no clear answer (in the absence of a satisfactory theory of quantum gravity) to the problem of the stability of a quantum superposition of two different gravitating states. We do not need to appeal to the contentious conceptual issues inherent in the measurement problem of quantum mechanics for some motivation for believing that such superpositions should be unstable.

In the above considerations, we have restricted attention to cases where the superposition involves just two different states, each of which individually has a well-defined spacetime geometry, and where the two amplitudes assigned to each of these two constituent states are of about the same size. This does not tell us, for example, what to do about the wave function of a single isolated proton, as it spreads throughout space in accordance with the Schrödinger equation. In the GRW scheme (Ghirardi et al. 1986) there would be a small probability for this state to reduce spontaneously to a more localized state. The present considerations, however, are not adequate to provide us with an expectation as to whether (or how frequently) such a spontaneous state reduction should occur, nor do they tell us what kind of state should be the result of such a spontaneous state reduction. We might think that the 'natural' states for the proton, having 'well-defined spacetime geometries', are those for which the proton is reasonably localized, say to a region of roughly its Compton wavelength. But if we adopt such a view, we are driven to consider that most wave functions are superpositions of a great many of these natural states, and with widely differing amplitudes. A more detailed theory is clearly needed if such questions are to be addressed adequately.

Despite these uncertainties, it is still possible, in many different circumstances, to estimate the expected order of magnitude of the rate of gravitationally induced state-vector reduction according to this scheme. For a single proton, we may expect that a superposed state of two separated spatial locations will decay to one or the other location in something of the order of a few million years. For a water speck $10^{-5}$ cm in radius, the timescale would be about an hour or so; for a speck $10^{-3}$ cm in radius, something like a millionth of a second. These results indeed seem reasonable, and if confirmed would supply a very plausible solution to the quantum measurement problem, but for the moment they appear to be rather beyond what can be experimentally tested.

It should be emphasized that none of the considerations of the present paper gives any clear indication of the mathematical nature of the theory that would be required to incorporate a plausible gravitationally induced spontaneous state-vector reduction. In all probability, such mathematical considerations would have to come from quite other directions. Indeed, this author's own expectations are that no fully satisfactory theory will be forthcoming until there is a revolution in the description of quantum phenomena that is of as great a magnitude as that which Einstein introduced (in the description of gravitational phenomena) with his general theory of relativity.

### Notes

I am grateful to Abhay Ashtekar, Jeeva Anandan, Joy Christian, Ted Newman, and Lee Smolin and others for some very helpful discussions. I am also grateful to NSF for support under research contract PHY 93-96246.

# 14     Why the quantum must yield to gravity

Joy Christian

## 14.1 Introduction: From Schrödinger's cat to Penrose's 'OR'

Quantum mechanics – one of our two most fundamental and successful theories –
is infested with a range of deep philosophical difficulties collectively known as the
measurement problem (Shimony 1963, Bell 1990). In a nutshell, the problem may
be stated as follows: If the orthodox formulation of quantum theory – which in
general allows attributions of only objectively indefinite properties or *potentialities* (Heisenberg 1958) to physical objects – is interpreted in compliance with what
is usually referred to as scientific realism, then one is faced with an irreconcil-
able incompatibility between the linearity of quantum dynamics – which governs
evolution of the network of potentialities – and the apparent definite or actual
properties of the physical objects of our 'macroscopic' world. Moreover, to date
no epistemic explanation of these potentialities (e.g. in terms of 'hidden variables') has been completely successful (Shimony 1989). Thus, on the one hand
there is overwhelming experimental evidence in favour of the quantum mechanical potentialities, supporting the view that they comprise a novel (i.e. classically
uncharted) metaphysical modality of Nature situated between logical possibility
and actuality (Shimony 1978, 1993a, pp. 140–62 and pp. 310–22), and on the
other hand there is phenomenologically compelling proliferation of actualities in
our everyday world, including even in the *microbiological* domain. The problem
then is that a universally agreeable mechanism for *transition* between these two
ontologically very different modalities – i.e. transition from the multiplicity of
potentialities to various specific actualities – is completely missing. As delineated,
this is clearly a very serious *physical* problem. What is more, as exemplified by
Shimony (1993, p. 56), the lack of a clear understanding of this apparent transition
in the world is also quite a 'dark cloud' for any reasonable programme of scientific
realism.

Not surprisingly, there exists a vast number of proposed solutions to the measurement problem in the literature (Christian 1996), some of which – the Copenhagen interpretation for example – being almost congenital to quantum mechanics. Among these proposed solutions there exists a somewhat dissident yet respectable tradition of ideas – going all the way back to Feynman's pioneering thoughts on the subject as early as in mid-1950s (Feynman 1957) – on a possible gravitational resolution of the problem. The basic tenet of these proposals can hardly be better motivated than in Feynman's own words. In his Lectures on Gravitation (Feynman 1995, pp. 12–13), he devotes a whole section to the issue, entitled 'On the philosophical problems in quantizing macroscopic objects', and contemplates on a possible breakdown of quantum mechanics:

> '... I would like to suggest that it is possible that quantum mechanics fails at large distances and for large objects. Now, mind you, I do not say that I think that quantum mechanics *does* fail at large distances, I only say that it is not inconsistent with what we do know. If this failure of quantum mechanics is connected with gravity, we might speculatively expect this to happen for masses such that $GM^2/\hbar c = 1$, of $M$ near $10^{-5}$ grams, which corresponds to some $10^{18}$ particles'.

Indeed, if quantum mechanics *does* fail near the Planck mass, as that is the mass scale Feynman is referring to here, then – at last – we can put the annoying problem of measurement to its final rest (see Fig. 14.1 for the meanings of the constants $G$, $\hbar$, and $c$). The judiciously employed tool in practice, the infamous Projection Postulate often referred to as the *reduction of quantum state* – which in orthodox formulations of the theory is taken as one of the unexplained basic postulates to resolve the tension between the linearity of quantum dynamics and the plethora of physical objects with apparent definite properties – may then be understood as an *objective* physical phenomenon. From the physical viewpoint such a resolution of the measurement problem would be quite satisfactory, since it would render the proliferation of diverse philosophical opinions on the matter to nothing more than a curious episode in the history of physics. For those who are not lured by pseudo-solutions such as the 'decohering histories' approaches (Kent 1997) and/or 'many worlds' approaches (Kent 1990), a resolution of the issue by 'objective reduction' ('OR', to use Penrose's ingenious pun) comes across as a very attractive option, provided of course that that is indeed the path Nature has chosen to follow (cf. Christian 1999).

Motivated by Feynman's inspiring words quoted above, there have been several concrete theoretical proposals of varied sophistication and predilections on how the breakdown of quantum mechanics might come about such that quantum superpositions are maintained only for 'small enough' objects, whereas reduction of the quantum state is objectively induced by gravity for 'sufficiently large' objects (Károlyházy 1966, Komar 1969, Kibble 1981, Diósi 1984, 1987, 1989, Károlyházy et al. 1986, Ellis et al. 1989, Ghirardi et al. 1990, Christian 1994, Percival 1995, Jones 1995, Pearle and Squires 1996, Frenkel 1997, Fivel 1997). Unfortunately, most of these proposals employ dubious or *ad hoc* notions such as 'quantum fluctuations of spacetime' (e.g. Percival 1995) and/or 'spontaneous localization of the wave function' (e.g. Ghirardi et al. 1990). Since the final 'theory of everything' or 'quantum gravity' is quite far from enjoying any concrete realization (Rovelli 1998), such crude notions

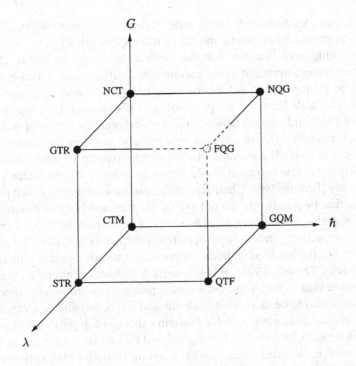

Fig. 14.1. The great dimensional monolith of physics indicating the fundamental role played by the three universal constants $G$ (the Newton's gravitational constant), $\hbar$ (the Planck's constant of quanta divided by $2\pi$), and $\lambda \equiv 1/c$ (the 'causality constant' (Ehlers 1981), where $c$ is the absolute upper bound on the speed of causal influences) in various basic theories. These theories, appearing at the eight vertices of the cube, are: CTM = Classical Theory of Mechanics, STR = Special Theory of Relativity, GTR = General Theory of Relativity, NCT = Newton–Cartan Theory, NQG = Newton–Cartan Quantum Gravity (constructed in Christian 1997), GQM = Galilean–relativistic Quantum Mechanics, QTF = Quantum Theory of (relativistic) Fields, and FQG = the elusive Full-blown Quantum Gravity. Note that FQG must reduce to QTF, GTR, and NQG in the respective limits $G \rightarrow 0$, $\hbar \rightarrow 0$, and $\lambda \rightarrow 0$ (Kuchař 1980, Christian 1997, Penrose 1997, pp. 90–2).

cannot be relied upon when discussing issues as fundamental as the measurement problem. In fact, these notions are not just unreliable, but, without the context of a consistent 'quantum theory of gravity', they are also quite meaningless. For this reason, in this chapter I shall concentrate exclusively on Penrose's proposal of quantum state reduction (1979, 1981, 1984, 1986, 1987, 1989, pp. 367–71, 1993, 1994, 1994a, pp. 339–46, 1996, 1997, 1998), since his is a minimalist approach in which he refrains from employing any ill-understood (or oxymoronic) notions such as 'quantum fluctuations of spacetime'. Rather, he argues from the first principles, exploiting the profound and fundamental conflict (sometimes manifested in the guise of the so-called 'problem of time') between the principle of general covariance of general relativity and the principle of superposition of quantum mechanics, to deduce a heuristic mechanism of gravity-induced quantum state reduction. Stated differently, instead of prematurely proposing a crude *theory* of quantum state reduction, he merely provides a rationale for the mass scale at which quantum mechanics must give way to gravitational effects, and hence to a superior theory.

Let me emphasize further that the motivations based on rather contentious conceptual issues inwrought in the measurement problem are *not* an essential pre-requisite to Penrose's proposal for the breakdown of quantum superpositions at a 'macroscopic' scale. Instead, his proposal can be viewed as a strategy not only to tackle the profound tension between the foundational principles of our two most fundamental physical theories, general relativity and quantum mechanics, but also to provide simultaneously a possible window of opportunity to go beyond the confining principles of these two great theories in order to arrive at even greater enveloping 'final' theory (Penrose 1984, Christian 1999a). Such a final theory, which presumably would neither be purely quantal nor purely gravitational but fundamentally different and superior, would then have to reduce to quantum mechanics and general relativity, respectively, in some appropriate approximations, as depicted in Fig. 14.1. Clearly, unlike the lopsided orthodox approaches towards a putative 'quantum theory of gravity' (Rovelli 1998), this is a fairly 'evenhanded' approach – as Penrose himself often puts it. For, in the orthodox approaches, quantum superpositions are indeed presumed to be sacrosanct at all physical scales, but only at a very high price of some radical compromises with Einstein's theory of gravity: for example, at a price of having to fix both the topological and differential structures of spacetime *a priori*, as in the 'loop quantum gravity' program (Rovelli 1998). Or – even worse – at a price of having to assume some *non-dynamical* causal structure as a fixed arena for dynamical processes, as in the currently voguish 'M-theory' program (Banks 1998, 1998a, Polchinski 1998). Either of the compromises are, of course, out-and-out anathema to the very essence of general relativity (Einstein 1994, p. 155, Stachel 1994, Isham 1994, Sorkin 1997).

In passing, let me also point out another significant feature of Penrose's proposal which, from a certain philosophical perspective (namely the 'process' perspective, Whitehead 1929), puts it in a class of very attractive proposals. Unlike some other approaches to the philosophical problems of quantum theory, his approach (and for that matter almost *all* approaches appealing to the 'objective reduction') implicitly takes *temporal transience* in the world – the incessant fading away of the specious present into the indubitable past – not as a merely phenomenological appearance, but as a *bona fide* ontological attribute of the world, in a manner, for example, espoused by Shimony (1998). For, clearly, any gravity-induced or other intrinsic mechanism, which purports to actualize – as a *real* physical process – a genuine multiplicity of quantum mechanical potentialities to a specific choice among them, evidently captures transiency, and thereby not only goes beyond the symmetric temporality of quantum theory, but also acknowledges the temporal transience as a fundamental and objective attribute of the physical world (Shimony 1998). (For anticipatory views on 'becoming' along this line, see also Eddington 1929, Bondi 1952, Reichenbach 1956, Whitrow 1961.) A possibility of an empirical test confirming the objectivity of this facet of the world via Penrose's approach is by itself sufficient for me to endorse his efforts wholeheartedly. But his approach has even more to offer. It is generally believed that the classical general relativistic notion of spacetime is meaningful only at scales well above the Planck regime, and that near the Planck scale the usual classical structure of spacetime emerges purely phenomenologically via a phase transition or symmetry breaking phenomenon (Isham 1994). Accordingly, one may incline to think that 'the concept of "spacetime" is not a fundamental one at all, but only something that applies in a "phenomenological" sense when the universe is not probed too closely' (Isham 1997). However, if the emergence of spacetime near the Planck scale is a byproduct of the actualization of quantum mechanical potentialities – via Penrose's or any related mechanism – then the general relativistic spacetime, along with its distinctively *dynamical* causal structure, comes into being not as a coarse-grained phenomenological construct, but as a genuine ontological attribute of the world, in close analogy with the special case of temporal transience. In other words, such an ontological coming into being of spacetime near the Planck scale would capture the 'becoming' not merely as temporal transience, which is a rather 'Newtonian' notion, but as a much wider, dynamical, spatio-temporal sense paralleling general relativity. (This will become clearer in Section 14.3 where I discuss Penrose's mechanism, which is tailor-made to actualize specific spacetime geometries out of 'superpositions' of such geometries.) This gratifying possibility leaves no shred of doubt that the idea of 'objective reduction' in general, and its variant proposed by Penrose in particular, is worth investigating seriously, both theoretically and experimentally.[1]

Since the principle of general covariance is at the heart of Penrose's proposal, I begin in the next section with a closer look at the physical meaning of this fundamental principle, drawing lessons from Einstein's struggle to come to terms with it by finding a resolution of his famous 'hole argument'. Even the reader familiar with this episode in the history of general relativity is urged to go through the discussion offered here, since the subtleties of the principle of general covariance provides the basis for both Penrose's central thesis as well as my own partial criticism of it. Next, after alluding to the inadequacies of the orthodox quantum measurement theory, I review Penrose's proposal in greater detail in Section 14.3, with a special attention to the experiment he has proposed to corroborate his quantitative prediction of the breakdown of quantum mechanics near a specific mass scale (Section 14.3.5). (As an aside, I also propose a new geometrical measure of gravity-induced deviation from quantum mechanics in Section 14.3.4.) Since Penrose's proposed experiment is entirely within the non-relativistic domain, in the subsequent section, 14.4.1, I provide an orthodox analysis of it strictly within this domain, thereby setting the venue for my partial criticism of his proposal in Section 14.4.2. The main conclusion here is that, since there remains no residue of the conflict between the principles of superposition and general covariance in the *strictly* Newtonian limit (and this happens to be a rather subtle limit), Penrose's formula for the 'decay-time' of quantum superpositions produces triviality in this limit, retaining the standard quantum coherence intact. Finally, in Section 14.4.3, before making some concluding remarks in Section 14.5, I suggest that an appropriate experiment which could in principle corroborate Penrose's predicted effect is not the one he has proposed, but a Leggett-type SQUID (Superconducting QUantum Interference Device) or BEC (Bose–Einstein Condensate) experiment involving superpositions of mass distributions in *relative rotation*. As a bonus, this latter analysis brings out one of the distinctive features of

Penrose's scheme, rendering it empirically distinguishable from all of the other (*ad hoc*) quantum state reduction theories involving gravity (e.g. Ghirardi et al. 1990).

### 14.2 How spatio-temporal events lost their individuality

Between 1913 and 1915 Einstein (1914) put forward several versions of an argument, later termed by him the 'hole argument' ('Lochbetrachtung'), to reject what is known as the principle of general covariance, which he himself had elevated earlier as a criterion for selecting the field equations of any reasonable theory of gravitation. It is only after two years of struggle to arrive at the correct field equations with no avail that he was led to reconsider general covariance, despite the hole argument, and realized the full significance and potency of the principle it enjoys today. In particular – and this is also of utmost significance for our purposes here – he realized that the hole argument and the principle of general covariance can peacefully coexist if, and only if, the mathematical individuation of the points of a spacetime manifold is physically meaningless. In other words, he realized that a bare spacetime manifold without some 'individuating field' (Stachel 1993) such as a specific metric tensor field defined on it is a highly fictitious mathematical entity without any direct physical content.

Although the physical meaninglessness of a mathematical individuation of space-time points – as a result of general covariance – is central to Penrose's proposal of quantum state reduction, he does not invoke the historical episode of the hole argument to motivate this non-trivial aspect of the principle. And justifiably so. After all, the non-triviality of the principle of general covariance (i.e. the freedom under *active* diffeomorphisms of spacetime) is one of the first things one learns about while learning general relativity. For example, Hawking and Ellis (1973) begin their seminal treatise on the large-scale structure of spacetime by simply taking a mathematical model of spacetime to be the entire equivalence class of copies of a 4-manifold, equipped, respectively, with Lorentzian metric fields related by active diffeomorphisms of the manifold, without even mentioning the hole argument. However, as we shall see, it is the hole argument – an argument capable of misleading even Einstein for two years – that demands such an identification in the first place. Therefore, and especially considering the great deal of persistent confusion surrounding the physical meaning of the principle of general covariance in the literature (as surveyed in Norton, 1993), for our purposes it would be worthwhile to take a closer look at the hole argument, and thereby appreciate what is at the heart of Penrose's proposal of quantum state reduction. For more details on the physical meaning of general covariance the reader is referred to Stachel's incisive analysis (1993) of it in the modern differential geometric language; it is the general viewpoint espoused in this reference that I shall be mostly following here (but see also Rovelli, 1991, and section 6 of Anandan, 1997, for analogous viewpoints).

Without further ado, here is Einstein's hole argument. As depicted in Fig. 14.2, suppose that the matter distribution encoded in a stress–energy tensor $T^{\mu\nu}$ is precisely known everywhere on a spacetime $\mathcal{M}$ outside of some hole $\mathcal{H} \subset \mathcal{M}$ – i.e. outside of an open subspace of the manifold $\mathcal{M}$. (Throughout this essay I shall be using Penrose's abstract index notation; see Wald 1984.) Further, let there be

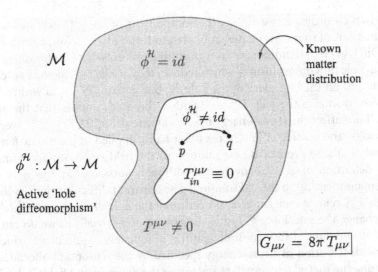

Fig. 14.2. Einstein's hole argument. If the field equations of a gravitational theory are generally covariant, then, inside a matter-free region of some known matter distribution, they appear to generate an infinite number of inequivalent solutions related by active diffeomorphisms of the underlying spacetime manifold.

no physical structure defined within $\mathcal{H}$ except a gravitational field represented by a Lorentzian metric tensor field $g^{\mu\nu}$; i.e. let the stress–energy vanish identically inside the hole: $T_{in}^{\mu\nu} \equiv 0$. Now suppose that the field equations of the gravitational theory under consideration are generally covariant. By definition, this means that if a tensor field $\mathcal{X}$ on the manifold $\mathcal{M}$ is a solution of the set of field equations, then the pushed-forward tensor field $\phi_* \mathcal{X}$ of $\mathcal{X}$ is also a solution of the same set of equations for any *active* diffeomorphism $\phi : \mathcal{M} \to \mathcal{M}$ of the manifold $\mathcal{M}$ onto itself. The set of such diffeomorphisms of $\mathcal{M}$ forms a group, which is usually denoted by Diff$(\mathcal{M})$. It is crucially important here to distinguish between this genuine group Diff$(\mathcal{M})$ of *global* diffeomorphisms of $\mathcal{M}$ and the pseudo-group of transformations between overlapping pairs of local co-ordinate charts. The elements of the latter group are sometimes referred to as passive diffeomorphisms because they can only produce trivial transformations by merely relabelling or renaming the points of a manifold. Admittance of only tensorial objects on $\mathcal{M}$ in any spacetime theory is sufficient to guarantee compatibility with this pseudo-group of passive diffeomorphisms. On the other hand, the elements of the genuine group Diff$(\mathcal{M})$ of active diffeomorphisms are smooth homeomorphisms $\phi : \mathcal{M} \to \mathcal{M}$ that can literally take each point $p$ of $\mathcal{M}$ into some other point $q := \phi(p)$ of $\mathcal{M}$ and thereby deform, for example, a doughnut-shaped manifold into its coffee-mug-shaped copy (Nakahara 1990, p. 54). Returning to the definition of general covariance, if a metric tensor field $g^{\mu\nu}(x)$ is a solution of the generally covariant field equations at any point $x$ of $\mathcal{M}$ in an adapted local co-ordinate system, then so is the corresponding pushed-forward tensor field $(\phi_* g)^{\mu\nu}(x)$ at the *same* point $x$ in the *same* co-ordinate system. Note that, in general, $g^{\mu\nu}$ and $(\phi_* g)^{\mu\nu}$ will be functionally different from each other in a given co-ordinate system; i.e. the components of $(\phi_* g)^{\mu\nu}$ will involve different functions of the co-ordinates compared to those of $g^{\mu\nu}$. Now, since a choice of $\phi \in \mathrm{Diff}(\mathcal{M})$ is by definition arbitrary, nothing prevents us from choosing a smooth $\phi^{\mathcal{H}}$ – a 'hole diffeomorphism' – which reduces to $\phi^{\mathcal{H}} = id$ (i.e. identity) everywhere outside and on the boundary of the hole $\mathcal{H}$, but remains $\phi^{\mathcal{H}} \neq id$ within $\mathcal{H}$. Such a choice, owing to the fact $T^{\mu\nu} \equiv 0$ within the hole, implies that the action of $\phi^{\mathcal{H}}$ will not affect the stress–energy tensor anywhere: $(\phi_*^{\mathcal{H}} T)^{\mu\nu} = T^{\mu\nu}$ everywhere, both inside and outside of $\mathcal{H}$. On the other hand, applied to the metric tensor $g^{\mu\nu}$, $\phi^{\mathcal{H}}$ will of course produce a new solution of the field equations according to the above definition of general covariance, although outside of $\mathcal{H}$ this new solution will remain identical to the old solution. The apparent difficulty, then, is that, even though $T^{\mu\nu}$ remains unchanged, our choice of the hole diffeomorphism $\phi^{\mathcal{H}}$ allows us to change the solution $g^{\mu\nu}$ inside the hole as non-trivially as we do not like, in a blatant violation of the physically natural uniqueness requirement, which states that the distribution of stress–energy specified by the tensor $T^{\mu\nu}$ should *uniquely* determine the metric tensor $g^{\mu\nu}$ representing the gravitational field. Indeed, under the diffeomorphism $\phi^{\mathcal{H}}$, identical matter fields $T^{\mu\nu}$ seem to lead to non-trivially different gravitational fields inside the hole, such as $g^{\mu\nu}$ and $(\phi_*^{\mathcal{H}} g)^{\mu\nu}$, since $\phi^{\mathcal{H}}$ is not an identity there. What is worse, even though nothing has been allowed to change outside or on the boundary of the hole $\mathcal{H}$, nothing seems to prevent the gravitational field $(\phi_*^{\mathcal{H}} g)^{\mu\nu}$ from being completely different for each one of the *infinitely many* inequivalent diffeomorphisms $\phi^{\mathcal{H}} \in \mathrm{Diff}(\mathcal{M})$ that can be carried out inside $\mathcal{H}$.

As mentioned above, Einstein's initial reaction to this dilemma was to abandon general covariance for the sake of uniqueness requirement, and he maintained this position for over two years. Of course, to a modern general relativist a resolution of the apparent problem is quite obvious: The tacit assumption in the hole argument that the mathematically different tensor fields $g^{\mu\nu}$ and $(\phi_*^{\mathcal{H}} g)^{\mu\nu}$ are also *physically* different – i.e. correspond to different physical realities – is clearly not justified. The two expressions $g^{\mu\nu}$ and $(\phi_*^{\mathcal{H}} g)^{\mu\nu}$, no matter how non-trivially they differ mathematically, must represent one and the same gravitational field *physically*. Thus, as Wald puts it (1984), 'diffeomorphisms comprise [nothing but a] gauge freedom of any theory formulated in terms of tensor fields on a spacetime manifold'. Accordingly, in formal analogy with the familiar gauge freedom of the gauge field theories, modern general relativists take a gravitational field to physically correspond to an entire equivalence class of metric tensor fields, related by arbitrary diffeomorphisms of the spacetime manifold, and not just to one of the members of this class.

The analogy with the gauge freedom of the gauge field theories, however, has only a limited appeal when it comes to general relativity. To see the difference, recall, for example, that electromagnetic gauge transformations – the prototype of all gauge transformations – occur at a *fixed* spacetime point: The vector potential $A_\mu(x)$ defined at a point $x$ of $\mathcal{M}$ is physically equivalent to the vector potential $A_\mu(x) + \partial_\mu f(x)$ defined at the same point $x$ of $\mathcal{M}$, for all scalar functions $f(x)$. Although mathematically different, both $A_\mu(x)$ and $A_\mu(x) + \partial_\mu f(x)$ correspond to one and the same physical electromagnetic field configuration $F_{\mu\nu}(x)$, which again depends locally on the same point $x$ of $\mathcal{M}$. On the other hand, as stressed above, in general relativity diffeomorphisms $\phi \in \text{Diff}(\mathcal{M})$ map one spacetime point, say $p$, to another spacetime point, say $q := \phi(p)$. Therefore, if the tensor fields $g^{\mu\nu}(p)$ and $(\phi_* g)^{\mu\nu}(q)$ are to be identified as representing one and the same gravitational field configuration, implying that they cannot be physically distinguishable by any means, then the two points $p$ and $q$ must also be physically indistinguishable, and, consequently, they must renounce their individuality. For, if the points of $\mathcal{M}$ did possess any ontologically significant individual identity of their own, then a point $p$ of $\mathcal{M}$ could be set apart from a point $q$ of $\mathcal{M}$, and that would be sufficient to distinguish the quantity $g^{\mu\nu}(p)$ from the quantity $(\phi_* g)^{\mu\nu}(q)$, contradicting the initial assertion.

As Einstein eventually realized, the conclusion is inescapable: The points of a spacetime manifold $\mathcal{M}$ have no direct ontological significance. A point in a bare spacetime manifold is not distinguishable from any other point – and, indeed, does not even become a point (i.e. an *event*) with physical meaning – unless and until a specific metric tensor field is *dynamically* determined on the manifold. In fact, in general relativity a bare manifold not only lacks this local property, but the entire *global* topological structure of spacetime is also determined only a posteriori via a metric tensor field (Einstein 1994, Stachel 1994, Isham 1994, Sorkin 1997). Since a dynamical metric tensor field on a manifold dynamizes the underlying topology of the manifold, in general relativity the topology of spacetime is also not an absolute element that 'affects without being affected'. Thus, strictly speaking, the bare manifold does not even become 'spacetime' with physical meaning until both the global and local spatio-temporal structures are dynamically determined along with a metric. Further, since spacetime points aquire their individuality in no other way but as a byproduct of a solution of Einstein's field equations, in general relativity 'here' and 'now' cannot be part of a physical question, but can only be part of the *answer* to a question, as Stachel so aptly puts it (1994). The concepts 'here' and 'now' – and hence the entire notion of local causality – acquire ontological meaning only a posteriori, as a part of the answer to a physical question. Anticipating the issue discussed in the next section, this state of affairs is in sharp contrast to what one can ask in quantum theory, which – due to its axiomatically non-dynamical causal structure – allows 'here' and 'now' to be part of a question. Indeed, in quantum mechanics, as we shall see, *a priori* individuation of spatio-temporal events is an essential prerequisite to any meaningful notion of time-evolution.

At a risk of repetition, let me recapitulate the central point of this section in a single sentence:

> In Einstein's theory of gravity, general covariance – i.e. invariance of physical laws under the action of the group $\text{Diff}(\mathcal{M})$ *of active* diffeomorphisms – expressly forbids *a priori* individuation of the points of a spacetime manifold as spatio-temporal events.

Although unfairly under-appreciated (especially within approaches towards 'quantum gravity' through 'string' or 'M' type theories, practically all of which being guilty of presupposing one form or another of blatantly unjustified non-dynamical background structure (Rovelli 1997, Banks 1998, 1998a, Polchinski 1998, Smolin 1999)), this is one of the most fundamental metaphysical tenets of general relativity.

In this respect, contrary to what is often asserted following Kretschmann (1917), the principle of general covariance is far from being physically vacuous. For instance, the potency of general covariance is strikingly manifest in the following circumstance: if $A_\mu$ is a vector field on a general relativistic manifold $\mathcal{M}$, then, unlike the situation in electromagnetism discussed above, the value $A_\mu(x)$ at a particular point $x \in \mathcal{M}$ has no invariant physical meaning. This is because the point $x$ can be *actively* transposed around by the action of the diffeomorphism group $\mathrm{Diff}(\mathcal{M})$, robbing it of any individuality of its own.

Of course, individuation of spacetime points *can* be achieved by a *fixed* 'gauge choice' – that is to say, by specification of a particular metric tensor field $g^{\mu\nu}$ out of the entire equivalence class of fields $g^{\mu\nu}$ related by gauge transformations, but that would be at odds with general covariance. In fact, if there are any non-dynamical structures present, such as the globally specified Minkowski metric tensor field $\eta^{\mu\nu}$ of special relativity, then the impact of general covariance is severely mitigated. This is because the non-dynamical Minkowski metric tensor field, for example, can be used to introduce a family of global inertial co-ordinate systems (or 'inertial individuating fields' (Stachel 1993)) that can be transformed into each other by the (extended) Poincaré group of isometries of the metric: $\pounds_x \eta^{\mu\nu} = 0$, where $\pounds_x$ denotes the Lie derivative, with the Killing vector field $x^\alpha$ being a generator of the Poincaré group of transformations (Wald 1984). These inertial co-ordinates in turn can be used to set apart a point $q$ from a point $p$ of a manifold, bestowing *a priori* spatio-temporal individuality to the points of the manifold (Wald 1984). For this reason, Stachel (1993) and Wald (1984), among others, strengthen the statement of general covariance by a condition – explicitly added to the usual requirement of tensorial form for the law-like equations of physics – that *there should not be any preferred individuating fields in spacetime other than, or independent of, the dynamically determined metric tensor field $g^{\mu\nu}$*. Here, preferred or background fields are understood to be the ones which affect the dynamical objects of a theory, but without being affected by them in return. They thereby provide non-dynamical backdrops for the dynamical processes. I shall return below to this issue of the background structure in spacetime.

In the light of this discussion, and in response to the lack of consensus on the meaning of general covariance in the literature (Norton 1993), let me end this section by proposing a litmus test for general covariance – formulated at the level of theory as a whole – which captures its true physical and metaphysical essence.

> **A litmus test for general covariance:** A given theory may qualify to be called generally covariant if and only if the points of the spacetime 4-manifold, or a more general N-manifold, belonging to any model of the theory do not possess physically meaningful *a priori* individuality of their own.

(A model of a theory is a set of dynamical variables constituting a particular solution to the dynamical equations of the theory, and may, in general, also contain non-dynamical structures.) Admittedly, this is not a very practical elucidation of the principle, but it does exclude theories which are not truly generally covariant in the sense discussed above. In particular, it excludes *all* of the 'string' or 'M' type theories known to date, since they all presuppose individuation-condoning background structure of one form or another. (For a recent attempt to overcome this potentially detrimental deficiency of M-theory see Smolin 1999.)

## 14.3 Penrose's mechanism for the objective state reduction

Even if we tentatively ignore the issue of individuation of spatio-temporal events, there exists a further concern regarding the notion of definite events in the quantum domain. In quantum theories (barring a few approaches to 'quantum gravity') one usually takes spacetime to be a fixed continuum whose constituents are the 'events' at points of space at instants of time. What is implicit in this assumption is the classicality or definiteness of the events. However, according to quantum mechanics, in general the notions such as 'here' and 'now' could have only *indefinite* or *potential* meaning. Further, if the conventional quantum framework is interpreted as universally applicable, objective (i.e. non-anthropocentric), and complete (Einstein et al. 1935), then, as pointed out above in the Introduction (Section 14.1), the linear nature of quantum dynamics gives rise to some serious conceptual difficulties collectively known as 'the measurement problem'. These difficulties make the notion of definite or actual events in the quantum world quite problematic, if not entirely meaningless (Jauch 1968, Haag 1990, 1992, Shimony 1993a). In particular, they render the orthodox quantum theory of measurement inadequate to explain the prolific occurrences of actual events in the 'macroscopic' domain, such as a formation of a droplet in a cloud chamber, or a blackening of a silver grain on a photographic plate. This absurdity, occurring as a direct consequence of the linearity of quantum dynamics, is well dramatized by Schrödinger in his notorious *gedanken*-experiment involving a poor cat, which ends up in a limbo between definite states of being alive and dead.

In conventional quantum mechanics this blatant contradiction with the apparent phenomenological facts about the occurrence of actual events is evaded by invoking an *ad hoc* postulate – the Projection Postulate, which in its simplest form is usually attributed to von Neumann. However, von Neumann's Projection Postulate is only a necessary but not sufficient condition for an unequivocal understanding of the occurrence of definite events. Further, even if we accept this *ad hoc* postulate unreservedly, the process of specific actualization out of the compendium of quantum mechanical potentialities remains completely obscure. Consequently, what is desperately needed is an unequivocal *physical* understanding underlying the non-unitary transition (in the standard notation; cf. Christian 1998)

$$|\Psi\rangle \equiv \sum_{j=1}^{N} \lambda_j |\psi_j\rangle \otimes |\varphi_j\rangle \longrightarrow |\psi_k\rangle \otimes |\varphi_k\rangle. \tag{14.1}$$

As discussed in Section 14.1, despite a multitude of attempts with varied sophistication and predilections, no universally acceptable explanation – physical or otherwise – of this mysterious transition is as yet in sight. In what follows we shall see that Penrose's scheme provides precisely the much desired physical explanation for the transition, and compellingly so.

### 14.3.1 Motivation via a concrete example

To illustrate Penrose's proposal within a concrete scenario, let us apply the usual measurement procedure to a model interaction, within the non-relativistic domain, in a specific representation – the co-ordinate representation. Let us begin by assuming a global inertial co-ordinate system whose origin is affixed at the centre of the Earth, and let $\Sigma^S$ and $\Sigma^A$ be two quantum mechanical systems constituting a *closed* composite system $\Sigma = \Sigma^S + \Sigma^A$ with their physical states represented by the rays corresponding to normalized vectors in the Hilbert spaces $\mathcal{H}^S$, $\mathcal{H}^A$, and $\mathcal{H}^\Sigma = \mathcal{H}^S \otimes \mathcal{H}^A$, respectively. Suppose now one wants to obtain the value of a dynamical variable corresponding to some property of the system $\Sigma^S$ by means of the system $\Sigma^A$, which serves as a measuring apparatus. To this end, using the global co-ordinate system, let a non-degenerate indicator variable $Q^A$ defined by $Q^A|\varphi_j\rangle = q_j|\varphi_j\rangle$ represent the location $q$ of the system $\Sigma^A$, which, say, has mass $M$, and let a non-degenerate dynamical variable $\Omega^S$ defined by $\Omega^S|\psi_j\rangle = \omega_j|\psi_j\rangle$ be a time-independent function of co-ordinate $x$ and its conjugate momentum $-i\hbar\,\partial/\partial x$ of the system $\Sigma^S$ exclusively. Further, let the mass $M$ (i.e. the apparatus system $\Sigma^A$) be localized initially ($t < t_a$) at $q_1$, and let the measurement, which is to be achieved by moving the mass from $q_1$ to some other location, consist in the fact that if the value of $\Omega^S(x, -i\hbar\,\partial/\partial x)$ is $\omega_1$ then the location of the mass remains unchanged at $q_1$ whereas if it is $\omega_{j\neq1}$ the mass is displaced from $q_1$ to a new location $q_{j\neq1} \equiv q_1 + \omega_{j\neq1}$. An interaction Hamiltonian which precisely accounts for such a process according to the conventional Schrödinger equation

$$i\hbar\,\frac{\partial}{\partial t}\,|\Psi(t)\rangle = H(t)\,|\Psi(t)\rangle \tag{14.2}$$

exists (cf. Christian 1998), and leads to the following set of evolutions

$$\Psi(t < t_a) = \psi_1\,\delta(q - q_1) \xrightarrow{H_{int}} \psi_1\,\delta(q - q_1) \quad \text{(location unchanged)} \tag{14.3}$$

but

$$\psi_{j\neq1}\,\delta(q - q_1) \xrightarrow{H_{int}} \psi_{j\neq1}\,\delta(q - q_{j\neq1}) \quad \text{(location shifted)}, \tag{14.4}$$

with $\langle q|\varphi_j\rangle = \delta(q - q_j)$. More generally, if the initial state of the quantum system $\Sigma^S$ is a superposition state represented by

$$\sum_{j=1}^{N}\lambda_j\,\psi_j, \qquad \sum_{j=1}^{N}|\lambda_j|^2 = 1, \tag{14.5}$$

then we have the Schrödinger's cat-type entanglement exhibiting superposition of the location-states of the mass at various positions:

$$\left[\sum_{j=1}^{N}\lambda_j\psi_j\right]\delta(q - q_1) \xrightarrow{H_{int}} \sum_{j=1}^{N}\lambda_j\psi_j\,\delta(q - q_j) \equiv \sum_{j=1}^{N}\lambda_j\psi_j\varphi_j. \tag{14.6}$$

In particular, if initially we have

$$\Psi(t < t_a) = (\lambda_2 \psi_2 + \lambda_3 \psi_3)\, \delta(q - q_1), \qquad |\lambda_2|^2 + |\lambda_3|^2 = 1, \qquad (14.7)$$

then, after the interaction $(t > t_b)$,

$$\Psi(t > t_b) = \lambda_2 \psi_2\, \delta(q - q_2) + \lambda_3 \psi_3\, \delta(q - q_3), \qquad (14.8)$$

and the location of the mass will be indefinite between the two positions $q_2$ and $q_3$. This of course is a perfectly respectable quantum mechanical state for the mass $M$ to be in, unless it is a 'macroscopic' object and the two locations are macroscopically distinct. In that case the indefiniteness in the location of the mass dictated by the linearity of quantum dynamics stands in a blatant contradiction with the evident phenomenology of such objects.

### 14.3.2 The *raison d'être* of state reduction

Recognizing this contradiction, Penrose, among others, has tirelessly argued that gravitation must be directly responsible for an *objective* resolution of this fundamental anomaly of quantum theory. He contends that, since the self-gravity of the mass must also participate in such superpositions, what is actually involved here, in accordance with the principles of Einstein's theory of gravity, is a superposition of two entirely *different* spacetime geometries; and, when the two geometries are sufficiently different from each other, the unitary quantum mechanical description of the situation – i.e. the linear superposition of a 'macroscopic' mass prescribed by eqn. 14.8 – must break down[2] (or, rather, 'decay'), allowing nature to choose between one or the other of the two geometries.

To understand this claim, let me bring to the surface some of the assumptions regarding spacetime structure underlying the time-evolution dictated by eqn. 14.2, which brought us to the state 14.8 in question. Recall that I began this section with an assumption of a globally specified inertial frame of reference affixed at the centre of the Earth. Actually, this is a bit too strong an assumption. Since the Schrödinger equation (eqn. 14.2) is invariant under Galilean transformations, all one needs is a *family* of such global inertial frames, each member of which is related to another by a Galilean transformation

$$t \longrightarrow t' = t + constant \qquad (14.9)$$

$$x^a \longrightarrow x'^a = O^a_b x^b + v^a t + constant \quad (a, b = 1, 2, 3), \qquad (14.10)$$

where $O^a_b \in SO(3)$ is a time-independent orthonormal rotation matrix (with Einstein's summation convention for like indices), and $\mathbf{v} \in \Re^3$ is a time-independent spatial velocity. Now, as discussed at the end of Section 14.2 above, existence of a global inertial frame grants *a priori* individuality to spacetime points – a point $p_1$ of a spacetime manifold can be set apart from a point $p_2$ using such inertial co-ordinates (Wald 1984, p. 6). Consequently, in the present scenario the concepts 'here' and 'now' have *a priori* meaning, and they can be taken as a part of any physical question (cf. Section 14.2). In particular, it is meaningful to take location $q_1$ of the mass $M$ to be a part of the initial state (eqn. 14.7), since it can be set apart from any other location, such as the location $q_2$ or $q_3$ in the final state (eqn. 14.8). If individuation of spatio-temporal events was not possible, then of course all of the locations, $q_1$, $q_2$, $q_3$, etc., would have been identified with each other as one and the same location, and it would not have been meaningful to take $q_1$ as a distinct initial location of the mass (as elaborated in Section 14.2 above, such an identification of *all* spacetime points is indeed what general covariance demands in full general relativity). Now, continuing to ignore gravity for the moment, but anticipating Penrose's reasoning when gravity *is* included, let us pretend, for the sake of argument, that the two components of the superposition in eqn. 14.8 correspond to two *different* (flat) spacetime geometries. Accordingly, let us take two separate inertial co-ordinate systems, one for each spacetime but related by the transformation 14.10, for separately describing the evolution of each of the two components of the superposition, with the initial location of the mass $M$ being $q_1$ as prescribed in eqn. 14.7 – i.e. assume for the moment that each component of the superposition is evolving on its own, as it were, under the Schrödinger equation (eqn. 14.2). Then, for the final superposed state 14.8 to be meaningful, a crucially important question would be: are these two time-evolutions corresponding to the two different spacetimes compatible with each other? In particular: is the time-translation operator '$\partial/\partial t$' in eqn. 14.2 the same for the two superposed evolutions – one displacing the mass $M$ from $q_1$ to $q_2$ and the other displacing it from $q_1$ to $q_3$? Unless the two time-translation operators in the two co-ordinate systems are equivalent in some sense, we do not have a meaningful quantum gestation of the superposition 14.8. Now, since we are in the Galilean-relativistic domain, the two inertial frames assigned to the two spacetimes must be related by the transformation 14.10, which, upon using the chain rule (and setting $O \equiv \mathbb{1}$ for simplicity), yields

$$\frac{\partial}{\partial x'} = \frac{\partial}{\partial x^a}, \quad \text{but} \quad \frac{\partial}{\partial t'} = \frac{\partial}{\partial t} - v^a \frac{\partial}{\partial x^a}. \tag{14.11}$$

Thus, the time-translation operators are *not* the same for the two spacetimes (cf. Penrose 1996, pp. 592–3). As a result, in general, unless **v** identically vanishes everywhere, the two superposed time-evolutions are *not* compatible with each other (see Section 14.4.2, however, for a more careful analysis). The difficulty arises for the following reason. Although in this Galilean-relativistic domain the individuality of spacetime points in a given spacetime is rather easy to achieve, when it comes to two entirely *different* spacetimes there still remains an ambiguity in registering the fact that the location, say $q_2$, of the mass in one spacetime is 'distinct' from its location, say $q_3$, in the other spacetime. On the other hand, the location $q_2$ must be unequivocally distinguishable from the location $q_3$ for the notion of superposition of the kind 14.8 to have any unambiguous physical meaning. Now, in order to meaningfully set apart a location $q_2$ in *one* spacetime from a location $q_3$ in *another*, a point-by-point identification of the two spacetimes is clearly necessary. But such a *pointwise* identification is quite ambiguous for the two spacetimes under consideration, as can be readily seen from eqn. 14.10, unless the arbitrarily chosen relative spatial velocity **v** is set to identically vanish everywhere (i.e. not just locally). Of course, in the present scenario, since we have ignored gravity, nothing prevents

318

us from setting $\mathbf{v} \equiv 0$ everywhere – i.e. by simply taking non-rotating co-ordinate systems with constant spatial distance between them – and the apparent difficulty completely disappears, yielding

$$\frac{\partial}{\partial t'} \equiv \frac{\partial}{\partial t}. \tag{14.12}$$

Therefore, as long as gravity is ignored, there is nothing wrong with the quantum mechanical time-evolution leading to the superposed state 14.8 from the initial state 14.7, since all of the hidden assumptions exposed in this paragraph are justified.

The situation becomes dramatically obscure, however, when one attempts to incorporate gravity in the above scenario in full accordance with the principles of general relativity.[3] To appreciate the central difficulty, let us try to parallel considerations of the previous paragraph with due respect to the ubiquitous general-relativistic features of spacetime.[4] To begin with, once gravity is included, even the initial state 14.7 becomes meaningless because any location such as $q_1$ loses its *a priori* meaning. Recall from Section 14.2 that in general relativity, since neither global topological structure of spacetime nor local individuality of spatio-temporal events has any meaning until a specific metric tensor field is dynamically determined, the concepts 'here' and 'now' can only be part of the *answer* to a physical question. On the other hand, the initial state 14.7 specifying the initial location $q_1$ of the mass $M$ is part of the question itself regarding the evolution of the mass. Thus, from the general-relativistic viewpoint – which clearly is the correct viewpoint for a 'large enough' mass – the statement 14.7 is entirely meaningless. In practice, however, for the non-relativistic situation under consideration, the much more massive Earth comes to rescue, since it can be used to serve as an *external* frame of reference providing prior – albeit approximate – individuation of spacetime points (Rovelli 1991). For the sake of argument, let us be content with such an approximate specification of the initial location $q_1$ of the mass, and ask: what role would the general-relativistic features of spacetime play in the evolution of this mass either from $q_1$ to $q_2$ or from $q_1$ to $q_3$, when these two evolutions are viewed separately – i.e. purely 'classically'? Now, since the self-gravity of the mass must also be taken into account here, and since each of the two evolutions would incorporate the self-gravitational effects in its own distinct manner to determine its own overall *a posteriori* spacetime geometry, to a good degree of classical approximation there will be essentially two *distinct* spacetime geometries associated with these two evolutions. Actually, as in the case of initial location $q_1$, the two final locations, $q_2$ and $q_3$, would also acquire physical meaning only *a posteriori* via the two resulting metric tensor fields – say $g_2^{\mu\nu}$ and $g_3^{\mu\nu}$, respectively, since the individuation of the points of each of these two spacetimes becomes meaningful only *a posteriori* by means of these metric tensor fields. It is of paramount importance here to note that, in general, the metric tensor fields $g_2^{\mu\nu}$ and $g_3^{\mu\nu}$ would represent two strictly *separate* spacetimes with their own distinct global topological and local causal structures. To dramatize this fact by means of a rather extreme example, note that one of the two components of the superposition leading to eqn. 14.8 might, in principle, end up having evolved into something like a highly singular Kerr–Newman spacetime, whereas the other one might end up having evolved into something like a non-singular Robertson–Walker spacetime. This observation is crucial to Penrose's argument because, as we did in the previous paragraph for the non-gravitational case, we must now ask whether it is meaningful to set apart one location of the mass, say $q_2$, from another, say $q_3$, in order for a superposition such as 14.8 to have any unambiguous physical meaning. And as before, we immediately see that in order to be able to distinguish the two locations of the mass – i.e. to register the fact that the mass has actually been displaced from the initial location $q_1$ to a final location, say $q_2$, and not to any other location, say $q_3$ – a point-by-point identification of the two spacetimes is essential. However, in the present general-relativistic picture such a pointwise identification is *utterly meaningless*, especially when the two geometries under consideration are 'significantly' different from each other. As a direct consequence of the principle of general covariance, there is simply no meaningful way to make a pointwise identification between two such distinct spacetimes in general relativity. Since the theory makes no *a priori* assumption as to what the spacetime manifold is and allows the Lorentzian metric tensor field to be any solution of Einstein's field equations, the entire causal structure associated with a general-relativistic spacetime is dynamical and not predetermined (cf. Section 14.2). In other words, unlike in special relativity and the case considered in the previous paragraph, there is simply no isometry group underlying the structure of general relativity which could allow existence of a preferred family of inertial reference frames that may be used, first, to individuate the points of each spacetime, and then to identify one spacetime with another *point-by-point*. Furthermore, the lack of an isometry group means that, in general, there are simply no Killing vector fields of any kind in a general-relativistic spacetime, let alone a time-like Killing vector field analogous to the time-translation operator '$\partial/\partial t$' of the non-gravitational case considered above (cf. eqn. 14.11). Therefore, in order to continue our argument, we have to make a further assumption: We have to assume, at least, that the spacetimes under consideration are actually two reasonably well-defined 'stationary' spacetimes with two time-like Killing vector fields corresponding to the time-symmetries of the two metric tensor fields $g_2^{\mu\nu}$ and $g_3^{\mu\nu}$, respectively. These Killing vector fields, we hope, would generate time-translations needed to describe the time-evolution analogous to the one provided by the operator '$\partial/\partial t$' in the non-gravitational case. However, even this drastic assumption hardly puts an end to the difficulties involved in the notion of time-evolution leading to a superposition such as 14.8. One immediate difficulty is that these two Killing vector fields generating the time-evolution are *completely different* for the two components of the superposition under consideration. Since they correspond to the time-symmetries of two essentially distinct spacetimes, they could hardly be the same. As a result, the two Killing vector fields represent two completely different causal structures, and hence, if we insist on implementing them, the final state corresponding to eqn. 14.8 would involve some oxymoronic notion such as 'superposition of two distinct causalities'. Incidentally, this problem notoriously reappears in different guises in various approaches to 'quantum gravity', and it is sometimes referred to as the 'problem of time' (Kuchař 1991, 1992, Isham 1993). In summary, for a 'large enough' mass $M$, the final superposed state such as 14.8 is fundamentally and hopelessly meaningless.

### 14.3.3 Phenomenology of the objective state reduction

In the previous two paragraphs we saw two extreme cases. In the first of the two we saw that, as long as gravity is ignored, the notion of quantum superposition is quite unambiguous, thanks to the availability of *a priori* and *exact* pointwise identification between the two 'spacetimes' into which a mass $M$ could evolve. However, since the ubiquitous gravitational effects cannot be ignored for a 'large enough' mass, in the last paragraph we saw that a notion of superposition within two general-relativistic spacetimes is completely meaningless. Thus, *a priori* and *exact* pointwise identification of distinct spacetimes – although expressly forbidden by the principle of general covariance – turns out to be an essential prerequisite for the notion of superposition. In other words, the superposition principle is not as fundamental a principle as the adherents of orthodox quantum mechanics would have us believe; it makes sense only when the other most important principle – the principle of general covariance – is severely mitigated. By contrast, of course, as a result of formidable difficulties encountered in attempts to construct a Diff$(\mathcal{M})$-invariant quantum field theory (Rovelli 1998), it is not so unpopular to assert that *active* general covariance may be truly meaningful only at the classical general-relativistic level – i.e. when the superposition principle is practically neutralized.

To bridge this gulf between our two most basic principles at least phenomeno-logically, Penrose invites us to contemplate an intermediate physical situation for which the notion of quantum superposition is at best approximately meaningful. In a nutshell, his strategy is to consider first a 'superposition' such as 14.8 with gravity included, but retaining at least some *approximate* meaning to an *a priori* pointwise identification of the two spacetimes (corresponding to the two components of the superposition). Then, after putting a practical measure on this approximation, he uses this measure to obtain a heuristic formula for the collapse time of this superposi-tion. Here is how this works: Consider two well-defined quantum states represented by $|\Psi_2\rangle$ and $|\Psi_3\rangle$ (analogous to the states in 14.4), each *stationary* on its own and possessing the same energy $E$:

$$i\hbar\frac{\partial}{\partial t}|\Psi_2\rangle = E|\Psi_2\rangle, \qquad i\hbar\frac{\partial}{\partial t}|\Psi_3\rangle = E|\Psi_3\rangle. \tag{14.13}$$

In standard quantum mechanics, when gravitational effects are ignored, linearity dictates that *any* superposition of these two stationary states such as

$$|\mathcal{X}\rangle = \lambda_2|\Psi_2\rangle + \lambda_3|\Psi_3\rangle \tag{14.14}$$

(cf. eqn. 14.8) must also be stationary, with the same energy $E$:

$$i\hbar\frac{\partial}{\partial t}|\mathcal{X}\rangle = E|\mathcal{X}\rangle. \tag{14.15}$$

Thus, quantum linearity necessitates a complete degeneracy of energy for super-positions of the two original states. However, when the gravitational fields of two different mass distributions are incorporated in the representations $|\Psi_2\rangle$ and $|\Psi_3\rangle$ of these states, a crucial question arises: will the state $|\mathcal{X}\rangle$ still remain station-ary with energy $E$? Of course, when gravity is taken into account, each of the two component states would correspond to two entirely *different* spacetimes with a good degree of classical approximation, whether or not we assume that they are reasonably well-defined stationary spacetimes. Consequently, as discussed above, the time-translation operators such as '$\partial/\partial t$' corresponding to the action of the time-like Killing vector fields of these two spacetimes would be completely different from each other in general. They could only be the same if there were an unequivocal pointwise correspondence between the two spacetimes. Let us assume, however, that these two Killing vector fields are not too different from each other for the physical situation under consideration. In that case, there would be a slight – but essential – ill-definedness in the action of the operator '$\partial/\partial t$' when it is employed to generate a superposed state such as 14.14, and this ill-definedness would be without doubt reflected in the energy $E$ of this state. One can use this ill-definedness in energy, $\Delta E$, as a measure of *instability* of the state 14.14, and postulate the life-time of such a 'stationary' superposition – analogous to the half-life of an unstable particle – to be

$$\tau = \frac{\hbar}{\Delta E}, \tag{14.16}$$

with two decay modes being the individual states $|\Psi_2\rangle$ and $|\Psi_3\rangle$ with relative probabilities $|\lambda_2|^2$ and $|\lambda_3|^2$, respectively. Clearly, when there is an exact pointwise identification between the two spacetimes, $\Delta E \to 0$, and the collapse of the superposition never happens. On the other hand, when such an identification is ambiguous or impossible, inducing much larger ill-definedness in the energy, the collapse is almost instantaneous.

A noteworthy feature of the above formula is that it is independent of the speed of light $c$, implying that it remains valid even in the non-relativistic domain (cf. Fig. 14.1 and Penrose 1994, p. 339, 1996, p. 592). Further, in such a Newtonian approximation, the ill-definedness $\Delta E$ (for an essentially static situation) turns out to be proportional to the gravitational self-energy of the *difference* between the mass distributions belonging to the two components of the superposition (Penrose 1996). Remarkably, numerical estimates (Penrose 1994, 1996) based on such Newtonian models for life-times of superpositions turn out to be strikingly realistic. For instance, the life-time of superposition for a proton works out to be of the order of a few million years, whereas a water droplet – depending on its size – is expected to be able to maintain superposition only for a fraction of a second. Thus, the boundary near which the reduction time is of the order of seconds is precisely the phenomenological quantum-classical boundary of our corroborative experience.[5]

An important issue in any quantum measurement theory is the 'preferred basis problem' (cf. section 3 of Christian 1998). The difficulty is that, without some further criterion, one does not know which states from the general compendium of possibilities are to be regarded as the 'basic' (or 'stable' or 'stationary') states and which are to be regarded as essentially unstable 'superpositions of basic states' – the states which are to reduce into the basic ones. Penrose's suggestion is to regard – within Newtonian approximation – the stationary solutions of what he calls the Schrödinger–Newton equation as the basic states (Penrose 1998, Moroz et al. 1998, Tod and Moroz 1999). (I shall elaborate on this equation, which I have independently studied in Christian (1997), in the next section.)

### 14.3.4 A different measure of deviation from quantum mechanics

As an aside, let me propose in this subsection a slightly different measure for the lack of exact pointwise identification between the two spacetimes under consideration. In close analogy with the above assumption of stationarity, let us assume that there exists a displacement isometry in each of the two spacetimes, embodied in the Killing vector fields $x_2$ and $x_3$ respectively – i.e. let $£_{x_2} g_2^{\mu\nu} = 0 = £_{x_3} g_3^{\mu\nu}$, where $£_x$ denotes the Lie derivative with Killing vector fields $x_2^\alpha$ and $x_3^\alpha$ as the generators of the displacement symmetry. Further, as before, let us assume that at least some approximate pointwise identification between these two spacetimes is meaningful. As a visual aid, one may think of two nearly congruent co-ordinate grids, one assigned to each spacetime. Then, à la Penrose, I propose a measure of incongruence between these two spacetimes to be the dimensionless parameter $d^\sigma d_\sigma$, taking values between zero and unity, $0 \leq d^\sigma d_\sigma \leq 1$, with

$$d^\sigma := x_2^\alpha \nabla_\alpha x_3^\sigma - x_3^\alpha \nabla_\alpha x_2^\sigma. \tag{14.17}$$

As it stands, this quantity is mathematically ill-defined since the Killing vectors $x_2^\alpha$ and $x_3^\alpha$ describe the same displacement symmetry in two quite distinct spacetimes. However, if we reinterpret these two vectors as describing two slightly different symmetries in one and the same spacetime, then the vector field $d^\sigma$ is geometrically well-defined, and it is nothing but the commutator Killing vector field (Misner et al. 1973, p. 654) corresponding to the two linearly independent vectors $x_2^\alpha$ and $x_3^\alpha$. In other words, $d^\sigma$ then is simply a measure of incongruence between the two co-ordinates adapted to simultaneously describe symmetries corresponding to both $x_2^\alpha$ and $x_3^\alpha$ within this single spacetime. This measure can now be used to postulate a gravity-induced deviation from the orthodox quantum commutation relation for the position and momentum of the mass $M$:

$$[Q, P] = i\hbar \left\{ 1 - d^\sigma d_\sigma \right\}. \tag{14.18}$$

Clearly, when there is an exact pointwise correspondence between the two space-times – i.e. when the Killing vector fields $x_2^\alpha$ and $x_3^\alpha$ are strictly identified and $d^\sigma d_\sigma \equiv 0$, we recover the standard quantum mechanical commutation relation between the position and momentum of the mass. On the other hand, when – for a 'large enough' mass – the quantity $d^\sigma d_\sigma$ reaches order unity, the mass exhibits essentially classical behaviour. Thus, the parameter $d^\sigma d_\sigma$ provides a good measure of ill-definedness in the canonical commutation relation due to a Penrose-type incongruence, but now between the *displacement* symmetries of the two spacetimes.

### 14.3.5 Penrose's proposed experiment

Finally, let me end this section by describing a variant of a *realizable* experiment proposed by Penrose to corroborate the contended 'macroscopic' breakdown of quantum mechanics (1998). The present version of the experiment due to Hardy (1998) is – arguably – somewhat simpler to perform. There are many practical problems in both Penrose's original proposal and Hardy's more clever version of it (contamination due to the ubiquitous decoherence effects being the most

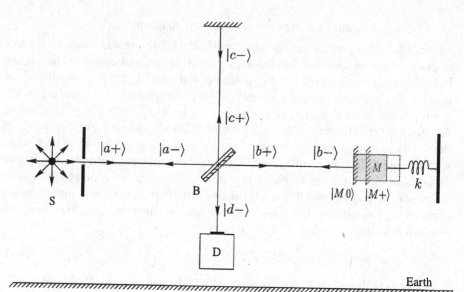

Fig. 14.3. Hardy's version of Penrose's proposed experiment. In an interferometric arrangement, a beam-splitter, B, is placed in the 'path' of an incident photon emanating form a source S. A horizontally movable mass $M$ is attached to the wall opposite to S by means of a restoring device with a spring constant $k$. There are two reflecting mirrors – one of them affixed on the mass, and the other one at the end of the vertical arm of the interferometer, both being at an exactly equal distance from the beam-splitter. The Earth provides a frame of reference, and the final destination of interest for the photon is the detector D.

intractable of all problems), but such practical problems will not concern us here (cf. Penrose 1998, Hardy 1998). Further, the use of a photon in the described experiment is for convenience only; in practice it may be replaced by any neutral particle, such as an ultracold atom of a suitable kind.

The basic experimental set-up is described in Fig. 14.3. The system consists of two objects: a 'photon' and a 'macroscopic' object of mass $M$, which in Penrose's version is a small Mössbauer crystal with about $10^{15}$ nuclei. The objective of the arrangement is to render the 'macroscopic' mass in a superposition of two macroscopically distinct positions, as in the state 14.8 above. The '+' or '−' sign in the photon states (such as $|a+\rangle$ or $|c-\rangle$), respectively, indicates a forward or backward motion along a given 'path'. For now, we simply look at the arrangement in a purely orthodox, quantum mechanical fashion. Then, the following transformations of the photon states due to a beam-splitter may be adopted from quantum optics:

$$|a\pm\rangle \longleftrightarrow \frac{1}{\sqrt{2}}\{|b\pm\rangle + |c\pm\rangle\} \qquad (14.19)$$

$$|d\pm\rangle \longleftrightarrow \frac{1}{\sqrt{2}}\{|c\pm\rangle - |b\pm\rangle\}, \qquad (14.20)$$

with inverse relations being

$$|c\pm\rangle \longleftrightarrow \frac{1}{\sqrt{2}} \{|a\pm\rangle + |d\pm\rangle\} \qquad (14.21)$$

$$|b\pm\rangle \longleftrightarrow \frac{1}{\sqrt{2}} \{|a\pm\rangle - |d\pm\rangle\}. \qquad (14.22)$$

If the initial state of the incident photon is taken to be $|a+\rangle$, and the initial (or unmoved) state of the mass $M$ is denoted by $|M0\rangle$, then the initial state of the *closed* composite system is the product state

$$|a+\rangle \otimes |M0\rangle. \qquad (14.23)$$

As the photon passes through the beam-splitter, this composite initial state evolves into

$$\frac{1}{\sqrt{2}} \{|b+\rangle + |c+\rangle\} \otimes |M0\rangle. \qquad (14.24)$$

Now, in the absence of the beam-splitter, if the photon happens to be in the horizontal 'path', then it would reflect off the mirror affixed on the mass, giving it a minute momentum in the '+' direction. On the other hand, if the photon is arranged to be in the vertical 'path', then it would simply reflect off the second mirror at the end of that path, without affecting the mass. The net result of these two alternatives in the presence of the beam-splitter, viewed quantum mechanically, is encoded in the state

$$\frac{1}{\sqrt{2}} \{|b-\rangle \otimes |M+\rangle + |c-\rangle \otimes |M0\rangle\}. \qquad (14.25)$$

Since each of the two options in this superposition would lead the photon back towards the beam-splitter, the composite state 14.25 – as the photon passes again through the beam-splitter – will evolve into

$$\frac{1}{2} \big[ \{|a-\rangle - |d-\rangle\} \otimes |M+\rangle + \{|a-\rangle + |d-\rangle\} \otimes |M0\rangle \big]. \qquad (14.26)$$

Now, our goal here is to generate a Penrose-type superposition of the mass $M$. Therefore, at this stage we isolate only those substates for which the photon could be detected by the detector D. Thus selected from 14.26, we obtain

$$\frac{1}{\sqrt{2}} \{|M0\rangle - |M+\rangle\} \qquad (14.27)$$

for the state of the mass, isolating it in the desired, spatially distinct, 'macroscopic' superposition. After some minute lapse of time, say $\Delta t$, the spring will bring the mass back to its original position with its momentum reversed, and thereby transform the above state into

$$\frac{1}{\sqrt{2}} \{|M0\rangle - |M-\rangle\}, \qquad (14.28)$$

where $|M-\rangle$ is the new state of $M$ with its momentum in the '−' direction (not shown in the figure).

At this precise moment, in order to bring about decisive statistics, we send *another* photon from S into the interferometer which, upon passing through the beam-splitter, will produce the product state

$$\tfrac{1}{2}\left\{|b+\rangle + |c+\rangle\right\} \otimes \left\{|M0\rangle - |M-\rangle\right\}. \tag{14.29}$$

Just as before, the four terms of this state will now evolve on their own, and, after a recoil of the photon from the two mirrors, the composite state will become

$$\tfrac{1}{2}\left[|b-\rangle \otimes |M+\rangle + |c-\rangle \otimes |M0\rangle - |b-\rangle \otimes |M0\rangle - |c-\rangle \otimes |M-\rangle\right]. \tag{14.30}$$

It is crucial to note here that, in the third term, the momentum of the mass has been reduced to zero by the interaction so that both the second and third terms have the same state $|M0\rangle$ for the mass. Finally, the evolution of the photon back through the beam-splitter will render the composite system to be in the state

$$\frac{1}{2\sqrt{2}}\left\{|a-\rangle - |d-\rangle\right\} \otimes |M+\rangle + \frac{1}{\sqrt{2}}|d-\rangle \otimes |M0\rangle$$

$$- \frac{1}{2\sqrt{2}}\left\{|a-\rangle + |d-\rangle\right\} \otimes |M-\rangle. \tag{14.31}$$

Thus, quantum mechanics predicts that the probability of detecting a photon in the detector D is 75%.

On the other hand, if the 'macroscopic' superposition of the mass such as 14.27 has undergone a Penrose-type process of state reduction, then the state of the mass just before the second photon is sent in would not be 14.28 but a *proper* mixture of $|M0\rangle$ and $|M-\rangle$. As a result, instead of 14.29, the overall disjoint state after the photon has passed through the beam-splitter would simply be

$$\frac{1}{\sqrt{2}}\left\{|b+\rangle + |c+\rangle\right\} \otimes |M0\rangle \quad \text{or} \quad \frac{1}{\sqrt{2}}\left\{|b+\rangle + |c+\rangle\right\} \otimes |M-\rangle, \tag{14.32}$$

without any quantum coherence between the two alternatives. As the photon is reflected off the two mirrors and passed again through the beam-splitter, these two 'classical' alternatives – instead of 14.31 – would evolve independently into the final disjoint state

$$\tfrac{1}{2}\left[\left\{|a-\rangle - |d-\rangle\right\} \otimes |M+\rangle + \left\{|a-\rangle + |d-\rangle\right\} \otimes |M0\rangle\right] \tag{14.33}$$

or

$$\tfrac{1}{2}\left[\left\{|a-\rangle - |d-\rangle\right\} \otimes |M0\rangle + \left\{|a-\rangle + |d-\rangle\right\} \otimes |M-\rangle\right]. \tag{14.34}$$

Consequently, if Penrose's proposal is on the right track, then, after the photon passes through the beam-splitter second time around, it would go to the detector only

50% of the time and not 75% of the time as quantum mechanics predicts. Practical difficulties aside (Penrose 1998, Hardy 1998), this is certainly a refutable proposition (especially because the commonly held belief concerning decoherence, cf. Kay 1998 – i.e. a belief that a strong coupling to the environment inevitably destroys the observability of quantum effects between macroscopically distinct states – is quite misplaced, as emphasised by Leggett 1998).

## 14.4 A closer look at Penrose's proposal within Newton–Cartan framework

My main goal in this section is, first, to put forward a delicate argument that demonstrates why Penrose's experiment – as it stands – is not adequate to corroborate the signatures of his proposed scheme of gravity-induced quantum state reduction, and then to discuss briefly a couple of decisive experiments which *would* be able to corroborate the putative breakdown of quantum mechanics along his line of reasoning.

### 14.4.1 An orthodox analysis within strictly Newtonian domain

In order to set the stage for my argument, let us first ask whether one can provide an orthodox quantum mechanical analysis of the physics underlying Penrose's proposed experiment. As it turns out, one can indeed provide such an orthodox treatment. Here, I shall outline one such treatment, which will not only direct us towards pinpointing where and for what reasons Penrose's approach differs from the orthodox approach, but will also allow us to explore more decisive experiments compared to the one he has proposed.

Clearly, to respond to Penrose's overall conceptual scheme in orthodox manner one would require a full-blown and consistent quantum theory of gravity, which, as we know, is not yet in sight (Rovelli 1998). If we concentrate, however, not on his overall conceptual scheme but simply on his proposed experiment, then we only require a *non-relativistic* quantum theory of gravity (recall from the last section that the formula 14.16 does not depend on the speed of light). And, fortunately, such a theory *does* exist. Recently, I have been able to demonstrate (Christian 1997) that the covariantly described Newtonian gravity – the so-called Newton–Cartan gravity which duly respects Einstein's principle of equivalence – interacting with Galilean-relativistic matter (Schrödinger fields) exists as an *exactly soluble* system, both classically and quantum mechanically (cf. Fig. 14.1). The significance of the resulting manifestly covariant unitary quantum field theory of gravity lies in the fact that it is the Newton–Cartan theory of gravity, and not the original Newtonian theory of gravity, that is the true Galilean-relativistic limit form of Einstein's theory of gravity. In fact, an alternative, historically counterfactual but logically more appropriate, formulation of general relativity is simply Newton–Cartan theory of gravity 'plus' the light-cone structure of the special theory of relativity. Newton's original theory in such a 'generally-covariant' Newton–Cartan framework emerges in an adscititiously chosen local inertial frame (modulo a crucially important additional restriction on the curvature tensor, as we shall see).

To begin the analysis, let us first look at the classical Newton–Cartan theory. (For further details and extensive references, consult section II of Christian 1997.) Cartan's spacetime reformulation of the classical Newtonian theory of gravity can be motivated in exact analogy with Einstein's theory of gravity. The analogy works because the universal equality of the inertial and the passive gravitational masses is independent of the relativization of time, and hence is equally valid at the Galilean-relativistic level. As a result, it is possible to parallel Einstein's theory and reconstrue the trajectories of (only) gravitationally affected particles as geodesics of a unique, 'non-flat' connection $\Gamma$ satisfying

$$\frac{d^2x^i}{dt^2} + \Gamma^i_{jk}\frac{dx^j}{dt}\frac{dx^k}{dt} = 0 \tag{14.35}$$

in a co-ordinate basis, such that

$$\Gamma^\mu_{\nu\lambda} \equiv \overset{v}{\Gamma}{}^\mu_{\nu\lambda} + \overset{v}{\Theta}{}^\mu_{\nu\lambda} := \overset{v}{\Gamma}{}^\mu_{\nu\lambda} + h^{\mu\alpha}\overset{v}{\nabla}_\alpha\overset{v}{\Phi}t_{\nu\lambda}, \tag{14.36}$$

with $\overset{v}{\Phi}$ representing the Newtonian gravitational potential relative to the freely falling observer field $v$, $\overset{v}{\Gamma}{}^\mu_{\nu\lambda}$ representing the coefficients of the corresponding 'flat' connection (i.e. one whose coefficients can be made to vanish in a suitably chosen linear co-ordinate system), and $\overset{v}{\Theta}{}^\mu_{\nu\lambda} := h^{\mu\alpha}\overset{v}{\nabla}_\alpha\overset{v}{\Phi}t_{\nu\lambda}$ representing the traceless gravitational field tensor associated with the Newtonian potential. Here $h^{\mu\nu}$ and $t_{\mu\nu}$, respectively, are the degenerate and mutually orthogonal spatial and temporal metrics with signatures $(0 + + +)$ and $(+ 0\ 0\ 0)$, representing the immutable chronogeometrical structure of the Newton–Cartan spacetime. They may be viewed as the '$c \to \infty$' limits of the Lorentzian metric: $h^{\mu\nu} = \lim_{c\to\infty}(g^{\mu\nu}/c^2)$ and $t_{\mu\nu} = \lim_{c\to\infty} g_{\mu\nu}$. The conceptual superiority of this geometrization of Newtonian gravity is reflected in the trading of the two 'gauge-dependent' quantities $\overset{v}{\Gamma}$ and $\overset{v}{\Theta}$ in favour of their gauge-independent sum $\Gamma$. Physically, it is the 'curved' connection $\Gamma$ rather than any 'flat' connection $\overset{v}{\Gamma}$ that can be determined by local experiments. Neither the potential $\overset{v}{\Phi}$ nor the 'flat' connection $\Gamma$ has an independent existence; they exist only relative to an arbitrary choice of a local inertial frame. It is worth noting that, unlike in both special and general theories of relativity, where the chronogeometrical structure of spacetime *uniquely* determines its inertio-gravitational structure, in Newton–Cartan theory these two structures are independently specified, subject only to the compatibility conditions $\nabla_\alpha h^{\beta\gamma} = 0$ and $\nabla_\alpha t_{\beta\gamma} = 0$. In fact, the connection $\Gamma$, as a solution of these compatibility conditions, is not unique unless a symmetry such as $R^\alpha{}_{\beta\cdot\delta}{}^\gamma = R^\gamma{}_{\delta\cdot\beta}{}^\alpha$ of the curvature tensor – capturing the 'curl-freeness' of the Newtonian gravitational filed – is assumed (here the indices are raised by the degenerate spatial metric $h^{\mu\nu}$). Further, although the two metric fields are immutable or non-dynamical in the sense that their Lie derivatives vanish identically,

$$\pounds_x t_{\mu\nu} \equiv 0 \quad \text{and} \quad \pounds_x h^{\mu\nu} \equiv 0, \tag{14.37}$$

the connection field remains dynamical, $\pounds_x\Gamma^\gamma_{\alpha\beta} \neq 0$, since it is determined by the evolving distributions of matter. The generators $x = (t, x^a)$ of the 'isometry' group defined by the conditions 14.37, represented in an arbitrary reference frame, take the form (cf. eqn. 14.10)

$$t' = t + constant \qquad (14.38)$$

$$\mathbf{x}'^a = O_b^a(t)\mathbf{x}^b + c^a(t) \quad (a, b = 1, 2, 3), \qquad (14.39)$$

where $O_b^a(t) \in SO(3)$ forms an orthonormal rotation matrix for each value of $t$ (with Einstein's summation convention for like indices), and $\mathbf{c}(t) \in \Re^3$ is an arbitrary *time-dependent* vector function. Physically, these transformations connect different observers in arbitrary (accelerating and rotating) relative motion.

With these physical motivations, the complete geometric set of gravitational field equations of the classical Newton–Cartan theory can be written as:

$$h^{\alpha\beta}t_{\beta\gamma} = 0, \quad \nabla_\alpha h^{\beta\gamma} = 0, \quad \nabla_\alpha t_{\beta\gamma} = 0, \quad \partial_{[\alpha}t_{\beta]\gamma} = 0, \qquad (14.40)$$

$$R^\alpha{}_{\beta \cdot \delta}{}^\gamma = R^\gamma{}_{\delta \cdot \beta}{}^\alpha, \qquad (14.41)$$

and

$$R_{\mu\nu} + \Lambda t_{\mu\nu} = 4\pi G M_{\mu\nu}, \qquad (14.42)$$

where the first four equations specify the degenerate 'metric' structure and a set of torsion-free connections on the spacetime manifold $\mathcal{M}$, the fifth one picks out the Newton–Cartan connection from this set of generic possibilities, and the last one, with mass-momentum tensor $M_{\mu\nu} := \lim_{c\to\infty} T_{\mu\nu}$, relates spacetime geometry to matter in analogy with Einstein's field equations. Alternatively, one can recover this entire set of field equations from Einstein's theory in the '$c \to \infty$' limit (Künzle 1976, Ehlers 1981, 1986, 1991).

The only other field equation that is compatible with the structure 14.42 (Dixon 1975), but which *cannot* be recovered in the '$c \to \infty$' limit of Einstein's field equations, is

$$R^{\alpha\lambda}{}_{\cdot \gamma\delta} = 0 \qquad (14.43)$$

(where, again, the index is raised by the degenerate spatial metric $h^{\lambda\sigma}$). It asserts the existence of absolute rotation in accordance with Newton's famous 'bucket experiment', and turns out to be of central importance for my argument against Penrose's experiment (cf. Section 14.4.2). Without this extra field equation, however, there does not even exist a classical Lagrangian density for the Newton–Cartan system, let alone a Hamiltonian density or an unambiguous phase space. Despite many diligent attempts to construct a consistent Lagrangian density, the goal remains largely elusive, thanks to the intractable geometrical obstruction resulting from the degenerate 'metric' structure of the Newton–Cartan spacetime.

If, however, we take the condition 14.43 as an extraneously imposed but necessary field equation on the Newton–Cartan structure, then, after some tedious manipulations (cf. Christian 1997), we can obtain an unequivocal constraint-free phase space for the classical Newton–Cartan system coupled with Galilean-relativistic matter (Schrödinger fields). What is more, the restriction 14.43 also permits the existence of a family of local inertial frames in the Newton–Cartan structure (cf. Section 14.4.2). Given such a local frame the inertial and gravitational parts of the Newton–Cartan connection-field can be unambiguously separated, as in the eqn. 14.36 above, and a non-rotating linear co-ordinate system may be introduced. Then, with some gauge choices appropriate for the Earth-nucleus system of Penrose's experiment (recall that Penrose's experiment involves displacements of some $10^{15}$ nuclei), the relevant action functional (i.e. eqn. 4.3 of Christian 1997) takes the simplified form

$$
\mathcal{I} = \int dt \int d\mathbf{x} \left[ \frac{1}{8\pi G} \Phi \nabla^2 \Phi + \frac{\hbar^2}{2m} \delta^{ab} \partial_a \psi \, \partial_b \bar{\psi} \right.
$$
$$
\left. + i \frac{\hbar}{2} (\psi \, \partial_t \bar{\psi} - \bar{\psi} \, \partial_t \psi) + m \bar{\psi} \psi \Phi \right],
\tag{14.44}
$$

where $\psi = \psi(\mathbf{x}_{CM}, \mathbf{x})$ is a complex Schrödinger field representing the composite Earth-nucleus system, $m$ is the reduced mass for the system, all spatial derivatives are with respect to the relative co-ordinate $\mathbf{x}$, and from now on the explicit reference to observer $\nu$ on the top of the scalar Newtonian potential $\Phi(\mathbf{x})$ is omitted. Evidently, the convenient inertial frame I have chosen here is the $CM$-frame in which kinetic energy of the centre-of-mass vanishes identically. In addition, one may also choose $\mathbf{x}_{CM} \equiv 0$ without loss of generality so that $\psi = \psi(\mathbf{x})$. Since the dynamics of the Earth-nucleus system is entirely encapsulated in the function $\psi(\mathbf{x})$, it is sufficient to focus only on this $\mathbf{x}$-dependence of $\psi$ and ignore the free motion of the centre-of-mass. Needless to say that, since $m_{Earth} \gg m_{nucleus}$, to an excellent approximation $m = m_{nucleus}$, and effectively the $CM$-frame *is* the laboratory-frame located at the centre of the Earth.

Extremization of the functional 14.44 with respect to variations of $\Phi(\mathbf{x})$ immediately yields the Newton–Poisson equation

$$
\nabla^2 \Phi(\mathbf{x}) = -4\pi G m \bar{\psi}(\mathbf{x}) \psi(\mathbf{x}),
\tag{14.45}
$$

which describes the manner in which a quantum mechanically treated particle bearing mass $m$ gives rise to a 'quantized' gravitational potential $\Phi(\mathbf{x})$, thereby capturing the essence of Newtonian quantum gravity. On the other hand, extremization of the action with respect to variations of the matter field $\bar{\psi}(\mathbf{x})$ leads to the familiar Schrödinger equation for a quantum particle of mass $m$ in the presence of an external field $\Phi(\mathbf{x})$:

$$
i\hbar \frac{\partial}{\partial t} \psi(\mathbf{x}, t) = \left[ -\frac{\hbar^2}{2m} \nabla^2 - m\Phi(\mathbf{x}) \right] \psi(\mathbf{x}, t).
\tag{14.46}
$$

The last two equations may be reinterpreted as describing the evolution of a *single* particle of mass $m$ interacting with its own Newtonian gravitational field. Then these coupled equations constitute a *non-linear* system, which can be easily seen as such by first solving eqn. 14.45 for the potential $\Phi(\mathbf{x})$, giving

$$
\Phi(\mathbf{x}) = Gm \int d\mathbf{x}' \frac{\bar{\psi}(\mathbf{x}') \, \psi(\mathbf{x}')}{|\mathbf{x} - \mathbf{x}'|},
\tag{14.47}
$$

and then – by substituting this solution into eqn. 14.46 – obtaining the integro-differential equation (cf. eqn. 5.18 of Christian 1997)

$$ i\hbar \frac{\partial}{\partial t}\psi(\mathbf{x}, t) = -\frac{\hbar^2}{2m}\nabla^2\psi(\mathbf{x}, t) - Gm^2 \int d\mathbf{x}' \frac{\bar{\psi}(\mathbf{x}', t)\,\psi(\mathbf{x}', t)}{|\mathbf{x} - \mathbf{x}'|}\psi(\mathbf{x}, t). $$

$$(14.48)$$

As alluded to at the end of Section 14.3.3, Penrose has christened this equation the 'Schrödinger–Newton equation', and regards the stationary solutions of it as the 'basic states' into which the quantum superpositions must reduce, within this Newtonian approximation of the full 'quantum gravity'.

As it stands, this equation is evidently a non-linear equation describing a self-interacting quantum particle. However, if we promote $\psi$ to a 'second-quantized' field operator $\hat{\psi}$ satisfying (Christian 1997)

$$ \left[\hat{\psi}(\mathbf{x}), \hat{\psi}^\dagger(\mathbf{x}')\right] = \hat{\mathbb{1}}\delta(\mathbf{x} - \mathbf{x}') \tag{14.49} $$

at equal times, then this equation corresponds to a *linear* system of many identical (bosonic) particles bearing mass $m$ in the Heisenberg picture, with $\hat{\psi}$ acting as an annihilation operator in the corresponding Fock space. In particular, the properly normal-ordered Hamiltonian operator for the system now reads

$$ \hat{H} = \hat{H}_O + \hat{H}_I, \tag{14.50} $$

with

$$ \hat{H}_O := \int d\mathbf{x}\,\hat{\psi}^\dagger(\mathbf{x})\left[-\frac{\hbar^2}{2m}\nabla^2\right]\hat{\psi}(\mathbf{x}) \tag{14.51} $$

and

$$ \hat{H}_I := -\frac{1}{2}Gm^2 \int d\mathbf{x} \int d\mathbf{x}' \frac{\hat{\psi}^\dagger(\mathbf{x}')\,\hat{\psi}^\dagger(\mathbf{x})\,\hat{\psi}(\mathbf{x})\,\hat{\psi}(\mathbf{x}')}{|\mathbf{x} - \mathbf{x}'|}, \tag{14.52} $$

which upon substitution into the Heisenberg equation of motion

$$ i\hbar \frac{\partial}{\partial t}\hat{\psi}(\mathbf{x}, t) = \left[\hat{\psi}(\mathbf{x}, t), \hat{H}\right] \tag{14.53} $$

yields an operator equation corresponding to eqn. 14.48. It is easy to show (Schweber 1961, p. 144) that the action of the Hamiltonian operator $\hat{H}$ on a multi-particle state $|\Psi\rangle$ is given by $\langle \mathbf{x}_1\mathbf{x}_2 \cdots \mathbf{x}_n|\hat{H}|\Psi\rangle =$

$$ \left[ -\frac{\hbar^2}{2m}\sum_{i=1}^{n}\nabla_i^2 - \frac{1}{2}Gm^2 \sum_{\substack{i,j=1 \\ i\neq j}}^{n} \frac{1}{|\mathbf{x}_i - \mathbf{x}_j|} \right] \langle \mathbf{x}_1\mathbf{x}_2 \cdots \mathbf{x}_n|\Psi\rangle, \tag{14.54} $$

which is indeed the correct action of the multi-particle Hamiltonian with gravitational pair-interactions. Put differently, since the Hamiltonian equation (eqn. 14.52) annihilates any single-particle state, the particles no longer gravitationally self-interact. Thus, *in a local inertial frame*, the Newton–Cartan–Schrödinger system (Christian 1997) reduces, formally, to the very first quantum field theory constructed by Jordan and Klein (1927).

### 14.4.2 The inadequacy of Penrose's proposed experiment

As noted above, the orthodox analysis carried out in the previous subsection is contingent upon the extraneously imposed field equation

$$R^{\alpha\lambda} \cdot_{\gamma\delta} = 0, \qquad (14.55)$$

(eqn. 14.43), without which the existence of even a classical Lagrangian density for the Newton–Cartan system seems impossible (cf. Christian 1997: subsection II C, subsection IV A, footnote 6). More significantly for our purposes, unless this extra condition prohibiting rotational holonomy is imposed on the curvature tensor, it is not possible to recover the Newton–Poisson equation (eqn. 14.45),

$$\nabla^2 \Phi(\mathbf{x}) = -4\pi G\rho(\mathbf{x}), \qquad (14.56)$$

from the usual set of Newton–Cartan field equations (eqns. 14.40–14.42) (which are obtained in the '$c \to \infty$' limit of Einstein's theory) without any unphysical global assumption. Thus, eqn. 14.55 embodies an essential *discontinuity* in the '$c \to \infty$' limit between the gravitational theories of Einstein and Newton, and without it the Schrödinger–Newton equation (eqn. 14.46) is *not* meaningful.

Let us look at this state of affairs more closely (cf. Misner et al. 1973, pp. 294–5, Ehlers 1981, 1986, 1991, 1997). The only non-zero components of the connection-field corresponding to the set of field equations (eqns. 14.40–14.42) (and the co-ordinate transformations 14.39) are

$$\Gamma^a_{00} =: -g^a \quad \text{and} \quad \Gamma^b_{0a} = O^b_c \dot{O}^c_a := h^{bc} \varepsilon_{acd} \Omega^d. \qquad (14.57)$$

With respect to a co-ordinate system, the spatial vector fields $\mathbf{g}(\mathbf{x}, t)$ and $\mathbf{\Omega}(\mathbf{x}, t)$ play the part of gravitational acceleration and Coriolis angular velocity, respectively, and the field equations (eqns. 14.40–14.42) reduce to the set

$$\nabla \cdot \mathbf{\Omega} = 0, \qquad \nabla \times \mathbf{g} + 2\dot{\mathbf{\Omega}} = 0, \qquad (14.58)$$

$$\nabla \times \mathbf{\Omega} = 0, \qquad \nabla \cdot \mathbf{g} - 2\Omega^2 = 4\pi G\rho, \qquad (14.59)$$

where $\mathbf{g}$ and $\mathbf{\Omega}$ in general depend on both $\mathbf{x}$ and $t$ (and I have set $\Lambda = 0$ for simplicity). It is clear from this set that the recovery of the Newton–Poisson equation – and hence the reduction to the *strictly Newtonian* theory – is possible if and only if a co-ordinate system exists with respect to which $\mathbf{\Omega} = 0$ holds. This can be achieved if $\mathbf{\Omega}$ is spatially constant – i.e. depends on time only. And this is precisely what is ensured by the extra field equation (eqn. 14.55), which asserts that the parallel-transport of spacelike vectors is path-independent. Given this condition, the co-ordinate system can be further specialized to a non-rotating one, with $\Gamma^b_{0a} = 0$, and the connection coefficients can be decomposed as in eqn. 14.36, with $\mathbf{g} := -\nabla\Phi$.

This entire procedure, of course, may be sidestepped if we admit only asymptotically flat spacetimes. With such a global boundary condition, the restriction 14.55 on the curvature tensor becomes redundant (Künzle 1972, Dixon 1975). However, physical evidence clearly suggests that we are *not* living in an 'island universe' (cf. Penrose 1996, pp. 593–4) – i.e. universe is not 'an island of matter surrounded by emptiness' (Misner et al. 1973, p. 295). Therefore, a better procedure of recovering the Newtonian theory from Einstein's theory is not to impose such a strong and unphysical global boundary condition, but, instead, to require that only the weaker condition on the curvature tensor, 14.55, is satisfied. For, this weaker condition is quite sufficient to recover the usual version of Newton's theory with gravitation as a force field on a flat, non-dynamical, *a priori* spacetime structure, and guarantees existence of a class of inertial co-ordinate systems *not* rotating with respect to each other; i.e. the condition suppresses time-dependence of the rotation matrix $O^a_b(t)$ (as a result of the restriction $\Gamma^b_{0a} = 0$), and reduces the transformation law 14.39 to

$$t \longrightarrow t' = t + constant \tag{14.60}$$

$$x^a \longrightarrow x'^a = O^a_b x^b + c^a(t) \quad (a, b = 1, 2, 3). \tag{14.61}$$

Note that, unlike the asymptotic-flatness condition $\lim_{|\mathbf{x}|\to\infty} \Phi(\mathbf{x}) = 0$, the weaker condition 14.55 *does not* suppress the arbitrary time-dependence of the function $c^a(t)$ – i.e. 14.55 does not reduce $c^a(t)$ to $v^a \times t$ as in the Galilean transformation 14.10 above. Consequently, the gravitational potential $\Phi$ in the resultant Newtonian theory remains non-unique (Misner et al. 1973, p. 295), and, under the diffeomorphism corresponding to the transformation 14.61, transforms (actively) as

$$\Phi(\mathbf{x}) \longrightarrow \Phi'(\mathbf{x}) = \Phi(\mathbf{x}) - \ddot{\mathbf{c}} \cdot \mathbf{x}. \tag{14.62}$$

Let us now go back to Penrose's hypothesis on the mechanism underlying quantum state reduction discussed in Section 14.3.2 above, and retrace the steps of that subsection within the present strictly Newtonian scenario. As before, although here $h^{\mu\nu}$ and $t_{\mu\nu}$ would serve as 'individuating fields' (cf. Section 14.2) allowing pointwise identification between two different spacetimes, due to the transformation law 14.61 there would appear to be an ambiguity in the notion of time-translation operator analogous to eqn. 14.11,

$$\frac{\partial}{\partial x'^a} = \frac{\partial}{\partial x^a} \quad \text{but} \quad \frac{\partial}{\partial t} \longrightarrow \frac{\partial}{\partial t'} = \frac{\partial}{\partial t} - \dot{c}^a(t)\frac{\partial}{\partial x^a}, \tag{14.63}$$

when superpositions involving two such different spacetimes are considered. However, I submit that this 'ambiguity' in the present – essentially Newtonian – case is entirely innocuous. For, in the strictly Newtonian theory being discussed here, where a 'spacetime' now is simply a flat structure 'plus' a gravitational potential $\Phi(\mathbf{x})$ as in eqn. 14.36, one must consider 14.63 *together* with the transformation 14.62. But the Schrödinger–Newton equation (eqn. 14.46) – which is the appropriate equation

Joy Christian

here – happens to be covariant under such a concurrent transformation,[6] and retains the original form

$$ i\hbar \frac{\partial}{\partial t'} \psi'(\mathbf{x}, t) = \left[ -\frac{\hbar^2}{2m} \nabla'^2 - m\Phi'(\mathbf{x}) \right] \psi'(\mathbf{x}, t) \tag{14.64} $$

(Rosen 1972, cf. also Christian 1997) with the following (active) transformation of its solution (if it exists):

$$ \psi(\mathbf{x}, t) \longrightarrow \psi'(\mathbf{x}, t) = e^{if(\mathbf{x}, t)} \, \psi(\mathbf{x}, t). \tag{14.65} $$

What is more (cf. Kuchař 1980, 1991), due to the inverse relation between transformations on the function space and transformations 14.61 on co-ordinates, eqn. 14.65 implies

$$ \psi'(\mathbf{x}', t') = \psi(\mathbf{x}, t). \tag{14.66} $$

That is to say, the new solution of the Schrödinger–Newton equation expressed in the new co-ordinate system is *exactly* equal to the old solution expressed in the old co-ordinate system. The new value of the $\psi$-field, as measured at the transformed spacetime point, is numerically the same as its old value measured at the original spacetime point. Now consider a superposition involving two entirely different strictly-Newtonian 'spacetimes' in the co-ordinate representation analogous to the 'superposition' 14.14 discussed in Section 14.3,

$$ \langle \mathbf{x} | \mathcal{X}(t) \rangle = \lambda_2 \Psi_2(\mathbf{x}, t) + \lambda_3 \Psi'_3(\mathbf{x}', t'), \tag{14.67} $$

where unprimed co-ordinates correspond to one spacetime and the primed co-ordinates to another.[7] *Prima facie*, in accordance with the reasonings of Section 14.3, such a superposition should be as unstable as eqn. 14.14. However, in the present strictly Newtonian case, thanks to the relationship 14.66, the physical state represented by 14.67 is equivalent to the superposed state

$$ \langle \mathbf{x} | \mathcal{X}(t) \rangle = \lambda_2 \Psi_2(\mathbf{x}, t) + \lambda_3 \Psi_3(\mathbf{x}, t). \tag{14.68} $$

And there is, of course, nothing unstable about such a superposition in this strictly Newtonian domain. Consequently, for such a superposition, $\Delta E \equiv 0$, and hence its life-time $\tau \sim \infty$ (cf. eqn. 14.16).

Thus, as long as restriction 14.55 on the curvature tensor is satisfied – i.e. as long as it is possible to choose a co-ordinate system with respect to which $\Gamma^b_{0a} = 0$ holds for each spacetime, the Penrose-type instability in quantum superpositions is non-existent, a conclusion not inconsistent with the results of Christian (1997). Put differently, given $\Gamma^b_{0a} = 0$, the Penrose-type obstruction to stability of superpositions is sufficiently mitigated to sustain stable quantum superpositions. In physical terms, since 14.55 postulates the existence of 'absolute rotation', the superposition 14.67 is perfectly Penrose-stable as long as there is no *relative* rotation involved between its two components. On the other hand, if there *is* a relative rotation between the two components of 14.67 so that $\Gamma^b_{0a} = 0$ does not hold for both spacetimes, then it is not possible to analyse the physical system in terms of the strictly Newtonian limit of Einstein's theory, and, as a result, the 'superposition' 14.67 would be Penrose-unstable. Unfortunately, neither in Penrose's original experiment (1998), nor in the version discussed in Section 14.3.5, is there any relative rotation between two components of the superposed mass distributions. In other words, in both cases $\Gamma^b_{0a} = 0$ holds everywhere, and hence no Penrose-type instability should be expected in the outcome of these experiments. (Incidentally, among the known solutions of Einstein's field equations, the only known solution which has a genuinely Newton–Cartan limit – i.e. in which $\Omega$ is not spatially constant, entailing that it cannot be reduced to the strictly Newtonian case with $\Gamma^b_{0a} = 0$ – is the NUT spacetime (Ehlers 1997).)

### 14.4.3 More adequate experiments involving relative rotations

It is clear from the discussion above that, in order to detect Penrose-type instability in superpositions, what we must look for is a physical system for which the components $\Gamma^b_{0a}$ of the connection field, *in addition* to the components $\Gamma^a_{00}$, are meaningfully non-zero. Most conveniently, there exists extensive theoretical and experimental work on just the kind of physical systems we require.

The first among these systems involves 'macroscopic' superpositions of two screening currents in r.f.-SQUID rings, first proposed by Leggett almost two decades ago (Leggett 1980, 1984, 1998, Leggett and Garg 1985). An r.f.-SQUID ring consists of a loop of superconducting material interrupted by a thin Josephson tunnel junction. A persistent screening current may be generated around the loop in response to an externally applied magnetic flux, which obeys an equation of motion similar to that of a particle moving in a one dimensional double-well potential. The thus-generated current in the ring would be equal in magnitude in both wells, but opposite in direction. If dissipation in the junction and decoherence due to environment are negligible, then the orthodox quantum analysis predicts coherent oscillations between the two distinct flux states, and, as a result, a coherent superposition between a large number of electrons flowing around the ring in opposite directions – clockwise or counterclockwise – is expected to exist, generating a physical situation analogous to the one in eqns. 14.14 or 14.67 above. Most importantly for our purposes, since there would be relative rotation involved between the currents in the two possible states, owing to the Lense–Thirring fields (Lense and Thirring 1918, Ciufolini et al. 1998) of these currents, the connection components $\Gamma^b_{0a}$, in addition to the components $\Gamma^a_{00}$, will be non-zero. And this will unambiguously give rise to a Penrose-type instability at an appropriate mass scale – say roughly around $10^{21}$ electrons. The number of electrons in the SQUID ring in an actual experiment currently under scrutiny in Italy (Castellano et al. 1996) is only of the order of $10^{15}$, but there is no reason for a *theoretical* upperbound on this number.

It should be noted that Penrose himself has briefly considered the possibility of a Leggett-type experiment to test his proposal (1994a, p. 343). Recently, Anandan (1998) has generalized Penrose's expression for $\Delta E$ to arbitrary connection fields (cf. footnote 5), which allows him to consider connection components other than $\Gamma^a_{00}$, in particular the components $\Gamma^b_{0a}$, and suggest a quantitative test of Penrose's ansatz via Leggett's experiment. What is novel in my own endorsement of this suggestion is the realization that Leggett-type experiments belong to a class of experiments – namely, the class involving $\Gamma_{0a}^{b} \neq 0$ – which is the only class available within the non-relativistic domain to unequivocally test Penrose's proposal.

A second more exotic physical system belonging to this class of experiments is a superposition of two vortex states of an ultracold Bose–Einstein Condensate (BEC), currently being studied by Cirac's group in Austria among others (Cirac et al. 1998, Dum et al. 1998, Butts and Rokhsar 1999). Again, owing to the Lense–Thirring fields of such a slowly whirling BEC (clockwise or counterclockwise), a Penrose-type instability can in principle be detected at an appropriate mass scale.

Finally, let me point out that the analysis of this section has opened up an exciting new possibility of empirically distinguishing Penrose's scheme from other *ad hoc* theories of gravity-induced state reduction (e.g. Ghirardi et al. 1990), with the locus of differentiation being the connection components $\Gamma_{0a}^{b}$. There is nothing intrinsic in such *ad hoc* theories that could stop a state from reducing when these connection components are zero – e.g. for the experiment described in Section 14.3.5 these theories predict reduction at an appropriate scale, whereas Penrose's scheme, for the reasons explicated above, *does not*.

### 14.5 Concluding remarks

It should be clear that my (partial) reservations against Penrose's proposed experiment has significance only in the *strictly* Newtonian domain. The classical world, of course, is not governed by Galilean-relativistic geometries, but by general-relativistic geometries. Accordingly, the true domain of the discussion under consideration must be the domain of full 'quantum gravity'. And, reflecting on this domain, I completely share Penrose's sentiments that 'our present picture of physical reality, particularly in relation to the nature of *time*, is due for a grand shake up' (1989, p. 371). (Similar sentiments, arrived at from quite a different direction, are also expressed by Shimony 1998.) The incompatibility between the fundamental principles of our two most basic theories – general relativity and quantum mechanics – is so severe that the unflinching orthodox view maintaining a *status quo* for quantum superpositions – including at such a special scale as the Planck scale – is truly baffling. As brought out in several of the chapters in this book, and elaborated on by myself in Section 14.3, the conflict between the two foundational theories has primarily to do with the axiomatically presupposed fixed causal structure underlying quantum dynamics, and the meaninglessness of such a fixed, non-dynamical, background causal structure in the general-relativistic picture of the world. The orthodox response to the conflict is to hold the fundamental principles of quantum mechanics absolutely sacrosanct at the price of severe compromises with those of Einstein's theory of gravity. For example, Banks, one of the pioneers of the currently popular M-theory program, has proclaimed (1998a): '... it seems quite clear that the fundamental rules of [M-theory] will seem outlandish to anyone with a background in ... general relativity. ... At the moment it appears that the only things which may remain unscathed are the fundamental principles of quantum mechanics'. In contrast, representing the view of a growing minority, Penrose has argued for a physically more meaningful *evenhanded* approach in which even the superposition principle is not held beyond reproach at all scales. It certainly requires an extraordinary leap of faith in quantum mechanics (a leap, to be precise, of some *seventeen* orders of magnitude in the mass scale!) to maintain that the Gordian knot – the conflict between our two most basic theories – can be cut without compromising the superposition principle in some manner. My own feeling, heightened by Penrose's tenacious line of reasoning, is that such a faith in quantum mechanics could turn out to be fundamentally misplaced, as so tellingly made plain by Leggett (1998):

> 'Imagine going back to the year 1895 and telling one's colleagues that classical mechanics would break down when the product of energy and time reached a value of order $10^{-34}$ joule seconds. They would no doubt respond gently but firmly that any such idea must be complete nonsense, since it is totally obvious that the structure of classical mechanics cannot tolerate any such characteristic scale!'

Indeed, one often comes across similar sentiments with regard to the beautiful internal coherence of quantum formalism. However, considering the extraordinary specialness of the Planck scale, I sincerely hope that our 'quantum' colleagues are far less complacent than their 'classical' counterparts while harbouring the 'dreams of a final theory'.

## Notes

I am truly grateful to my mentor Abner Shimony for his kind and generous financial support without which this work would not have been possible. I am also grateful to Roger Penrose for discussions on his ideas about gravity-induced state reduction, Lucien Hardy for kindly letting me use his own version of Penrose's proposed experiment before publication, Ashwin Srinivasan for his expert help in casting figures in TeX, and Jeeva Anandan, Julian Barbour, Harvey Brown, Roger Penrose, and Paul Tod for their comments on parts of the manuscript.

1. It should be noted that Penrose's views on 'becoming' are rather different from the stance I have taken here (1979, 1989, 1994a). In the rest of this essay I have tried to remain as faithful to his writings as possible. For recent discussions on 'becoming', other than the paper by Shimony cited above, see, for example, Zeilicovici (1986), Saunders (1996), and Magnon (1997).
2. It is worth emphasizing here that, as far as I can infer from his writings, Penrose is *not* committed to any of the existing proposals of nonlinear (e.g. Weinberg 1989a) and/or stochastic (e.g. Pearle 1993) modifications of quantum dynamics (neither am I for that matter). Such proposals have their own technical and/or interpretational problems, and are far from being completely satisfactory. As discussed in Section 14.1, Penrose's proposal, by contrast, is truly minimalist. Rather than prematurely proposing a *theory* of quantum state reduction, he simply puts forward a rationale why his heuristic scheme for the actualization potentialities must inevitably be a built-in feature of the sought-for 'final theory'.
3. It is worth noting here that the conventional 'quantum gravity' treatments are of no help in the conceptual issues under consideration. Indeed, as Penrose points out (1996, p. 589), the conventional attitude is to treat superpositions of different spacetimes in merely formal fashion, in terms of complex functions on the space of 3- or 4-geometries, with no pretence at conceptual investigation of the physics that takes place *within* such a formal superposition.
4. Within our non-relativistic domain, a more appropriate spacetime framework is of course that of Newton–Cartan theory (Christian 1997). This framework will be taken up in a later more specialized discussion, but for now, for conceptual clarity, I rather not deviate from the subtleties of the full general-relativistic picture of spacetime.
5. It should be noted that, independently of Penrose, Diósi has also proposed the same formula 14.16 for the collapse time (1989), but he arrives at it from a rather different

direction. Penrose's scheme should also be contrasted (Penrose 1996) with the 'semi-classical approaches' to 'quantum gravity' (e.g. Kibble 1981), which are well-known to be inconsistent (Eppley and Hannah 1977, Wald 1984, pp. 382–3, Anandan 1994). Recently, Anandan (1998) has generalized Penrose's Newtonian expression for $\Delta E$ to a similar expression for an arbitrary superposition of relativistic, but weak, gravitational fields, obtained in the gravitational analogue of the Coulomb gauge in a linearized approximation applied to the Lorentzian metric tensor field (cf. Section 14.4.3 for further comments).

6. Better still: under simultaneous gauge transformations 14.62, 14.63, and 14.65, the Lagrangian density of the action 14.44 remains invariant except for a change in the spatial boundary term, which of course does not contribute to the Euler–Lagrange equations 14.45 and 14.46. Thus, the entire Schrödinger–Newton theory is unaffected by these transformations, implying that it is independent of a particular choice of reference frame represented by $\partial/\partial t$ out of the whole family given in 14.63. It should be noted, however, that here, as in any such demonstration of covariance, all variations $\delta\Phi$ of the Newtonian potential are assumed to vanish identically at the spatial boundary (and this is perhaps a contentious requirement in the present context).

7. Of course, since the Schrödinger–Newton equation is a non-linear equation, its more adequate (orthodox) quantum mechanical treatment is the one given by eqns. 14.49–14.54 of Section 14.4.1. My purpose here, however, is simply to parallel Penrose's argument of instability in quantum superpositions near the Planck mass.

# References

Aharonov, Y. and Albert, D. (1981), 'Can We Make Sense of the Measurement Process in Relativistic Quantum Mechanics?', *Physical Review D 24 2*: 359–70.

Aharonov, Y. and Bohm, D. (1959), 'Significance of Electromagnetic Potentials in Quantum Theory', *Physical Review 115*: 485–91.

Aharonov, Y. and Vaidman, L. (1993), 'Measurement of the Schrödinger Wave of a Single Particle', *Physics Letters A 178*: 38–42

Anandan, J. (1991), 'A Geometric Approach to Quantum Mechanics', *Foundations of Physics 21*: 1265–84.

Anandan, J. (1992), 'The Geometric Phase', *Nature 360*: 307–13.

Anandan, J. (1994), 'Interference of Geometries in Quantum Gravity', *General Relativity and Gravitation 26*: 125–33.

Anandan, J. (1996), 'Gravitational Phase Operator and Cosmic Strings', *Physical Review D 53*: 779–86.

Anandan, J. (1997), 'Classical and Quantum Physical Geometry', in *Potentiality, Entanglement and Passion-at-a-Distance: Quantum Mechanical Studies for Abner Shimony, Vol. 2*, R. S. Cohen, M. Horn, and J. Stachel (eds.). Dordrecht: Kluwer Academic, 31–52.

Anandan, J. (1998), 'Quantum Measurement Problem and the Gravitational Field', in *The Geometric Universe: Science, Geometry, and the Work of Roger Penrose*, S. A. Huggett, L. J. Mason, K. P. Tod, S. T. Tsou, and N. M. J. Woodhouse (eds.). Oxford, England: Oxford University Press, 357–68. Also available as preprint gr-qc/9808033.

Arnowitt, R., Deser, S., and Misner, C. W. (1962), 'The Dynamics of General Relativity', in *Gravitation: An Introduction to Current Research*, L. Witten (ed.). New York: Wiley.

Ashtekar, A. (1986), 'New Variables for Classical and Quantum Gravity', *Physical Review Letters 57*: 2244–7.

Ashtekar, A. (1987), 'New Hamiltonian Formulation of General Relativity', *Physical Review D 36*: 1587–602.

Ashtekar, A. (1995), 'Mathematical Problems of Non-Perturbative Quantum General Relativity', in *Gravitation and Quantizations*, B. Julia and J. Zinn-Justin (eds.). Amsterdam: Elsevier, 181–283.

339

Ashtekar, A. (1998), 'Geometric Issues in Quantum Gravity', in *The Geometric Universe: Science, Geometry, and the Work of Roger Penrose*, S. A. Huggett, L. J. Mason, K. P. Tod, S. T. Tsou, and N. M. J. Woodhouse (eds.). New York: Oxford University Press, 173–94.

Ashtekar, A. (1999), 'Quantum Mechanics of Geometry', preprint available as gr-qc/9901023.

Ashtekar, A. and Lewandowski, J. (1997), 'Quantum Theory of Gravity I: Area Operators', *Classical and Quantum Gravity 14*: A55–81.

Ashtekar, A. and Lewandowski, J. (1997a), 'Quantum Theory of Geometry II: Volume Operators', preprint available as gr-qc/9711031.

Ashtekar, A. and Stachel, J. (eds.) (1991), *Conceptual Problems of Quantum Gravity*. Boston: Birkhäuser.

Ashtekar, A. and Tate, R. (1994), 'An Algebraic Extension of Dirac Quantization: Examples', *Journal of Mathematical Physics 35*: 6434–70.

Ashtekar, A., Rovelli, C., and Smolin, L. (1992), 'Weaving a Classical Metric with Quantum Threads', *Physical Review Letters 69*: 237.

Atiyah, M. (1990), *The Geometry and Physics of Knots*. Cambridge, England: Cambridge University Press.

Baadhio, R. and Kauffman, L. (1993), *Quantum Topology*. Singapore: World Scientific Press.

Baez, J. (1996), 'Spin Networks in Nonperturbative Quantum Gravity', in *The Interface of Knots and Physics*, L. Kauffman (ed.). Providence: American Mathematical Society, 167–203.

Baez, J. (1997), 'An Introduction to *n*-Categories', in *7th Conference on Category Theory and Computer Science Lecture Notes in Computer Science 1290*, E. Moggi and G. Rosolini (eds.). Berlin: Springer Verlag, 1–33.

Baez, J. (1998), 'Spin Foam Models', *Classical and Quantum Gravity 15*: 1827–58.

Baez, J. and Dolan, J. (1995), 'Higher-Dimensional Algebra and Topological Quantum Field Theory', *Journal of Mathematical Physics 36*: 6073–105.

Baez, J. and Dolan, J. (1998), 'Categorification', in *Higher Category Theory*, E. Getzler and M. Kapranov (eds.). Providence: American Mathematical Society, 1–36.

Baez, J. and Munian, P. (1994), *Gauge Fields, Knots, and Gravity*. Singapore: World Scientific.

Banks, T. (1985), 'TCP, Quantum Gravity, the Cosmological Constant, and All That', *Nuclear Physics B 249*: 332.

Banks, T. (1998), 'Matrix Theory', *Nuclear Physics B – Proceedings Supplements 67*: 180–224.

Banks, T. (1998a), 'The State of Matrix Theory', *Nuclear Physics B – Proceedings Supplements 68*: 261–7.

Banks, T., Susskind, L., and Peskin, M. E. (1984), *Nuclear Physics B 244*: 125.

Barbour, J. (1982), 'Relational Concepts of Space and Time', *British Journal for the Philosophy of Science 33*: 251–74.

Barbour, J. (1989), *Absolute or Relative Motion? A Study from the Machian Point of View of the Discovery and the Structure of Dynamical Theories*. Cambridge: Cambridge University Press.

Barbour, J. (1994), 'The Timelessness of Quantum Gravity: I. The Evidence from the Classical Theory', *Classical and Quantum Gravity 11*: 2853–73.

Barbour, J. (1994a), 'The Timelessness of Quantum Gravity: II. The Appearance of Dynamics in Static Configurations', *Classical and Quantum Gravity 11*: 2875–97.

Barbour, J. (1999), *The End of Time: The Next Revolution in Physics*. London: Weidenfeld and Nicolson.

Barbour, J. (1999a), 'The Development of Machian Themes in the Twentieth Century', in *The Arguments of Time*, J. Butterfield (ed.). Oxford: Oxford University Press, 83–109.

Barbour, J. (in preparation), *Absolute or Relative Motion? Vol. 2. The Deep Structure of General Relativity*. New York: Oxford University Press.

Barbour, J. and Bertotti, B. (1982), 'Mach's Principle and the Structure of Dynamical Theories', *Proceedings of the Royal Society of London Series A382*: 295–306.

Barbour, J. and Pfister, H. (1995), *Mach's Principle: From Newton's Bucket to Quantum Gravity*. Boston: Birkhäuser.

Bardeen, J. M., Carter, B., and Hawking, S. W. (1973), *Communications in Mathematical Physics 31*: 161.

Barrett, J. (1995), 'Quantum Gravity as Topological Quantum Field Theory', *Journal of Mathematical Physics 36*: 6161–79.

Barrett, J. and Crane, L. (1998), 'Relativistic Spin Networks and Quantum Gravity', *Journal of Mathematical Physics 39*: 3296–302.

Bartnik, R. and Fodor, G. (1993), 'Proof of the Thin Sandwich Conjecture' *Physical Review D 48*: 3596–9.

Beig, R. (1994), 'The Classical Theory of Canonical General Relativity', in *Canonical Gravity: From Classical to Quantum*, J. Ehlers and H. Friedrich (eds.). New York: Springer-Verlag, 59–80.

Bekenstein, J. D. (1973), *Physical Review D 7*: 2333.

Bekenstein, J. D. (1974), 'The Quantum Mass Spectrum of a Kerr Black Hole', *Lettere a Nuovo Cimento 11*: 467–70.

Bekenstein, J. D. (1974a), *Physical Review D 9*: 3292.

Bekenstein, J. D. (1981), *Physical Review D 23*: 287.

Bell, J. S. (1976), 'How to Teach Special Relativity', *Progress in Scientific Culture, Vol. 1, No 2*. Reprinted in Bell (1987), 67–80.

Bell, J. S. (1987), *Speakable and Unspeakable in Quantum Mechanics*. Cambridge: Cambridge University Press.

Bell, J. S. (1990), 'Against Measurement', *Physics World 3*: 33–40.

Bell, J. S. (1992), 'George Francis FitzGerald', *Physics World 5*: 31–5. Based on a lecture given by Bell in 1989 at Trinity College, Dublin. Abridged by Denis Weaire.

Belot, G. (1998), 'Understanding Electromagnetism', *British Journal for the Philosophy of Science 49*: 531–55.

Belot, G. (1998a), 'Why General Relativity Does Need an Interpretation', *Philosophy of Science (Proceedings) 63*: S80–8.

Belot, G. (1999), 'Rehabilitating Relationalism', *International Studies in Philosophy of Science 13*: 35–52.

Belot, G. and Earman, J. (1999), 'From Metaphysics to Physics', in *From Physics to Philosophy*, J. Butterfield and C. Pagonis (eds.). New York: Cambridge University Press, 166–86.

Belot, G., Earman, J., and Ruetsche, L. (1999), 'The Hawking Information Loss Paradox: the Anatomy of Controversy', *British Journal for the Philosophy of Science 50*: 189–229.

Bialynicki-Birula, I. and Mycielski, J. (1976), *Annals of Physics 100*: 62.

Blaut, A. and Kowalski-Glikman, J. (1996), 'Quantum Potential Approach to Class of Cosmological Models', *Classical and Quantum Gravity 13*: 39–50.

Bohm, D. (1951), *Quantum Theory*. Prentice-Hall, Englewood Cliffs.

Bohm, D. (1952), 'A Suggested Interpretation of Quantum Theory in Terms of "Hidden" Variables, I and II', *Physical Review 85*: 166–93.

Bohm, D. and Hiley, B. (1994), *The Undivided Universe*. London: Routledge.

Bohr, N. and Rosenfeld, L. (1933), 'Zur frage der messbarkeit der elektromagnetischen feldgrossen', *Kgl. Danek Vidensk. Selsk. Math.-fys. Medd. 12*: 8. Reprinted as 'On the Question of the Measureability of Electromagnetic Field Quantities' in *Quantum Theory and Measurement*, Wheeler and Zurek (eds.). Princeton: Princeton University Press (1983).

Bondi, H. (1952), 'Relativity and Indeterminacy', *Nature 169*: 660.

Boulware, D. G. and Deser, S. (1975), 'Classical General Relativity Derived From Quantum Gravity', *Annals of Physics 89*: 193.

Brighouse, C. (1994), 'Spacetime and Holes', in *PSA 1994, Vol. 1*, D. Hull, M. Forbes, and R. Burian (eds.). East Lansing Michigan: Philosophy of Science Association, 117–25.

de Broglie, L. (1956), *Tentative d'Interpretation Causale et Nonlineaire de la Mechanique Ondulatoire*. Paris: Gauthier-Villars.

Brown, H. R. (1993), 'Correspondence, Invariance and Heuristics in the Emergence of Special Relativity', in *Correspondence, Invariance and Heuristics*, S. French and H. Kamminga (eds.). The Netherlands: Kluwer Academic Publishers, 227–60.

Brown, H. R. (1997), 'On the Role of Special Relativity in General Relativity', *International Studies in the Philosophy of Science 11*: 67–81.

Brown, H. R. (1999), 'The Origins of Length Contraction', manuscript.

Brown, H. R. and Maia, A. (1993), 'Light-Speed Constancy Versus Light-Speed Invariance in the Derivation of Relativistic Dynamics', *British Journal for the Philosophy of Science 44*: 381–407.

Brown, H. R. and Redhead, M. L. G. (1981), 'A Critique of the Disturbance Theory of Indeterminacy in Quantum Mechanics', *Foundations of Physics 11*: 1–20.

Brown, H. R. and Sypel R. (1995), 'On the Meaning of the Relativity Principle and Other Symmetries', *International Studies in the Philosophy of Science 9*: 235–53.

Brown, J. D. and York, J. W. (1989), 'Jacobi's Action and the Recovery of Time in General Relativity', *Physical Review D 40*: 3312.

Brown, R. (1992), 'Out of Line', *Royal Institution Proceedings 64*: 207–43.

Bub, J. (1997), *Interpreting the Quantum World*. Cambridge: Cambridge University Press.

Budden, T. (1997), 'A Star in the Minkowskian Sky: Anisotropic Special Relativity', *Studies in the History and Philosophy of Modern Physics 28B*: 325–61.

Butterfield, J. (1989), 'The Hole Truth', *British Journal for the Philosophy of Science 40*: 1–28.

Butterfield, J. (1995), 'Worlds, Minds and Quanta', *Aristotelian Society Supplementary Volume 69*: 113–58.

Butterfield, J. (1996), 'Whither the Minds?', *British Journal for the Philosophy of Science 47*: 200–21.

Butterfield, J. and Isham, C. J. (1998), 'A Topos Perspective on the Kochen-Specker Theorem: I. Quantum States as Generalised Valuations', *International Journal of Theoretical Physics 37*: 2669–733. Also available as quant-ph/980355.

Butterfield, J. and Isham, C. J. (1999), 'On the Emergence of Time in Quantum Gravity', in *The Arguments of Time*, J. Butterfield (ed.). Oxford: Oxford University Press, 111–68. Also available as gr-qc/9901024.

Butterfield, J. and Isham, C. J. (1999a), 'A Topos Perspective on the Kochen-Specker Theorem: II. Conceptual Aspects, and Classical Analogues', *International Journal of Theoretical Physics 38*: 827–59. Also available as quant-ph/9808067.

Butts, D. A. and Rokhsar, D. S. (1999), 'Predicted Signatures of Rotating Bose-Einstein Condensates', *Nature 397*: 327–9.

Callender, C. and Huggett, N. (2001), 'Why Quantize Gravity (or any other Field for that Matter)?', *Philosophy of Science* (Proceedings), forthcoming.

Callender, C. and Weingard, R. (1994), 'The Bohmian Model of Quantum Cosmology', *Philosophy of Science Association 1*: 218–27.

Callender, C. and Weingard, R. (1995), 'Bohmian Cosmology and the Quantum Smearing of the Initial Singularity', *Physics Letters A 208*: 59–61.

Callender, C. and Weingard, R. (1996), 'Time, Bohm's Theory, and Quantum Cosmology', *Philosophy of Science 63*: 470–4.

Callender, C. and Weingard, R. (2000), 'Topology Change and the Unity of Space', *Studies in the History and Philosophy of Modern Physics 31*: 227–46.

Carlip, S. (1998), *Quantum Gravity in 2 + 1 Dimensions*. Cambridge: Cambridge University Press.

Cartan, É. (1923), 'Sur les Variétés à Connexion Affine et la Théorie de la Relativité Generalisée', *Ann. École Norm. Sup. 40*: 325–412.

Cartan, É. (1924), 'Sur les Variétés à Connexion Affine et la Théorie de la Relativité Generalisée', *Ann. École Norm. Sup. 41*: 1.

Castellano, M. G., Leoni, R., Torrioli, G., Carelli, P., and Cosmelli, C. (1996), 'Development and Test of Josephson Devices for an Experiment of Macroscopic Quantum Coherence', *Il Nuovo Cimento D 19*: 1423–8.

Choquet-Bruhat, Y., DeWitt-Morrette, C., and Dillard-Bleick, M. (1982), *Analysis, Manifolds, and Physics*. Amsterdam: North-Holland.

Christian, J. (1994), 'On Definite Events in a Generally Covariant Quantum World', Oxford University preprint.

Christian, J. (1995), 'Definite Events in Newton-Cartan Quantum Gravity', Oxford University preprint.

Christian, J. (1996), 'The Plight of "I am" ', *Metaphysical Review 3*: 1–4. Also available as quant-ph/9702012.

Christian, J. (1997), 'Exactly Soluble Sector of Quantum Gravity', *Physical Review D 56*: 4844–77. Also available as gr-qc/9701013.

Christian, J. (1998), 'Why the Quantum Must Yield to Gravity', available as preprint gr-qc/9810078.

Christian, J. (1999), 'Potentiality, Entanglement and Passion-at-a-distance', *Studies in History and Philosophy of Modern Physics 30*(4): 561–7. Also available as quant-ph/9901008.

Christian, J. (1999a), 'Evenhanded Quantum Gravity vs. the World as a Hologram', in preparation.

Christodolou, D. (1970), *Physical Review Letters 25*: 1596.

Cirac, J. I., Lewenstein, M., Molmer, K., and Zoller, P. (1998), 'Quantum Superposition States of Bose-Einstein Condensates', *Physical Review A 57*: 1208–18.

Ciufolini, I. and Wheeler, J. A. (1995), *Gravitation and Inertia*. Princeton University Press, Princeton.

Ciufolini, I., Pavlis, E., Chieppa, F., Fernandes-Vieira, E., and Pérez-Mercader, J. (1998), 'Test of General Relativity and Measurement of the Lense-Thirring Effect with Two Earth Satellites', *Science 279*: 2100–3.

Connes, A. (1994), *Non Commutative Geometry*. New York: Academic Press.

Connes, A. and Rovelli, C. (1994), 'Von Neumann Algebra Automorphisms and Time Versus Thermodynamics Relation in General Covariant Quantum Theories', *Classical and Quantum Gravity 11*: 2899.

Cosgrove, R. (1996), 'Consistent Evolution and Different Time Slicings in Quantum Gravity', *Classical and Quantum Gravity 13*: 891–919.

Crane, L. (1991), '2d Physics and 3d Topology', *Communications in Mathematical Physics 135*: 615–40.

Daumer, M., Dürr, D., Goldstein, S., and Zanghì, N. (1997), 'Naive Realism About Operators', *Erkenntnis 45*: 379–97.

Daumer, M., Dürr, D., Goldstein, S., and Zanghì, N. (1999), 'On the Role of Operators in Quantum Theory', in preparation.

Davies, P., Fulling, S., and Unruh, W. G. (1976), *Physical Review D 13*: 2720.

Descartes, R. (1983 [1644]), *Principia Philosophiae*, Translated by V. R. Miller and R. P. Miller. Reidel: Dordrecht.

DeWitt, B. (1962), 'Definition of Commutators via the Uncertainty Principle', *Journal of Mathematical Physics 3*: 619–24.

DeWitt, B. S. (1967), 'Quantum Theory of Gravity. I. The Canonical Theory', *Physical Review 160*: 1113–48.

DeWitt, B. S. and Graham, R. D. (eds.) (1973), *The Many Worlds Interpretation of Quantum Mechanics*. Princeton: Princeton University Press.

Diósi, L. (1984), 'Gravitation and Quantum-Mechanical Localization of Macro-Objects', *Physics Letters A 105*: 199–202.

Diósi, L. (1987), 'A Universal Master Equation for the Gravitational Violation of Quantum Mechanics', *Physics Letters A 120*: 377–81.

Diósi, L. (1989), 'Models for Universal Reduction of Macroscopic Quantum Fluctuations', *Physical Review A 40*: 1165–74.

Dirac, P. A. M. (1930), *The Principles of Quantum Mechanics*. Oxford: Clarendon Press.

Dirac, P. A. M. (1964), *Lectures on Quantum Mechanics*. New York: Yeshiva University.

DiSalle, R. (1995), 'Spacetime Theory as Physical Geometry', *Erkenntnis 42*: 317–37.

Dixon, W. G. (1975), 'On the Uniqueness of the Newtonian Theory as a Geometric Theory of Gravitation', *Communications in Mathematical Physics 45*: 167–82.

Donoghue, J. (1998), 'Perturbative Dynamics of Quantum General Relativity', in *Proceedings of the Eighth Marcel Grossmann Conference on General Relativity*.

Duff, M. J. (1981), 'Inconsistency of Quantum Field Theory in a Curved Spacetime', in *Quantum Gravity 2: A Second Oxford Symposium*, C. J. Isham, R. Penrose, and D. Sciama (eds.). Oxford: Clarendon Press, 81–105.

Dum, R., Cirac, J. I., Lewenstein, M., and Zoller, P. (1998), 'Creation of Dark Solitons and Vortices in Bose-Einstein Condensates', *Physical Review Letters 80*: 2972–5.

Dürr, D., Goldstein, S., and Zanghì, N. (1992), 'Quantum Equilibrium and the Origin of Absolute Uncertainty', *Journal of Statistical Physics 67*: 843–907.

Dürr, D., Goldstein, S., and Zanghì, N. (1997), 'Bohmian Mechanics and the Meaning of the Wave Function', in *Experimental metaphysics: Quantum mechanical studies for Abner Shimony, Vol. 1*, R. S. Cohen, M. Horne, and J. Stachel (eds.). Boston Studies in the Philosophy of Science 193: Kluwer, 25–38.

Dziobek, O. (1888), *Die mathematischen Theorien der Planeten-Bewegungen*. Leipzig: Johann Ambrosius Barth. English translation: *Mathematical Theories of Planetary Motions*. New York: Dover (published around 1890).

Earman, J. (1989), *World Enough and Spacetime: Absolute vs. Relational Theories of Space and Time*. Cambridge: MIT Press.

Earman, J. (1995), *Bangs, Crunches, Whimpers and Shrieks*. Oxford: Oxford University Press.

Earman, J. (1995a), 'Recent Work on Time Travel', in *Time's Arrow Today*, S. Savitt (ed.). Cambridge: Cambridge University Press, 268–310.

Earman, J. and Norton, J. (1987), 'What Price Spacetime Substantivalism? The Hole Story', *British Journal for the Philosophy of Science 38*: 515–25.

Eddington, A. S. (1929), *The Nature of the Physical World*. Cambridge: Cambridge University Press.

Ehlers, J. (1973), 'Survey of General Relativity Theory', in *Proceedings of the Summer School on Relativity, Astrophysics and Cosmology*, W. Israel (ed.). Dordrecht, Netherlands: Reidel, 1–125.

Ehlers, J. (1981), 'Über den Newtonschen Grenzwert der Einsteinschen Gravitationstheorie', in *Grundlagenprobleme der modernen Physik.*, J. Nitsch, J. Pfarr, and E. W. Stachow (eds.). Mannheim: Bibliographisches Institut, 65–84.

Ehlers, J. (1986), 'On Limit Relations Between, and Approximative Explanations of, Physical Theories', in *Logic, Methodology and Philosophy of Science VII*, R. Barcan Marcus, G. J. W. Dorn, and P. Weingartner (eds.). Amsterdam: North-Holland, 387–403.

Ehlers, J. (1991), 'The Newtonian Limit of General Relativity', in *Classical Mechanics and Relativity: Relationship and Consistency*, G. Ferrarese (ed.). Naples: Bibliopolis, 95–106.

Ehlers, J. (1997), 'Examples of Newtonian Limits of Relativistic Spacetimes', *Classical and Quantum Gravity 14*: A119–26.

Ehrenberg, W. and Siday, R. E. (1949), 'The Refractive Index in Electron Optics and the Principles of Dynamics', *Proceedings of the Physical Society of London B62*: 8–21.

Einstein, A. (1905), 'Zur Elektrodynamik bewegter Körper', *Annalen der Physik 17*: 891–921. For English translation, see Einstein (1952).

Einstein, A. (1905a), 'Über einen die Erzeugung und Verwandlung des Lichtes betreffenden heuristichen Gesichtspunkt', *Annalen der Physik 17*: 132–48.

Einstein, A. (1914), 'Die Formale Grundlage der allgemeinen Relativitätstheorie', *Königlich Preussische Akademie der Wissenschaften (Berlin) Sitzungsberichte*: 1030–85.

Einstein, A. (1916), 'Die Grundlage der allgemeinen Relativitätstheorie', *Annalen der Physik 49*: 1030–85. For English translation, see Einstein (1952).

Einstein, A. (1918), 'Prinzipielles zur allgemeinen Relativitätstheorie', *Annalen der Physik 55*: 241–4.

Einstein A. (1918a), 'Comment on Weyl 1918', published at the end of Weyl (1918). Not included in the English translation.

Einstein, A. (1919), 'My Theory', *The Times, November 28*. London, p. 13. Reprinted as 'What Is the Theory of Relativity?' in Einstein (1982), 227–32.

Einstein, A. (1921), 'Geometrie und Erfahrung', *Erweiterte Fassung des Festvortrages gehalten an der preussischen Akademie*. Berlin: Springer. Translated by S. Bargmann as 'Geometry and Experience' in Einstein (1982), 232–46.

Einstein, A. (1928), 'À propos de 'La Déduction Relativiste' de M. Émile Meyerson', *Revue Philosophique 105*: 161–6.

Einstein, A. (1952), *The Principle of Relativity*. New York: Dover Publications, Inc.

Einstein, A. (1952a), Letter to M. von Laue, January 17, 1952. *Einstein Archive*, 16–168.

Einstein, A. (1961 [1917]), *Relativity. The Special and General Theory*, Translated by R. W. Lawson. New York: Three Rivers Press. Originally published as *Über die Spezielle und die Allgemeine Relativitäts-theorie, Gemeinverständlich* (Braunschwieg, Vieweg (Sammlung Vieweg, Heft 38)).

Einstein, A. (1969), 'Autobiographical Notes', in *Albert Einstein: Philosopher-Scientist, Vol. 1*, P. A. Schilpp (ed.). Illinois: Open Court, 1–94.

Einstein, A. (1982), *Ideas and Opinions*. New York: Crown Publishers, Inc.

Einstein, A. (1994), *Relativity, The Special and the General Theory*. London: Routledge.

Einstein, A., Podolsky, B., and Rosen, N. (1935), 'Can Quantum-Mechanical Description of Physical Reality be Considered Complete?', *Physical Review 47*: 777–80.

Ellis, J., Mavromatos, N. E., and Nanopoulos, D. V. (1999), 'Search for Quantum Gravity', preprint available as gr-qc/9905048.

Ellis, J., Mohanty, S., and Nanopoulos, D. V. (1989), 'Quantum Gravity and the Collapse of the Wavefunction', *Physics Letters B 221*: 113–19.

Eppley, K. and Hannah, E. (1977), 'The Necessity of Quantizing the Gravitational Field', *Foundations of Physics 7*: 51–68.

Everett, H. (1957), *Reviews of Modern Physics 29*: 454.

Faraday, M. (1991 [1859]), *Experimental Researches in Chemistry and Physics*. London: Taylor and Francis.

Feynman, R. (1957), a talk given at the Chapel Hill Conference, Chapel Hill, North Carolina, January 1957. [A report on the conference proceedings can be obtained from Wright Air Development Center, Air Research and Development Command, United States Air Force, Wright-Patterson Air Force Base, Ohio, USA: WADC Technical Report 57–216.]

Feynman, R. (1963), 'Lectures on Gravitation', *Acta Physica Polonica XXIV*: 697.

Feynman, R. P. (with Morinigo, F. B. and Wagner, W. G.) (1995), *Feynman Lectures on Gravitation*, B. Hatfield (ed.). Reading, Massachusetts: Addison-Wesley.

Field, H. (1985), 'Can We Dispense With Space-Time?', in *PSA 1984, Vol. 2*, P. Asquith and P. Kitcher (eds.). East Lansing Michigan: Philosophy of Science Association, 33–90.

Fine, A. (1986), *The Shaky Game: Einstein, Realism and the Quantum Theory*. Chicago: University of Chicago Press.

Finkelstein, D. R. (1996), *Quantum Relativity: A Synthesis of the Ideas of Einstein and Heisenberg*. New York: Springer-Verlag.

Fischer, A. and Moncrief, V. (1996), 'A Method of Reduction of Einstein's Equations of Evolution and a Natural Symplectic Structure on the Space of Gravitational Degrees of Freedom', *General Relativity and Gravitation 28*: 207–19.

Fivel, D. I. (1997), 'An Indication From the Magnitude of CP Violations That Gravitation is a Possible Cause of Wave-Function Collapse', preprint available as quant-ph/9710042.

Fradkin, E. S. and Tseytlin, A. A. (1981), 'Renormalizable Asymptotically Free Quantum Theory of Gravity', *Physics Letters 104B*: 377–81.

Freidel, L. and Krasnov, K. (1998), 'Spin Foam Models and the Classical Action Principle', preprint available as hep-th/9807092.

Frenkel, A. (1997), 'The Model of F. Károlyházy and the Desiderata of A. Shimony for a Modified Quantum Dynamics', in *Experimental Metaphysics: Quantum Mechanical Studies for Abner Shimony, Vol. 1*, R. S. Cohen, M. Horn, and J. Stachel (eds.). Dordrecht: Kluwer Academic, 39–59.

Friedman, M. (1983), *Foundations of Space-Time Theories*. Princeton: Princeton University Press.

Friedrichs, K. (1927), 'Eine invariante Formulierung des Newtonschen Gravitationsgesetzes und des Grenzüberganges vom Einsteinschen zum Newtonschen Gesetz', *Mathematische Annalen 98*: 566–75.

Fulling, S. A. (1989), *Aspects of Quantum Field Theory in Curved Space-Time*. Cambridge: Cambridge University Press.

Gell-Mann, M. and Hartle, J. B. (1993), *Physical Review D 47*: 3345.

Georgi, H. and Glashow, S. L. (1972), 'Unified Weak and Electromagnetic Interactions Without Neutral Currents', *Physical Review Letters 28*: 1494–7.

Geroch, R. (1972), 'Einstein Algebras', *Communications in Mathematical Physics 25*: 271–5.

Geroch, R. and Horowitz, G. (1979), 'Global Structure of Spacetimes', in *General Relativity: An Einstein Centenary Survey*, S. W. Hawking and W. Israel, (eds.). Cambridge: Cambridge University Press, 212–93.

Ghirardi, G. C., Grassi, R., and Rimini, A. (1990), 'Continuous Spontaneous Reduction Model Involving Gravity', *Physical Review A 42*: 1057–64.

Ghirardi, G. C., Rimini, A., and Weber, T. (1986), 'Unified Dynamics for Microscopic and Macroscopic Systems', *Physical Review D 34*: 470–91.

Goldberg, J. N., Lewandowski, J., and Stornaiolo, C. (1992), 'Degeneracy in Loop Variables', *Communications in Mathematical Physics 148*: 377–402.

Goldblatt, R. (1979), *Topoi, the Categorial Analysis of Logic*. New York: North-Holland.

Goldstein, S. (1998), 'Quantum Theory Without Observers—Part One', *Physics Today*, March 1998, 42–6.

Goldstein, S. (1998a), 'Quantum Theory Without Observers—Part Two', *Physics Today*, April 1998, 38–42.

Goldstein, S. and Teufel, S. (1999), in preparation.

Gotay, M., Isenberg, J., and Marsden, J. (1998), 'Momentum Maps and Classical Relativistic Fields: Part I: Covariant Field Theory', preprint available as physics/9801019.

Green, M. B. and Schwartz, J. H. (1984), 'Superstring Field Theory', *Nuclear Physics B 243*: 475–536.

Green, M. B., Schwarz, J. H., and Witten, E. (1987), *Superstring Theory, Vol. 1*. Cambridge: Cambridge University Press.

Griffiths, R. (1984), *Journal of Statistical Physics 36*: 219.

Gross, D. J. and Mende, P. F. (1988), 'String Theory Beyond the Planck Scale', *Nuclear Physics B 303*: 407–54.

Guichardet, A. (1984), 'On Rotation and Vibration Motions of Molecules', *Annales Institut Henri Poincaré 40*: 329–42.

Haag, R. (1990), 'Fundamental Irreversibility and the Concept of Events', *Communications in Mathematical Physics 132*: 245–51.

Haag, R. (1992), *Local Quantum Physics: Fields, Particles, Algebras*. Berlin: Springer-Verlag.

Hájiček, P. (1991), 'Comment on "Time in Quantum Gravity: An Hypothesis" ', *Physical Review D 44*: 1337–8.

Hájiček, P. (1994), 'Quantization of Systems with Constraints', in *Canonical Gravity: From Classical to Quantum*, J. Ehlers and H. Friedrich (eds.). New York: Springer-Verlag, 113–49.

Hájiček, P. and Isham, C. (1996), 'Perennials and the Group-Theoretic Quantization of a Parameterized Scalar Field on a Curved Background', *Journal of Mathematical Physics 37*: 3522–38.

Hájiček, P. and Isham, C. (1996a), 'The Symplectic Geometry of a Parameterized Scalar Field on a Curved Background', *Journal of Mathematical Physics 37*: 3505–21.

Hardy, L. (1998), 'An Experiment to Observe Gravitationally Induced Collapse', Oxford University preprint.

Hartle, J. B. (1995), 'Spacetime Quantum Mechanics and the Quantum Mechanics of Spacetime', in *Proceedings of the 1992 Les Houches School, Gravitation and Quantisation*, B. Julia and J. Zinn-Justin (eds.). Elsevier Science, 285–480.

Hartle, J. B. (1996), 'Time and Time Functions in Parameterized Non-Relativistic Quantum Mechanics', *Classical and Quantum Gravity 13*: 361–75.

Hartle, J. B. and Hawking, S. W. (1983), 'Wavefunction of the Universe', *Physical Review D 28*: 2960–75.

Hawking, S. W. (1974), *Nature 248*: 30.

Hawking, S. W. (1975), *Communications in Mathematical Physics 43*: 199.

Hawking, S. W. and Ellis, G. F. R. (1973), *The Large Scale Structure of Space-time*. Cambridge: Cambridge University Press.

Heath, T. L. (1897), *The Works of Archimedes*. Cambridge: Cambridge University Press.

Heisenberg, W. (1958), *Physics and Philosophy*. New York: Harper and Row.

Helling, R. and Nicolai, H. (1998), 'Supermembranes and (M)atrix Theory', preprint available as hep-th/9809103.

Henneaux, M. and Teitelboim, C. (1992), *Quantization of Gauge Systems*. Princeton: Princeton University Press.

Hojman, S. A., Kuchař, K., and Teitelboim, C. (1976), 'Geometrodynamics Regained', *Annals of Physics 96*: 88–135.

Holland, P. (1993), *The Quantum Theory of Motion*. Cambridge: Cambridge University Press.

't Hooft, G. (1993), 'Dimensional Reduction in Quantum Gravity', preprint available as gr-qc/9310026.

't Hooft, G. (1997), 'Distinguishing Causal Time from Minkowski Time and a Model for the Black Hole Quantum Eigenstates', preprint available as gr-qc/9711053.

Horowitz, G. (1997), 'Quantum States of Black Holes', preprint available as hep-th/9704072.

Horowitz, G. T. and Marolf, D. (1997), *Physical Review D 55*: 3654–63.

Horowitz, G. T. and Polchinski, J. (1997), *Physical Review D 55*: 6189.

Horowitz, G. T., Lykken, J., Rohm, R., and Strominger, A. (1986), 'A Purely Cubic Action for String Field Theory', NSF-ITP-86-61.

Huggett, N. (1999), *Space from Zeno to Einstein: Classic Readings with a Contemporary Commentary*. Cambridge, MA: MIT Press.

Isenberg, J. and Marsden, J. (1982), 'A Slice Theorem for the Space of Solutions of Einstein's Equations', *Physics Reports 89*: 179–222.

Isenberg, J. and Moncrief, V. (1996), 'A Set of Nonconstant Mean Curvature Solutions of the Einstein Constraint Equations on Closed Manifolds', *Classical and Quantum Gravity 13*: 1819–47.

Isham, C. J. (1990), 'An Introduction to General Topology and Quantum Topology', in *Physics, Geometry and Topology*, H. C. Lee (ed.). New York: Plenum Press, 129–90.

Isham, C. J. (1991), 'Conceptual and Geometrical Problems in Quantum Gravity', in *Recent Aspects of Quantum Fields*, H. Mitter and Gausterer (eds.). New York: Springer-Verlag, 123–229.

Isham, C. J. (1993), 'Canonical Quantum Gravity and the Problem of Time', in *Integrable Systems, Quantum Groups, and Quantum Field Theories*, L. A. Ibort and M. A. Rodriguez (eds.). Dordrecht: Kluwer, 157–288.

Isham, C. J. (1994), 'Prima Facie Questions in Quantum Gravity', in *Canonical Gravity: From Classical to Quantum (Lecture Notes in Physics 434)*, J. Ehlers and H. Friedrich (eds.). Berlin: Springer-Verlag, 1–21.

Isham, C. J. (1997), 'Structural Issues in Quantum Gravity', in *General Relativity and Gravitation: GR14*, E. Sorace, G. Longhi, L. Lusanna, and M. Francaviglia (eds.). Singapore: World Scientific, 167–209.

Isham, C. J. and Linden, N. (1994), 'Quantum Temporal Logic and Decoherence Functionals in the Histories Approach to Generalised Quantum Theory', *Journal of Mathematical Physics 35*: 5452–76.

Isham, C. J., Penrose, R., and Sciama, D. W. (1975), *Quantum Gravity: An Oxford Symposium*. Clarendon Press.

Isham, C. J., Penrose, R., and Sciama, D. W. (1981), *Quantum Gravity: A Second Oxford Symposium*. Clarendon Press.

Jacobson, T. (1991), *Physical Review D 44*: 1731.

Jacobson, T. (1995), 'Thermodynamics of Spacetime: The Einstein Equation of State', *Physical Review Letters 75*: 1260.

Jacobson, T. and Corley, S. (1996), *Physical Review D 54*: 1568.

Jauch, J. M. (1968), *Foundations of Quantum Mechanics*. Reading, Massachusetts: Addison-Wesley.

Jones, K. R. W. (1995), 'Newtonian Quantum Gravity', *Australian Journal of Physics 48*: 1055–81.

Jordan, P. and Klein, O. (1927), 'Zum Mehrkörperproblem der Quantentheorie', *Zeitschrift für Physik 45*: 751–65.

Jordan, T. (1969), 'Why $-i\nabla$ is the Momentum', *American Journal of Physics 43*: 1089–93.

Kaiser, D. (1992), 'More Roots of Complementarity: Kantian Aspects and Influences', *Studies in History and Philosophy of Science 23*: 213–39.

Kane, G. (1997), 'Superstring Theory is Testable, Even Supertestable', *Physics Today 50 2*: 40–2.

Károlyházy, F. (1966), 'Gravitation and Quantum Mechanics of Macroscopic Bodies', *Il Nuovo Cimento A 42*: 390–402.

Károlyházy, F. (1974), *Magyar Fizikai Polyoirat 12*: 24.

Károlyházy, F., Frenkel, A., and Lukács, B. (1986), in *Quantum Concepts in Space and Time*, R. Penrose and C. J. Isham (eds.). Oxford: Oxford University Press, 109–128.

Kauffman, L. (1993), *Knots and Physics*. Singapore: World Scientific Press.

Kay, B. S. (1998), 'Decoherence of Macroscopic Closed Systems within Newtonian Quantum Gravity', to appear in *Classical and Quantum Gravity*. Also available as hep-th/9810077.

Kenmoku, M., Kubotani, H., Takasugi, E., and Yamazaki, Y. (1997), 'de Broglie-Bohm Interpretation for the Wave Function of Quantum Black Holes', preprint available as gr-qc/9711039.

Kent, A. (1990), 'Against Many-Worlds Interpretations', *International Journal of Modern Physics A 5*: 1745–62. A Foreword has been added to the updated e-print version located at gr-qc/9703089.

Kent, A. (1997), 'Consistent Sets Yield Contrary Inferences in Quantum Theory', *Physical Review Letters 78*: 2874–7.

Kibble, T. W. B. (1981), 'Is a Semi-Classical Theory of Gravity Viable?', in *Quantum Gravity 2: A Second Oxford Symposium*, C. J. Isham, R. Penrose, and D. Sciama (eds.). Oxford: Clarendon Press, 63–80.

Komar, A. B. (1969), 'Qualitative Features of Quantized Gravitation', *International Journal of Theoretical Physics 2*: 157–60.

Kowalski-Glikman, J. and Vink, J. C. (1990), 'Gravity-matter Mini-super-space: Quantum Regime, Classical Regime and In Between', *Classical Quantum Gravity 7*: 901–18.

Kretschmann, E. (1917), 'Über die prinzipielle Bestimmbarkeit der berechtigten Bezugssysteme beliebiger Relativistätstheorien', *Annalen der Physik (Leipzig) 53*: 575–614.

Kuchař, K. (1972), 'A Bubble–Time Canonical Formalism for Geometrodynamics', *Journal of Mathematical Physics 13*: 768–81.

Kuchař, K. (1980), 'Gravitation, Geometry, and Nonrelativistic Quantum Theory', *Physical Review D 22*: 1285–99.

Kuchař, K. (1986), 'Canonical Geometrodynamics and General Covariance', *Foundations of Physics 16*: 193–208.

Kuchař, K. (1988), 'Canonical Quantization of Generally Covariant Systems', in *Highlights in Gravitation and Cosmology*, B. Iyer, A Kambhavi, J. Narlikar, and C. Vishveshvara (eds.). New York: Cambridge University Press, 93–120.

Kuchař, K. (1991), 'The Problem of Time in Canonical Quantization of Relativistic Systems', in *Conceptual Problems of Quantum Gravity*, A. Ashtekar and J. Stachel (eds.). Boston: Birkhäuser, 141–68.

Kuchař, K. (1992), 'Time and Interpretations of Quantum Gravity', in *Proceedings of the 4th Canadian Conference on General Relativity and Astrophysics*, G. Kunsatter, D. Vincent, and J. Williams (eds.). Singapore: World Scientific, 211–314.

Kuchař, K. (1993), 'Canonical Quantum Gravity', in *General Relativity and Gravitation 1992: Proceedings of the Thirteenth International Conference on General Relativity and Gravitation*, R. J. Gleiser, C. N. Kozameh, and O. M. Moreschi (eds.). Philadelphia: Institute of Physics Publishing, 119–50.

Kuchař, K. (1993a), 'Matter Time in Canonical Quantum Gravity', in *Directions in General Relativity*, B. Hu, M. Ryan, and C. Vishveshvara (eds.). New York: Cambridge University Press, 201–21.

Künzle, H. P. (1972), 'Galilei and Lorentz Structures on Space-time: Comparison of the Corresponding Geometry and Physics', *Annales Institut Henri Poincaré A 17*: 337–62.

Künzle, H. P. (1976), 'Covariant Newtonian Limit of Lorentz Space-Time', *General Relativity and Gravitation 7*: 445–57.

Lanczos, C. (1949), *The Variational Principles of Mechanics*. Toronto: University of Toronto Press. Also published by New York: Dover.

Landsman, N. (1995), 'Against the Wheeler–DeWitt Equation', *Classical and Quantum Gravity 12*: L119–23.

Landsman, N. (1998), *Mathematical Topics Between Classical and Quantum Mechanics*. New York: Springer-Verlag.

Lange, L. (1884), 'Über die wissenschaftliche Fassung des Galilei'schen Beharrungsgesetz', *Philosophische Studien 2*: 266–97.

Laue, M. von (1951), *Die Relativitätstheorie*, Band 1, Chapter 1. Braunschweig: Vieweg.

Lawrence, R. (1993), 'Triangulation, Categories and Extended Field Theories', in *Quantum Topology*, R. Baadhio and L. Kauffman (eds.). Singapore: World Scientific Press, 191–208.

Leggett, A. J. (1980), 'Macroscopic Quantum Systems and the Quantum Theory of Measurement', *Supplement of the Progress of Theoretical Physics 69*: 80–100.

Leggett, A. J. (1984), 'Schrödinger's Cat and her Laboratory Cousins', *Contemporary Physics 25*: 583–98.

Leggett, A. J. (1998), 'Macroscopic Realism: What Is It, and What Do We Know About It From Experiment?', in *Minnesota Studies in the Philosophy of Science, Vol. XVII, Quantum Measurement: Beyond Paradox*, R. A. Healey and G. Hellman (eds.). Minneapolis: University of Minnesota Press, 1–22.

Leggett, A. J. and Garg, A. (1985), 'Quantum Mechanics versus Macroscopic Realism: Is the Flux There When Nobody Looks?', *Physical Review Letters 54*: 857–60.

Lense, J. and Thirring, H. (1918), 'Über den Einfluss der Eigenrotation der Zentralkörper auf die Bewegung der Planeten und Monde nach der Einsteinschen Gravitationstheorie', *Physikalische Zeitschrift (Germany) 19*: 156–63. English translation by B. Mashhoon, F. W. Hehl, and D. S. Theiss, (1984), 'On the Gravitational Effects of Rotating Masses – The Thirring-Lense Papers', *General Relativity and Gravitation 16*: 711–25.

Littlejohn, R. and Reinsch, M. (1997), 'Gauge Fields in the Separation of Rotations and Internal Motions in the *n*-Body Problem', *Reviews of Modern Physics 69*: 213–75.

Lorentz, H. A. (1892), 'La Théorie Électromagnétique de Maxwell et Son Application aux Corps Mouvants', *Archives Néerlandaises des Sciences Exactes et Naturelles 25*: 363–542.

Lorentz, H. A. (1937 [1923]), 'The Determination of the Potentials in the General Theory of Relativity, With Some Remarks About the Measurement of Lengths and Intervals of Time and About the Theories of Weyl and Eddington', in his *Collected Papers, Vol. 5*, A. D. Fokker and P. Zeeman (eds.). The Hague: Nijhoff, 363–82. Originally published in *Proceedings of the Royal Academy of Amsterdam 29*: 383.

Lynden-Bell, D. (1995), 'A Relative Newtonian Mechanics', in *Mach's Principle: From Newton's Bucket to Quantum Gravity*, J. Barbour and H. Pfister (eds.). Boston: Birkhäuser, 172–8.

MacLane, S. (1988), *Categories for the Working Mathematician*. Berlin: Springer-Verlag.

Mach, E. (1960), *The Science of Mechanics: A Critical and Historical Account of Its Development*. LaSalle: Open Court.

Magnon, A. (1997), *Arrow of Time and Reality: In Search of a Conciliation*. Singapore: World Scientific.

Major, S. A. (1999), 'A Spin Network Primer', *American Journal of Physics 67.11*: 972–80.

Maldecena, J. and Strominger, A. (1997), *Physical Review D 55*: 861.

Markopoulou, F. and Smolin, L. (1998), 'Quantum Geometry With Intrinsic Local Causality', *Physical Review D 58*: 084032.

Marsden, J. E. (1981), *Lectures on Geometric Methods in Mathematical Physics*. Philadelphia: SIAM.

Maudlin, T. (1990), 'Substances and Spacetime: What Aristotle Would Have Said to Einstein', *Studies in History and Philosophy of Science 21*: 531–61.

Maudlin, T. (1994), *Quantum Non-locality and Relativity*. Oxford: Blackwell.

Mellor, D. H. (1993), 'The Unreality of Tense', in *The Philosophy of Time*, R. Le Poidevin and M. MacBeath (eds.). New York: Oxford University Press, 47–59.

Miller, A. I. (1981), *Albert Einstein's Special Theory of Relativity*. Reading, Massachusetts: Addison-Wesley.

Minkowski, H. (1908), 'Raum und Zeit', address delivered at the 80th Assembly of German Natural Scientists and Physicians, Cologne. English translation 'Space and Time' in *The Principle of Relativity*. New York: Dover.

Misner, C. W., Thorne, K. S., and Wheeler, J. A. (1973), *Gravitation*. New York: W. H. Freeman and Company.

Møller, C. (1962), 'The Energy-Momentum Complex in General Relativity and Related Problems', in *Les Theories Relativistes de la Gravitation*, A. Lichnerowicz and M. A. Tonnelat (eds.). Paris: CNRS, 2–29.

Moroz, I. M., Penrose, R., and Tod, P. (1998), 'Spherically-symmetric Solutions of the Schrödinger-Newton Equations', *Classical and Quantum Gravity 15*: 2733–42.

Münch-Berndl, K., Dürr, D., Goldstein, S., and Zanghì, N. (1999), 'Hypersurface Bohm-Dirac models', *Physical Review A 60*: 2729–36.

Nakahara, M. (1990), *Geometry, Topology and Physics*. Bristol: Adam Hilger.

Newton, I. (1962), *De Gravitatione et Aequipondio Fluidorum*, translation in *Unpublished Papers of Isaac Newton*, A. R. Hall and M. B. Hall (eds.). Cambridge: Cambridge University Press.

Norton, J. D. (1984), 'How Einstein Found His Field Equations: 1912–1915', *Historical Studies in the Physical Sciences 14*: 253–315. Reprinted in *Einstein and the History of General Relativity: Einstein Studies Vol. I*, D. Howard and J. Stachel (eds.). Boston: Birkhäuser, 101–59.

Norton, J. D. (1993), 'General Covariance and the Foundations of General Relativity: Eight Decades of Dispute', *Reports on Progress in Physics 56*: 791–858.

Omnès, R. (1992), *Reviews of Modern Physics 64*: 339.

Page, D. (1994), 'Black Hole Information', in *Proceedings of the 5th Conference on General Relativity and Relativistic Astrophysics*, R. Mann and R. McLenaghan (eds.). Singapore: World Scientific. Also available as hep-th/9305040.

Page, D. N. and Geilker, C. D. (1981), 'Indirect Evidence for Quantum Gravity', *Physical Review Letters 47*: 979–82.

Parfionov, G. N. and Zapatrin, R. R. (1995), 'Pointless Spaces in General Relativity', *International Journal of Theoretical Physics 34*: 717.

Pauli, W. (1981 [1921]), *Theory of Relativity*, translated by G. Field. New York: Dover Publications, Inc. Originally published as 'Relativitätstheorie', *Encyklopädie der matematischen Wissenschaften, Vol. V19*. Leibzig: B. G. Teubner.

Pearle, P. (1986), in *Quantum Concepts in Space and Time*, R. Penrose and C. J. Isham (eds.). Oxford: Oxford University Press, 84–108.

Pearle, P. (1989), 'Combining Stochastic Dynamical State-vector Reduction with Spontaneous Localization', *Physical Review A 39*: 2277–89.

Pearle, P. (1993), 'Ways to Describe Dynamical Statevector Reduction', *Physical Review A 48*: 913–23.

Pearle, P. and Squires, E. J. (1996), 'Gravity, Energy Conservation, and Parameter Values in Collapse Models', *Foundations of Physics 26*: 291–305. Also available as preprint quant-ph/9503019.

Penrose, R. (1979), 'Singularities and Time-Asymmetry', in *General Relativity: an Einstein Centenary*, S. W. Hawking and W. Israel (eds.). Cambridge: Cambridge University Press, 581–638.

Penrose, R. (1981), 'Time-Asymmetry and Quantum Gravity', in *Quantum Gravity 2: a Second Oxford Symposium*, C. J. Isham, R. Penrose, and D. W. Sciama (eds.). Oxford: Oxford University Press, 244–72.

Penrose, R. (1984), 'Donaldson's Moduli Space: a "Model" for Quantum Gravity?', in *Quantum Theory of Gravity*, S. M. Christensen (ed.). Bristol: Adam Hilger, 295–8.

Penrose, R. (1986), 'Gravity and State Vector Reduction', in *Quantum Concepts in Space and Time*, R. Penrose and C. J. Isham (eds.). Oxford: Oxford University Press, 129–6.

Penrose, R. (1987), 'Newton, Quantum Theory and Reality', in *300 Years of Gravity*, S. W. Hawking and W. Israel (eds.). Cambridge: Cambridge University Press, 17–49.

Penrose, R. (1989), *The Emperor's New Mind: Concerning Computers, Minds and the Laws of Physics*. Oxford: Oxford University Press.

Penrose, R. (1993), 'Gravity and Quantum Mechanics', in *General Relativity and Gravitation 13. Part 1: Plenary Lectures*, R. J. Gleiser, C. N. Kozameh, and O. M. Moreschi (eds.). Bristol/Philadelphia: Institute of Physics Publishing, 179–89.

Penrose, R. (1994), 'Non-Locality and Objectivity in Quantum State Reduction', in *Fundamental Aspects of Quantum Theory*, J. Anandan and J. L. Safko (eds.). Singapore: World Scientific, 238–46.

351

Penrose, R. (1994a), *Shadows of the Mind: An Approach to the Missing Science of Consciousness.* Oxford: Oxford University Press.

Penrose, R. (1996), 'On Gravity's Role in Quantum State Reduction', *General Relativity and Gravitation 28*: 581–600, reprinted in this volume.

Penrose, R. (1997), *The Large, the Small and the Human Mind.* Cambridge: Cambridge University Press.

Penrose, R. (1998), 'Quantum Computation, Entanglement, and State Reduction', *Philosophical Transactions of the Royal Society, London A 356*: 1927–39.

Percival, I. C. (1995), 'Quantum Spacetime Fluctuations and Primary State Diffusion', *Proceedings of the Royal Society, London A 451*: 503–13.

Peters, A., Chung, K. Y., and Chu, S. (1999), *Nature*: 849.

Plyushchay, M. and Razumov, A. (1996), 'Dirac Versus Reduced Phase Space Quantization for Systems Admitting No Gauge Conditions', *International Journal of Modern Physics A 11*: 1427–62.

Le Poidevin, R. and MacBeath, M. (eds.) (1993), *The Philosophy of Time.* New York: Oxford University Press.

Poincaré, H. (1902), *Science et Hypothèse.* Paris: Flammarion. English translation: *Science and Hypothesis.* London: Walter Scott Publ. Co.

Polchinski, J. (1998), *String Theory: Superstring Theory and Beyond.* Cambridge: Cambridge University Press.

Redhead, M. L. G. (1975), 'Symmetry in Intertheory Relations', *Synthese 32*: 77–112.

Redhead, M. L. G. (1983), 'Quantum Field Theory for Philosophers', in *PSA 1982, Vol. 2*, P. D. Asquith and R. N. Giere (eds.). East Lansing Michigan: Philosophy of Science Association, 57–99.

Reichenbach, H. (1956), *The Direction of Time.* Berkeley: University of California Press.

Reichenbach, H. (1958 [1927]), *The Philosophy of Space and Time (English translation).* New York: Dover.

Reisenberger, M. (1996), 'A Left-handed Simplicial Action for Euclidean General Relativity', preprint available as gr-qc/9609002.

Reisenberger, M. and Rovelli, C. (1997), ' "Sum over Surfaces" Form of Loop Quantum Gravity', *Physical Review D 56*: 3490–508. Also available as preprint gr-qc/9612035.

Ridderbos, K. (1999), 'The Loss of Coherence in Quantum Cosmology', *Studies in History and Philosophy of Modern Physics 30B*: 41–60.

Rosen, G. (1972), 'Galilean Invariance and the General Covariance of Nonrelativistic Laws', *American Journal of Physics 40*: 683–7.

Rosenfeld, L. (1963), 'On Quantization of Fields', *Nuclear Physics 40*: 353–6.

Rovelli, C. (1991), 'What is Observable in Classical and Quantum Gravity?', *Classical and Quantum Gravity 8*: 297–316.

Rovelli, C. (1991a), 'Is There Incompatibility Between the Way Time is Treated in General Relativity and in Standard Quantum Mechanics?', in *Conceptual Problems of Quantum Gravity*, A. Ashtekar and J. Stachel (eds.). Boston: Birkhäuser, 126–40.

Rovelli, C. (1991b), 'Quantum Evolving Constants. Reply to "Comment on 'Time in Quantum Gravity: An Hypothesis' " ', *Physical Review D 44*: 1339–41.

Rovelli, C. (1991c), 'Time in Quantum Gravity: An Hypothesis', *Physical Review D 43*: 442–56.

Rovelli, C. (1991d), 'Quantum Reference Systems', *Classical and Quantum Gravity 8*: 317–31.

Rovelli, C. (1991e), 'Quantum Mechanics Without Time: a Model', *Physical Review D 42*: 2638.

Rovelli, C. (1993), 'A Generally Covariant Quantum Field Theory and a Prediction on Quantum Measurements of Geometry', *Nuclear Physics B 405*: 797.

Rovelli, C. (1993a), 'Statistical Mechanics of Gravity and Thermodynamical Origin of Time', *Classical and Quantum Gravity 10*: 1549.

Rovelli, C. (1993b), 'The Statistical State of the Universe', *Classical and Quantum Gravity 10*: 1567.

Rovelli, C. (1995), 'Analysis of the Different Meaning of the Concept of Time in Different Physical Theories', *Il Nuovo Cimento 110B*: 81.

Rovelli, C. (1996), 'Relational Quantum Mechanics', *International Journal of Theoretical Physics* 35: 1637.

Rovelli, C. (1997), 'Halfway Through the Woods: Contemporary Research on Space and Time', in *The Cosmos of Science: Essays of Exploration*, J. Earman and J. Norton (eds.). Pittsburgh: University of Pittsburgh Press, 180–223.

Rovelli, C. (1998), 'Strings, Loops and the Others: a Critical Survey on the Present Approaches to Quantum Gravity', in *Gravitation and Relativity: At the Turn of the Millennium*, N. Dadhich and J. Narlikar (eds.). Poona University Press. Also available as preprint gr-qc/9803024.

Rovelli, C. (1998a), 'Incerto tempore, incertisque loci: Can We Compute the Exact Time at Which the Quantum Measurement Happens?', *Foundations of Physics 28*: 1031–43. Also available as preprint quant-ph/9802020.

Rovelli, C. (1998b), 'Loop Quantum Gravity', *Living Reviews in Relativity*, (refereed electronic journal), http://www.livingreviews.org/Articles/Volume1/1998-1rovelli. Also available as preprint gr-qc/9709008.

Rovelli, C. and Smolin, L. (1988), 'Knot Theory and Quantum Gravity', *Physical Review Letters* 61: 1155.

Rovelli, C. and Smolin, L. (1990), 'Loop Representation of Quantum General Relativity', *Nuclear Physics B 331*: 80–152.

Rovelli, C. and Smolin, L. (1995), 'Discreteness of Area and Volume in Quantum Gravity', *Nuclear Physics B 442*: 593–622. Erratum: *Nuclear Physics B 456*: 734.

Rovelli, C. and Smolin, L. (1995a), 'Spin Networks and Quantum Gravity', *Physical Review D 53*: 5743.

Rugh, S. E., Zinkernagel, H., and Cao, T. Y. (1999), 'The Casimir Effect and the Interpretation of the Vacuum', *Studies in History and Philosophy of Modern Physics 30B*: 111–39.

Ryckman, T. A. (1994), 'Weyl, Reichenbach and the Epistemology of Geometry', *Studies in History and Philosophy of Modern Physics 25B*: 831–70.

Saunders, S. (1996), 'Time, Quantum Mechanics, and Tense', *Synthese 107*: 19–53.

Schmid, R. (1987), *Infinite Dimensional Hamiltonian Systems*. Naples: Bibliopolis.

Schön, M. and Hájiček, P. (1990), 'Topology of Quadratic Super-Hamiltonians', *Classical and Quantum Gravity 7*: 861–70.

Schrödinger, E. (1985 [1950]), *Space-Time Structure* (Reprint). Cambridge, England: Cambridge University Press.

Schwarz, J. H. (ed.) (1985), *Superstrings: The First 15 Years of Superstring Theory, Vols. 1 and 2*. Singapore: World-Scientific.

Schweber, S. S. (1961), *An Introduction to Relativistic Quantum Field Theory*. Evanston, Illinois: Row, Peterson and Company.

Shapere, A. and Wilczek, F. (1987), 'Self-propulsion at Low Reynolds Number', *Physical Review Letters 58*: 2051–4.

Shapere, A. and Wilczek, F. (1989), 'Gauge Kinematics of Deformable Bodies', *American Journal of Physics 57*: 514–18.

Shimony, A. (1963), 'Role of the Observer in Quantum Theory', *American Journal of Physics 31*: 755–73.

Shimony, A. (1978), 'Metaphysical Problems in the Foundations of Quantum Mechanics', *International Philosophical Quarterly 18*: 3–17.

Shimony, A. (1989), 'Conceptual Foundations of Quantum Mechanics', in *The New Physics*, P. Davies (ed.). Cambridge: Cambridge University Press, 373–95.

Shimony, A. (1993), *Search for a Naturalistic World View, Vol. I*. Cambridge: Cambridge University Press.

Shimony, A. (1993a), *Search for a Naturalistic World View, Vol. II.* Cambridge: Cambridge University Press.

Shimony, A. (1998), 'Implications of Transience for Spacetime Structure', in *The Geometric Universe: Science, Geometry, and the Work of Roger Penrose*, S. A. Huggett, L. J. Mason, K. P. Tod, S. T. Tsou, and N. M. J. Woodhouse (eds.). Oxford: Oxford University Press, 161–72.

Shtanov, Y. (1996), 'On Pilot Wave Quantum Cosmology', *Physical Review D 54*: 2564–70.

Smolin, L. (1991), 'Space and Time in the Quantum Universe', in *Conceptual Problems of Quantum Gravity*, A. Ashtekar and J. Stachel (eds.). Boston: Birkhäuser, 228–91.

Smolin, L. (1995), 'The Bekenstein Bound, Topological Quantum Field Theory and Pluralistic Quantum Cosmology', preprint available as gr-qc/9508064.

Smolin, L. (1995a), 'Linking Topological Quantum Field Theory and Nonperturbative Quantum Gravity', preprint available as gr-qc/9505028.

Smolin, L. (1997), 'The Future of Spin Networks', preprint available as gr-qc/9702030.

Smolin, L. (1998), 'Strings as Perturbations of Evolving Spin-networks', hep-th/9801022.

Smolin, L. (1999), 'A Candidate for a Background Independent Formulation of M Theory', e-print hep-th/9903166.

Sorkin, R. D. (1991), 'Finitary Substitute for Continuous Topology', *International Journal of Theoretical Physics 30*: 923–47.

Sorkin, R. D. (1997), 'Forks in the Road, on the Way to Quantum Gravity', *International Journal of Theoretical Physics 36*: 2759–81.

Squires, E. (1990), *Physics Letters A 145*: 67.

Squires, E. (1992), 'A Quantum Solution to a Cosmological Mystery', *Physics Letters A 162*: 35–6.

Squires, E. (1992a), *Foundations of Physics Letters 5*: 279.

Stachel, J. (1989), 'Einstein's Search for General Covariance 1912–1915', in *Einstein Studies Vol. 1*, D. Howard and J. Stachel (eds.). Boston: Birkhäuser, 63–100.

Stachel, J. (1993), 'The Meaning of General Covariance: The Hole Story', in *Philosophical Problems of the Internal and External Worlds: Essays on the Philosophy of Adolf Grünbaum*, J. Earman, A. Janis, G. Massey, and N. Rescher (eds.). Pittsburgh, PA: University of Pittsburgh Press, 129–60.

Stachel, J. (1994), 'Changes in the Concepts of Space and Time Brought About by Relativity', in *Artifacts, Representations and Social Practice*, C. C. Gould and R. S. Cohen (eds.). Dordrecht: Kluwer Academic, 141–62.

Stein, H. (1999), 'Physics and Philosophy Meet: the Strange Case of Poincaré', forthcoming.

Stewart, J. (1991), *Advanced General Relativity.* Cambridge: Cambridge University Press.

Strominger, A. and Vaffa, C. (1996), *Physics Letters B 379*: 99.

Susskind, L. (1995), 'The World as a Hologram', *Journal of Mathematical Physics 36*: 6377–96.

Susskind, L. (1997), 'Black Holes and the Information Paradox', *Scientific American*, April.

Susskind, L. and Uglum, J. (1994), *Physical Review D 50*: 2700.

Tait, P. G. (1883), 'Note on Reference Frames', *Proceedings of the Royal Society of Edinburgh*, Session 1883–84, 743.

Teller, P. (1995), *An Interpretive Introduction to Quantum Field Theory.* Princeton: Princeton University Press.

Thomson, J. (1883), 'On the Law of Inertia: The Principle of Chronometry; and the Principle of Absolute Clinural Rest, and of Absolute Rotation', *Proceedings of the Royal Society of Edinburgh*, Session 1883–84, 568, 730.

Thorne, K. S. (1994), *Black Holes and Time Warps: Einstein's Outrageous Legacy.* New York: W. W. Norton.

Thorne, K., Zurek, W., and Price, R. (1986), 'The Thermal Atmosphere of a Black Hole', in *Black Holes: The Membrane Paradigm*, K. Thorne, R. H. Price, and D. A. Macdonald (eds.). New Haven: Yale University Press.

Tod, P. and Moroz, I. M. (1999), 'An Analytic Approach to the Schrödinger-Newton Equations', *Nonlinearity 12*: 201–16.

Torre, C. G. (1992), 'Is General Relativity an "Already Parameterized" Theory?', *Physical Review D 46*: R3231–4.

Torre, C. G. (1993), 'Gravitational Observables and Local Symmetries', *Physical Review D 48*: R2373–6.

Torre, C. G. (1994), 'The Problems of Time and Observables: Some Recent Mathematical Results', preprint available as gr-qc/9404029.

Torre, C. G. and Varadarajan, M. (1998), 'Functional Evolution of Free Quantum Fields', preprint available as hep-th/9811222.

Torretti, R. (1983), *Relativity and Geometry*. Oxford: Pergamon Press.

Trautman, A. (1965), in *Lectures on General Relativity (Brandeis 1964 Summer Inst. on Theoretical Physics), Vol. I*, by A. Trautman, F. A. E. Pirani, and H. Bondi. Englewood Cliffs: Prentice-Hall, 7.

Turaev, V. (1994), *Quantum Invariants of Knots and 3-Manifolds*. New York: W. de Gruyter.

Unruh, W. G. (1976), *Physical Review D 14*: 870.

Unruh, W. G. (1980), *Physical Review Letters 46*: 1351.

Unruh, W. G. (1984), 'Steps Towards a Quantum Theory of Gravity', in *Quantum Theory of Gravity: Essays in Honor of the 60th Birthday of Bryce S. De Witt*, S. M. Christensen (ed.). Bristol: Hilger, 234–42.

Unruh, W. G. (1988), 'Time and Quantum Gravity', in *Quantum Gravity*, M. A. Markov, V. A. Berezin, and V. P. Frolov (eds.). Singapore: World Scientific, 252–68.

Unruh, W. G. (1991), 'No Time and Quantum Gravity', in *Gravitation: A Banff Summer Institute*, R. Mann and P. Wesson (eds.). Singapore: World Scientific, 260–75.

Unruh, W. G. (1995), *Physical Review D 51*: 2827–38.

Unruh, W. G. and Wald, R. M. (1982), *Physical Review D 25*: 942.

Unruh, W. G. and Wald, R. M. (1983), *Physical Review D 27*: 2271.

Unruh, W. G. and Wald, R. M. (1989), 'Time and the Interpretation of Canonical Quantum Gravity', *Physical Review D 40*: 2598–614.

Unruh, W. G. and Wald, R. M. (1995), *Physical Review D 52*: 2176.

Valentini, A. (2000), *On the Pilot-Wave Theory of Classical, Quantum and Subquantum Physics*. Berlin: Springer-Verlag.

Vilenkin, A. (1982), 'Creation of Universes from Nothing', *Physics Letters B 117*: 25–8.

Vizgin, V. P. (1994 [1985]), *Unified Field Theories in the First Third of the 20th Century*, translated by J. B. Barbour. Originally published as *Edinye teorii polya v perevoi tret XXveka* (Moskva: Nauka). Basel: Birkhäuser Verlag.

Wald, R. M. (1984), *General Relativity*. Chicago: University of Chicago Press.

Wald, R. M. (1994), *Quantum Field Theory in Curved Spacetime and Black Hole Thermodynamics*. Chicago: University of Chicago Press.

Weinberg, S. (1983), 'Why the Renormalization Group is a Good Thing', in *Asymptotic Realms of Physics: Essays in Honour of Francis E. Low*, A. H. Guth, K. Huang, and R. L. Jaffee (eds.). Cambridge, MA: MIT Press, 1–19.

Weinberg, S. (1989), *Physical Review Letters 62*: 485.

Weinberg, S. (1989a), 'Testing Quantum Mechanics', *Annals of Physics (New York) 194*: 336–86.

Weingard, R. (1984), 'Grand Unified Gauge Theories and the Number of Elementary Particles', *Philosophy of Science 51*: 150–5.

Weinstein, S. (1996), 'Strange Couplings and Space-time Structure', *Philosophy of Science 63*: S63–70.

Weinstein, S. (1998), *Conceptual and Foundational Issues in the Quantization of Gravity*. Unpublished PhD dissertation, Northwestern University.

Weinstein, S. (1998a), 'Time, Gauge, and the Superposition Principle in Quantum Gravity', in *Proceedings of the Eighth Marcel Grossman Meeting*. Singapore: World Scientific. Also available as gr-qc/9711056.

Weinstein, S. (2001), 'Absolute Quantum Mechanics', forthcoming in *British Journal for the Philosophy of Science*.

Weyl, H. (1918), 'Gravitation und Elektrizität', *Sitzungsberichte der Königlich Preussischen Akademie der Wissenschaften zu Berlin*: 465–80. Translated by W. Perret and G. B. Jeffery as 'Gravitation and Electricity', in *The Principle of Relativity: A Collection of Original Memoirs on the Special and General Theory of Relativity*, A. Einstein, H.A. Lorentz, H. Minkowski, and H. Weyl. New York: Dover (1952), 201–16.

Weyl, H. (1918a), *Raum-Zeit-Materie*. Berlin: Springer.

Weyl, H. (1952 [1921]), *Space-Time-Matter, 4th edition*, translated by H. L. Brose. New York: Dover Publications, Inc. Originally published as *Raum-Zeit-Materie*. Berlin: Springer.

Wheeler, J. A. (1968), 'Superspace and the Nature of Quantum Geometrodynamics', in *Battelle Rencontres: 1967 Lectures in Mathematics and Physics*, C. DeWitt and J. A. Wheeler (eds.). New York: Benjamin, 242–307.

Whitehead, A. N. (1929), *Process and Reality*. London: Macmillan.

Whitrow, G. J. (1961), *The Natural Philosophy of Time*. Edinburgh: Nelson.

Wigner, E. P. (1961), 'Remarks on the Mind-Body Question' in *The Scientist Speculates*, I. J. Good (ed.). London: Heinemann. Reprinted in Wigner (1967), 171–84 and in J. A. Wheeler and W. H. Zurek (eds.) (1983), *Quantum Theory and Measurement*. Princeton: Princeton University Press, 168–81.

Wigner, E. P. (1967), *Symmetries and Reflections*. Bloomington: Indiana University Press.

Williams, D. C. (1951), 'The Myth of Passage', *Journal of Philosophy 48*: 457–72.

Witten, E. (1986), 'Non-Commutative Geometry and String Field Theory', *Nuclear Physics B 268*: 253–94.

Witten, E. (1997), 'Duality, Spacetime and Quantum Mechanics', *Physics Today 50 5*: 28–33.

Yang, C. N. (1984), 'Gauge Fields, Electromagnetism and the Bohm-Aharonov Effect', in *Foundations of Quantum Mechanics in the Light of New Technology*, S. Kamefuchi, H. Ezaira, Y. Murayama, M. Namiki, S. Nomura, Y. Ohnuki, and T. Yajima (eds.). Tokyo: Physical Society of Japan, 5–9.

Yang, C. N. (1986), 'Hermann Weyl's Contribution to Physics', in *Hermann Weyl 1885–1985*, K. Chandrasekharan (ed.). Berlin: Springer-Verlag, 7–21.

Zanstra, H. (1924), 'A Study of Relative Motion in Connection with Classical Mechanics', *Physical Review 23*: 528–45.

Zeilicovici, D. (1986), 'A (Dis)Solution of McTaggart's Paradox', *Ratio 28*: 175–95.

Zeilinger, A. (1997), in *Gravitation and Relativity: At the Turn of the Millennium*, N. Dadhich and J. Narlikar (eds.). Poona University Press.

Zurek, W. H. (1991), *Physics Today 44*: 36.

# Notes on contributors

**John C. Baez** received his B.A. from Princeton in 1982 and his Ph.D. from MIT in 1986. He now teaches in the Mathematics Department at the University of California, Riverside. He works on quantum gravity and *n*-category theory, and spends a lot of time explaining math and physics on the internet – see http://math.ucr.edu/home/baez/ for some of this.

**Julian B. Barbour** has worked on fundamental issues in physics for 35 years, and made important and original contributions to the theory of time and inertia. He is the author of *Absolute or Relative Motion?* (CUP) and co-editor of *Mach's Principle: From Newton's Bucket to Quantum Gravity* (Birkhauser). He has contributed to three recent major television documentaries on modern physics that have been shown world-wide.

**Gordon Belot** is Associate Professor of Philosophy at New York University. He has published a number of papers on philosophy of physics.

**Harvey R. Brown** studied physics in New Zealand, and philosophy of physics at London University. After teaching for some years in Brazil, he came to Oxford University in 1984, where he is now a Reader in Philosophy and a Fellow of Wolfson College.

**Jeremy Butterfield** is Senior Research Fellow, All Souls College, University of Oxford. His main research interests are the philosophy of quantum theory and relativity.

**Joy Christian** obtained his Ph.D. in Foundations of Quantum Mechanics in 1991 from Boston University under the supervision of Professor Abner Shimony.

357

He was then elected a Junior Research Fellow in Philosophy of Physics at the University of Oxford, where he has remained since 1991 as a member of the Sub-Faculty of Physics.

**John Earman** is University Professor of History and Philosophy of Science at the University of Pittsburgh.

**Sheldon Goldstein** is Professor of Mathematics and Physics at Rutgers University.

**Christopher Isham** is Professor of Theoretical Physics at Imperial College, London. His main research interests are foundational and mathematical issues in quantum gravity and quantum theory.

**Sir Roger Penrose** is a Professor at the Mathematical Institute, University of Oxford.

**Oliver Pooley** holds a fixed-term lectureship in philosophy at Exeter College, Oxford. He is completing a D.Phil. at Oxford University looking at space-time realism in the light of parity violation and geometrodynamical formulations of general relativity. He was formerly Hargeaves Senior Scholar at Oriel College, Oxford.

**Carlo Rovelli** is Professor in the Physics Department and in the Department of History and Philosophy of Science of the University of Pittsburgh, and Professor de Physique in the Université de la Méditerranée in Marseille, France. He received the Xanthopoulos award in 1995. His main research interests are in quantum gravity and fundamental spacetime physics.

**Stefan Teufel**: Mathematisches Insitut der Universität München.

**William G. Unruh**, a Professor of Physics at the University of British Columbia and the Founding Director and current Fellow if the Cosmology and Gravity Program of the Canadian Institute for Advanced Research, is interested in the wide range of research touching on the interaction between gravity and quantum theory. This has taken him as far afield as quantum computers, analysis of counterfactual arguments in quantum mechanics, and the initial conditions of the universe.

**Robert Weingard** was Professor of Philosophy at Rutgers University until his untimely death in 1996. His work on such topics as quantum mechanics, quantum field theory, Bohm's theory, relativity, the physics of the mind–body problem, cosmology, and string theory continues to inspire many of the contributors to this volume.

**Steven Weinstein** received his Ph.D. in Philosophy at Northwestern University in 1998. He was a Killam Postdoctoral Fellow in physics and philosophy at the University of British Columbia during 1998–1999, and is presently Visiting Assistant Professor in the Philosophy Department at Princeton University.

**Edward Witten** is a Professor in the School of Natural Sciences at the Institute for Advanced Study, Princeton.

# Index

# Index

diffeomorphisms, 61, 69
dimension compactification, 140–142
history, 113
infinitesimal parallel transport, 265–266
instant, 241, 242
manifold, 71, 73, 78, 81–82, 83, 84, 107
mirror asymmetry, 136
NUT, 335
in particle-physics, 69–70, 126
points, 26, 51–54, 81–83, 93, 106, 227–228, 247, 298–303, 310, 313, 320, 323
quantization, 77–83, 109, 110–115, 239
regions, 83, 86
string theory, 15–16, 71–72, 126, 136
superposition, 296
special relativity, 104, 107, 111, 258–261, 269
speed of light, 179, 180
two-way, 260, 262
spin-connection, 75
spin foams, 114, 194
spin networks, 110–111, 193, 194
'spins', 188
spontaneous localization, 10, 11, 51
SQUID rings, 335
standard model, 5, 177
state reduction, 290–304, 306, 307, 315–337
states, time-independent 113
statistical time, 115
stress–energy tensor, 4, 171
string field theory, 143–145
string theory, 4, 14–16, 20, 59, 65, 70–73, 84, 129–131, 132, 133, 136–137, 138–151, 276
background-free, 182–183
black holes, 168–170
covariant field theory, 143–145
dualities, 15, 16, 73, 134–136, 150–151
Green–Schwartz string, 140
heterotic string, 141–142
historical aspects, 14, 71, 150–151
massless modes, 139–140
minimum length, 134–136
non-perturbative, 15, 72–73
open and closed strings, 139, 140
perturbative, 15, 39, 59, 71–72, 109
spacetime in, 15–16, 71–72, 126, 136
string coupling constant, 168
substantivalism, 25, 223, 227–228, 248–249
superluminal signals, 6–9
superposition, 25, 104, 208–209, 293–297, 307–308, 316–322, 334–335, 337
superspace, 147–148, 226, 240
superstrings, *see* string theory
supersymmetry, 70–71, 130, 146–148
black holes, 168
symplectic form/geometry, 215, 249

*T*-duality, 73

tachyon, 140
theories, evolution of, 115–119
thermodynamical time, 114–115
thermodynamics of black holes, 152–153, 154–157, 164
tidal effects, 94
time, 22–24, 111–114, 230–231, 241
absolute vs. relative, 51–52, 106
Everettian view, 48
flow, 115
general relativity, 63–65, 111–112, 242–246
instants, 212
internal, 243–246, 247–249
Kantian view, 44–45
problem of, 23, 62–65, 98, 279, 280, 320
in quantum theory, 62–63
statistical, 115
thermodynamical, 114–115
translation, 296, 299, 301–302
time travel, 9
topological quantum field theory, 21, 62, 178, 183–187, 192
topological space, 53, 136
topology change, 21, 137, 184–185
topos theory, 83, 86
trace anomaly, 171
transcendental idealism, 44
trial equilocality relationship, 209–210
'trickle-down effects', 78, 82–83
Turaev–Viro model, 178, 187–189
two-way light speed, 260, 262

ultraviolet divergences, 57, 75
uncertainty principle, 104
underdetermination of theories, 38
unified field theory, 264–267
'unity of space', 22

*V – A* theory, 148–149
vanishing intrinsic angular momentum, 204
vector constraint, 224–225, 244
Veneziano's model, 14, 15, 151

Ward identities, 69
wave function, 28–29
weave, 111
Wheeler–DeWitt equation, 19, 23, 28, 74–75, 77, 237–238, 278
'when and where', 114
world sheet, 71
spinner, 140
worldlines and worldtubes, 129–130

Yang–Mills theory, 67, 68, 141, 144

Zeno, 24
zero temperature heat baths, 155

Printed in the United States
By Bookmasters